Current Progress in Thermal Engineering

Current Progress in Thermal Engineering

Edited by
Abigail Davis

STATES
ACADEMIC PRESS
www.statesacademicpress.com

Published by States Academic Press,
109 South 5th Street,
Brooklyn, NY 11249, USA

ISBN: 978-1-63989-133-7

Cataloging-in-Publication Data

Current progress in thermal engineering / edited by Abigail Davis.
p. cm.
Includes bibliographical references and index.
ISBN 978-1-63989-133-7
1. Heat engineering. 2. Thermodynamics. 3. Power (Mechanics) I. Davis, Abigail.
TJ260 .C87 2021
621.402 2--dc23

For information on all States Academic Press publications
visit our website at www.statesacademicpress.com

STATES
ACADEMIC PRESS

Contents

Preface

Thermal engineering is the branch of mechanical engineering that undertakes the study of controlling the heating and cooling processes in an enclosed or open atmosphere. It is mostly used by chemical and mechanical engineers. Thermal engineering encompasses the concepts related to the design, development, and demonstration of components, devices, equipment, technologies, and systems involving thermal processes. These are applied to the production, storage, utilization, and conservation of energy. Thermal engineering borrows concepts from various areas of study such as thermodynamics, fluid dynamics, fluid statics and heat transfer. This book is a compilation of chapters that discuss the most vital concepts and emerging trends in the field of thermal engineering. It picks up individual branches and explains their need and contribution to a growing economy. This book will provide comprehensive knowledge to the readers.

This book unites the global concepts and researches in an organized manner for a comprehensive understanding of the subject. It is a ripe text for all researchers, students, scientists or anyone else who is interested in acquiring a better knowledge of this dynamic field.

I extend my sincere thanks to the contributors for such eloquent research chapters. Finally, I thank my family for being a source of support and help.

Editor

ENERGY AND EXERGY ANALYSIS OF A VAPOR ABSORPTION REFRIGERATION SYSTEM IN AN INTERCITY BUS APPLICATION

N. Kurtulmuş[1,*], M. Bilgili[2], B. Şahin[3]

ABSTRACT

A Vapor Absorption Refrigeration (VAR) system driven by the exhaust gas waste heat received from the internal combustion engine of an intercity bus is modeled and analyzed for air-conditioning the intercity bus cabin under different operating parameters. Initially, the hourly comfort cooling load of the intercity bus is calculated for a cooling season spanning five months between May and October in Turkey. After determining the capacity of heat source sufficiency for air-conditioning the intercity bus, energy and exergy analyses of the VAR system are conducted, then designed and compared with the vapor compression refrigeration system in respect to the effect of fuel consumption. The results show that approximately 4,489 kg/year of fuel can be saved by using the VAR system driven by an exhaust gas waste heat in an intercity bus. The maximum coefficient of performance (COP) of the VAR system is obtained as 0.78 at 5 a.m. in May, and the maximum total exergy destruction for the VAR system is obtained as 15.25 kW at 4 p.m. in July. Lastly, the specific time is selected to investigate the effect of operating and environmental parameters on the VAR system.

Keywords: *Energy, Exergy, Intercity Bus Air-Conditioning, Vapor Absorption Refrigeration System, Water-Lithium Bromide*

INTRODUCTION

A large amount of waste heat is released to the ambient from the internal combustion engines in the automotive industry. There are several methods to utilize waste heat to meet the energy requirement of a refrigeration or air- conditioning process. Air-conditioning of the cabin, especially for vehicles with large cabins such as intercity buses, requires considerable mechanical energy input, ranging from 5 kW to10 kW. One of the methods that is suitable in terms of energy efficiency and economic feasibility is to utilize the waste heat of exhaust gas for the implementation of Vapor Absorption Refrigeration (VAR). The VAR system provides the required cooling by the evaporator component, which is also the case for a Vapor Compression Refrigeration (VCR) system. The VAR system requires heat energy to be driven because there is a physico-chemical process in the cycle, whereas a VCR system requires mechanical energy to drive the system. Commercial VAR systems generally use two different working solutions—ammonia-water (ammonia is refrigerant) and water-LiBr (water is refrigerant). While the VAR system using ammonia-water is preferred for refrigeration processes requiring low temperatures, the VAR system using water-LiBr is preferred for air-conditioning applications [1-4].

Several methods were applied for recovering waste heat from exhaust gases of the internal combustion engines [5, 6]. The researchers claimed that utilizing exhaust gas waste heat received from engines in the VAR systems could be an alternative for cooling applications [7, 8]. When using the VAR system driven by waste heat of exhaust gas, some applications require low temperature refrigeration such as the truck refrigeration system used for preserving the perishable food during the transportation. Koehler et al. [9] investigated a VAR system driven by waste heat received from the exhaust gases for truck refrigeration. For that refrigeration purpose, they designed and constructed a prototype of an absorption refrigeration system and finally tested it under various operational conditions. The system was also evaluated at different cases including city traffic, mountains and plain road. The results of their investigations indicated that the COP value of the prototype was around 0.27, which could be improved beyond that value, and that the system could be an alternative for long-distance plain road transportation. Another experimental research work were conducted on the application of a low-temperature refrigeration system by Horuz [10], who aimed to investigate the performance of a VAR system driven by waste heat received from exhaust gases under different conditions. Many other researchers [11, 12] studied VAR

This paper was recommended for publication in revised form by Regional Editor Tolga Taner

[1]Adana Science and Technology University, Faculty of Engineering and Natural Science, Mech. Eng. Dept., Adana/Turkey
[2]Cukurova University, Ceyhan Engineering Faculty, Mech. Eng. Dept., Adana, Turkey
[3]Cukurova University, Faculty of Engineering and Architecture, Mech. Eng. Dept., Adana, Turkey
**E-mail address: nkurtulmus@adanabtu.edu.tr*

systems driven by waste heat from exhaust gas for air-conditioning vehicle cabins. They all confirmed the feasibility of the VAR system application.

Kilic and Kaynaklı [13] conducted a second law-based thermodynamic analysis of a VAR system using water-LiBr as the working fluid pair. They found that the highest rate of exergy loss at the generator corresponds to the rate of 45.68% of total exergy loss, and the lowest rate of exergy loss occurs at the pump and corresponds to the rate of 0.0034% of total exergy loss. Effects of varying the generator temperature on the exergy loss rate of each component and on the exergy efficiency of the VAR system were investigated. Şencan et al. [14] performed studies on the exergy analysis of a single-effect VAR system using water-LiBr solution. They found that the highest exergy loss occurs through the absorber and the lowest exergy loss occurs through the evaporator. Talbi and Agnew [15] also conducted an exergy analysis on a single-effect VAR system using water-LiBr solution. Gomri [16] made a comparison between single-effect and double-effect VAR systems in terms of the second law of thermodynamics. The obtained results reveal that the maximum COP of the single-effect VAR system is 0.79 and that the maximum COP of the double-effect VAR system is 1.42. In addition, the maximum exergy efficiency of the single-effect VAR system is about 23.2%, whereas the maximum exergy efficiency of the double-effect VAR system is approximately 25.1%. Energy and exergy analyses were conducted by Kaynakli et al. [17] on the double-effect VAR system to determine the effects of heat source types. They chose three types of heat sources, including hot air, steam, and hot water, to drive the system. Their results reported that the maximum exergy destruction occurs through the double-effect VAR system when hot air is used as the heat source. A few of the researchers analyzed the VAR system on an hourly basis of cooling load and solar power for air-conditioning applications. Another research was conducted by Arora et al. [18] on double-effect VAR system. An exergy analysis of a solar-assisted VAR system was performed by Onan et al. [19].

In this study, energy and exergy analyses are conducted for an application of the VAR system for air-conditioning an intercity bus cabin with atmospheric data such as ambient temperature, solar radiation, and relative humidity on an hourly basis. The second objective is to investigate the effects of operating and environmental parameters on the system performance and exergy destruction of the system. Therefore, the advantages of VAR system application for air-conditioning purpose for an intercity bus cabin is presented at different conditions in terms of fuel saving, emissions and weight. To perform the analysis, components of a VAR system using water-LiBr solution as the working fluid are investigated in detail with respect to the variation of ambient parameters.

SYSTEM DESCRIPTIONS

Vapor Absorption Refrigeration System

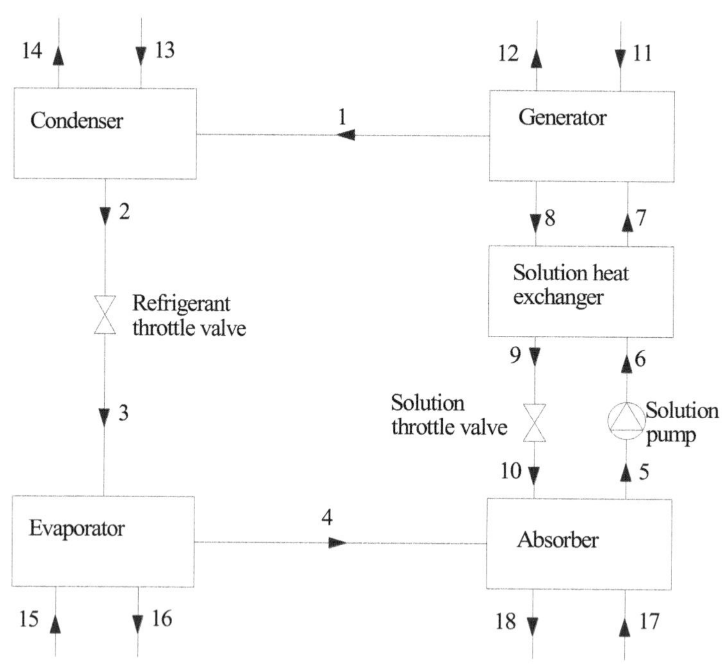

Figure 1. Schematic illustration of the vapor absorption refrigeration system

As seen in Figure 1, the single-effect VAR system consists of the evaporator, condenser, generator, absorber, solution heat exchanger, solution pump, solution throttle valve, and refrigerant throttle valve. The generator, solution heat exchanger, and condenser operate at high pressure, and the evaporator and absorber operate at low pressure. The generator receives heat from exhaust gas. The absorber and condenser are cooled with ambient air and the evaporator takes heat to provide refrigeration.

Operating Conditions of the VAR System

In the modeled system, the generator of the VAR system is coupled with the exhaust system of the intercity bus to utilize waste heat of exhaust gas received from the 6-cylinder, 4-stroke turbocharged diesel engine that drives the bus. The data shown in Table 1 are taken from Yilmaz [20] and Shu et al. [21]. The gas mixture method [21] is applied to determine the exhaust gas properties, including specific heat, enthalpy, and entropy. For example, enthalpy of exhaust gas is calculated as follows and X, is the mass fraction of each individual gas in the exhaust gas.

$$h_{exgas} = X_{CO_2}h_{CO_2} + X_{O_2}h_{O_2} + X_{H_2O}h_{H_2O} + X_{N_2}h_{N_2} \quad (1)$$

Considering the highest and lowest ambient temperatures (because ambient temperature determines the condenser outlet temperature, so does the generator pressure) in terms of the VAR system performance, the generator temperature is chosen as 110 °C. The exhaust gas mass composition (CO_2, N_2, O_2, H_2O) for different engine loads is shown in Table 1 [21].

Table 1. Data for the 6-cylinder, 4-stroke turbocharged diesel engine [20, 21]

Engine load (%)	Exhaust gas temperature (°C)	Exhaust gas mass flow rate (kg/h)	Engine power output (kW)	Engine efficiency (%)	CO_2 (%)	N_2 (%)	O_2 (%)	H_2O (%)
50	420	610.91	117.7	41.72	12.4	73.8	9	4.8
75	474	804.43	176.2	42.15	13.9	73.4	7.2	5.5
90	498	930.81	211.6	41.89	14.5	73.2	6.6	5.7
100	519	990.79	235.8	41.81	15.2	73	5.8	6

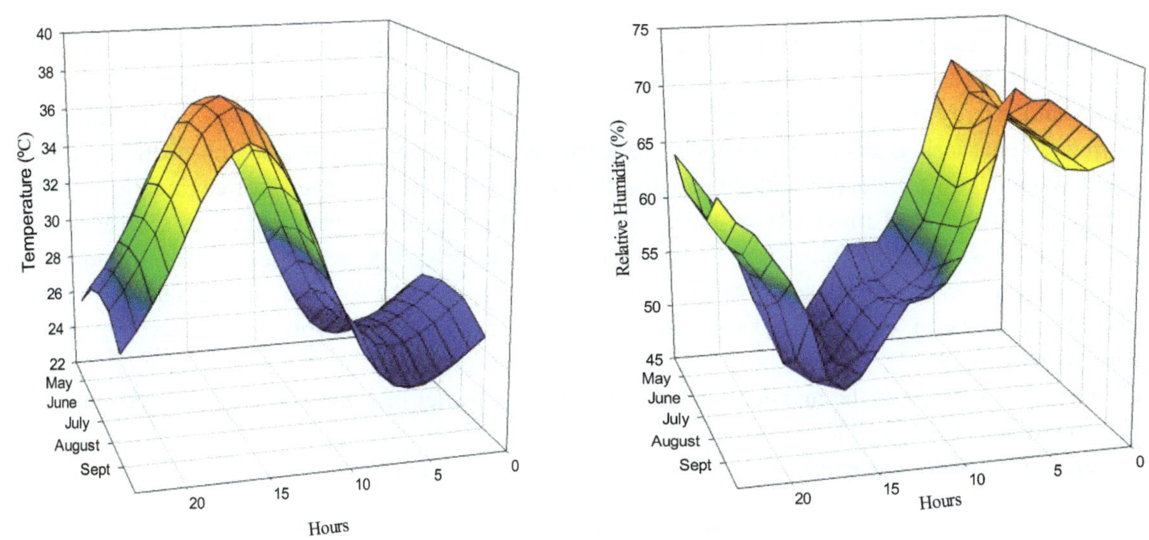

Figure 2. (a) Hourly environmental temperature distributions, (b) Hourly relative humidity distributions

The evaporator is used in a TS 45 type intercity bus cabin belonging to TEMSA Bus Company to meet the comfort cooling load calculated for Adana province in Turkey. For performing energy and exergy analyses, temperature and humidity in the bus cabin are taken as 24 °C and 50%, respectively. The other main components, such as the condenser and absorber, are integrated into the bus appropriately, and they release heat to the medium of outdoor air. Temperatures of the condenser and absorber are determined to be T_A+14 °C [22, 23]. A simulation was prepared using Engineering Equation Solver (EES) software [24] for applying the energy and exergy analyses of the single-effect VAR system. In the analyses, the properties of water/steam and the enthalpy of water-LiBr were obtained from the EES library. The density and specific heat of water-LiBr were obtained from Florides et al. [25] and the entropy of water-LiBr was obtained from Chua et al. [26].

Hourly environmental temperature and the relative humidity values taken from the Turkish State Meteorological Service are shown in Figures 2(a) and 2(b), respectively. As is shown, the environmental temperature reaches a maximum of 38 °C at 3 p.m. in July and August and a minimum 23.4 °C at 5 a.m. in May. Maximum relative humidity is 73% at 6 a.m. in September, and the minimum is 48% at 4 p.m. in June and at 5 p.m. in July.

Assumptions

The following assumptions were made [16, 17]:

- All system components are at steady-state conditions.
- Refrigerant leaving the condenser is saturated water at condenser pressure, and refrigerant leaving the evaporator is saturated vapor at evaporator pressure.
- Condenser pressure is equal to generator pressure, and evaporator pressure is equal to absorber pressure.
- There is no pressure drop in the heat exchangers or piping systems.
- There are no heat losses or gains in the various components and piping systems.
- Temperature and pressure at the reference state are 298 K and 101.325 kPa, respectively. Enthalpy and entropy of the working fluid at the reference state used for calculating the exergy of the VAR system are equal to the values of water at an environmental temperature and pressure of 25 °C and 101.325 kPa, respectively.
- The kinetic, chemical, and potential exergy of all streams of the VAR system are negligible.
- The solution pump efficiency is 0.9.
- The specific humidity is constant for heat exchangers in the VAR system.

Thermodynamic Analysis of the VAR System

Mass balance and energy conversion methods are implemented to determine and optimize the capacity of the components and the COP value of the VAR system. These methods provide any necessary information about the irreversibility of the components of the VAR system. Cengel and Boles [27] reported that entropy production and exergy destruction analyses could be utilized to set up principles for the performance of engineering systems such as absorbers, generators, condensers, and evaporators. Entropy production can be used as a quantitative degree of irreversibilities related to the processes. The performance of engineering systems is decayed by the existence of irreversibility during processes. An increase in irreversibilities increases entropy production. There is no irreversibility in a reversible process, but reversible processes, in fact, do not occur in nature. In addition, reversible processes can be considered to be theoretical limits for the related irreversible processes. For that reason, exergy analysis is substantially important for the performance of a vapor absorption refrigeration system [19, 21].

Exergy is defined as the maximum work potential of a matter or a form of energy with respect to its reference environment [13].

The mass balance equations for the system components are as follows:

$$\sum \dot{m}_i = \sum \dot{m}_e \tag{2}$$

$$\sum \dot{m}_i X_i = \sum \dot{m}_e X_e \tag{3}$$

The flow ratio, f, is a parameter that helps to calculate the heat capacities of the components of VAR system. The LiBr balance of the generator is used to obtain the flow ratio, which is defined as the ratio of the strong solution of mass flow rate, $\dot{m}_s$, to the refrigerant mass flow rate, $\dot{m}_r$.

$$f = \frac{\dot{m}_s}{\dot{m}_r} = \frac{X_7}{X_8 - X_7} \tag{4}$$

General energy and exergy balance equations are given as follows [23]:

$$E_i = E_o \tag{5}$$

$$\dot{E}x_i - \dot{E}x_o = \dot{E}x_{dest} \tag{6}$$

$$\dot{E}x = \dot{m}(ex) \tag{7}$$

$$ex = h - h_0 - T_0(s - s_0) \tag{8}$$

Heat capacity, $\dot{Q}$, and exergy destructions, $\dot{E}x_{dest}$, obtained from energy and exergy balances of each of the components of the VAR system illustrated in Figure 1 can be expressed as follows [14, 15]:

$$q_{CO} = \frac{\dot{Q}_{CO}}{\dot{m}_1} = h_2 - h_1 \tag{9}$$

$$\dot{E}x_{dest,CO} = \dot{m}_1(ex_1 - ex_2) + \dot{m}_{13}(ex_{13} - ex_{14}) \tag{10}$$

$$q_{EV} = \frac{\dot{Q}_{EV}}{\dot{m}_1} = h_4 - h_3 \tag{11}$$

$$\dot{E}x_{dest,EV} = \dot{m}_3(ex_3 - ex_4) + \dot{m}_{15}(ex_{15} - ex_{16}) \tag{12}$$

$$q_{GE} = \frac{\dot{Q}_{GE}}{\dot{m}_1} = h_1 + fh_8 - (f + 1)h_7 \tag{13}$$

$$\dot{E}x_{dest,GE} = \dot{m}_{11}(ex_{11} - ex_{12}) + \dot{m}_7 ex_7 - \dot{m}_8 ex_8 - \dot{m}_1 ex_1 \tag{14}$$

$$q_{AB} = \frac{\dot{Q}_{AB}}{\dot{m}_1} = (f + 1)h_5 - h_4 - fh_{10} \tag{15}$$

$$\dot{E}x_{dest,AB} = \dot{m}_{17}(ex_{17} - ex_{18}) + \dot{m}_4 ex_4 + \dot{m}_{10} ex_{10} - \dot{m}_5 ex_5 \tag{16}$$

$$q_{she} = \frac{\dot{Q}_{she}}{\dot{m}_1} = f(h_8 - h_9) = (f + 1)(h_7 - h_6) \tag{17}$$

$$\dot{E}x_{dest,SHE} = \dot{m}_8(ex_8 - ex_9) + \dot{m}_5(ex_6 - ex_7) \tag{18}$$

$$\dot{E}x_{dest,rtv} = \dot{m}_1 T_0(s_3 - s_2) \tag{19}$$

$$\dot{W}_p = \dot{m}_5(h_6 - h_5) = \frac{\dot{m}_5\vartheta_s(P_{CO}-P_{EV})}{n_p} \tag{20}$$

$$\dot{Ex}_{dest,p} = \dot{m}_5(ex_5 - ex_6) + \dot{W}_p \tag{21}$$

$$\dot{Ex}_{dest,tot} = \sum_{j=1}^{N} Ex_{dest,j} \tag{22}$$

$$COP_{VAR} = \frac{\dot{Q}_{EV}}{\dot{Q}_{GE}+W_p} \tag{23}$$

RESULTS AND DISCUSSION

To validate the present model, the simulation results were compared with the previous works available in the literature. The comparative heat capacities, $\dot{Q}$, coefficients of performance, COP, and exergy destructions, $\dot{Ex}_{dest}$, of the VAR system are presented in Table 2. For validations, the following input parameters were used: T_{GE} = 87.8 °C, T_{EV} = 7.2 °C, T_{CO} = T_{AB} = 37.8 °C, the effectiveness of solution heat exchanger, ϵ = 0.7, and refrigerant mass flow rate, $\dot{m}_r$ = 1 kg/s. As seen, the results obtained from the present simulation are in good agreement with the literature.

Table 2. Comparison of the analytical results with data given in the literature

	Anand and Kumar [28]	Arora and Kaushik [29]	Present study
$\dot{Q}_{GE}$ (kW)	3073.11	3095.698	3078.547
$\dot{Q}_{AB}$ (kW)	2922.39	2945.269	2932.376
$\dot{Q}_{CO}$ (kW)	2507.89	2505.91	2505.910
$\dot{Q}_{EV}$ (kW)	2357.17	2355.45	2355.450
$\dot{Q}_{SHE}$ (kW)	523.25	518.717	529.465
W_P (kW)	-	0.03143	0.03431
COP	0.76703	0.7609	0.7651
$Ex_{dest,GE}$ (kW)	-	55.568	54.639
$Ex_{dest,AB}$ (kW)	-	70.478	67.420
$Ex_{dest,CO}$ (kW)	-	6.606	6.606
$Ex_{dest,EV}$ (kW)	-	86.275	87.524
$Ex_{dest,SHE}$ (kW)	-	25.081	26.608
$Ex_{dest,RTV}$ (kW)	-	6.936	6.936
$Ex_{dest,TOT}$ (kW)	-	250.967	249.733
Ex_{INPUT} (kW)		538.637	535.622
Ex_{OUTPUT} (kW)		63.277	62.028
η		0.1175	0.1158

The heat capacity of evaporator, $\dot{Q}_{EV}$, is equal to the hourly comfort cooling load of the subject bus model. As seen in Figure 3(a), the minimum heat capacity of the evaporator, $\dot{Q}_{EV}$, is 13.17 kW at 6 a.m. in September, and the maximum heat capacity of the evaporator, $\dot{Q}_{EV}$, is 25.96 kW at 5 p.m. in July. As expected, the comfort cooling load at night hours is lower than in day hours. The trend of hourly exergy destruction, $\dot{Ex}_{dest,EV}$, values shown in Figure 3(b) is similar to that of the heat capacity of the evaporator. The highest exergy destruction, $\dot{Ex}_{dest,EV}$, occurring in the evaporator is 0.9631 kW at 5 p.m. in July, and the lowest exergy destruction, $\dot{Ex}_{dest,EV}$, occurring in the evaporator is 0.4885 kW at 6 a.m. in September.

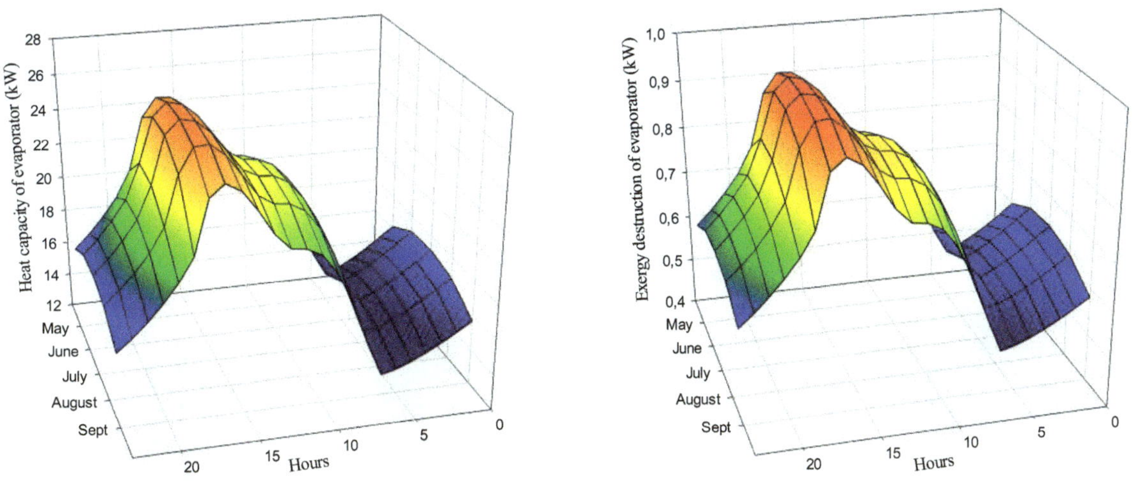

(a) (b)

Figure 3. (a) Hourly variation of heat capacity for the evaporator, (b) hourly variation of exergy destruction for the evaporator

The COP variation of the VAR system is shown in Figure 4(a). The maximum COP value of the VAR system is 0.778 at 5 a.m. in May, and the corresponding outdoor temperature, T_o, is 23.4 °C. The minimum COP value of the VAR system is 0.6628 at 3 p.m. in July and August, and the corresponding outdoor temperature is 38 °C. Figure 4(b) shows the hourly variation of total exergy destruction, $\dot{Ex}_{dest,TOT}$, for the VAR system. The lowest total exergy destruction, $\dot{Ex}_{dest,TOT}$ value is 8.69 kW at 6 a.m. in September and the corresponding outdoor temperature, T_o, and relative humidity, RH are 24.7 °C, 73%, respectively. The highest total exergy destruction, $\dot{Ex}_{dest,TOT}$ value is 15.25 kW at 4 p.m. in July and the corresponding outdoor temperature, T_o, and the relative humidity, RH, are 37.6 °C, 48%, respectively. The results indicate that the COP of the system decays in day hours and rises at night. On the other hand, the total exergy destruction increases in day hours and decays at night.

To present the benefits of a VAR system driven by exhaust gas waste heat received from the engine, the results for using a conventional VCR system driven by the engine by a belt drive should be studied for the purpose of comparison. The COP variation of the VCR system and the compressor (which is the main energy consumer in the basic VCR system) load variations on an hourly basis during the cooling season are presented in Figures 4(c) and 4(d). The temperatures of the evaporator and condenser for the VCR and VAR systems are assumed to be equal. The VCR system is assumed for the calculations to use R134a refrigerant as a working fluid. The calculated results reveal that the maximum coefficient of performance, COP, is equal to 3.9 and that the minimum COP is equal to 2.67, based on the outdoor temperatures. A similar trend is obtained for the VAR system. It is worth mentioning that the compressor of the VCR system consumes considerable energy, as seen in Figure 4(d). Specifically, this power input varies between 3.84 kW and 9.57 kW. When the compressor load is at maximum, it consumes approximately 9% of the power received from the engine when the engine works 50% load. Figure 4(e) shows the hourly variation of fuel consumption of the VCR system. First, the specific fuel consumption by the engine is calculated as follows [30]:

$$BSFC = \frac{1}{n * 0.0119531} \tag{24}$$

At equation (24), lower heating value of diesel fuel is taken as 0.0119531 kWh/g [30] and engine efficiency is taken from Table 1.

Then the fuel consumption is calculated by using brake-specific fuel consumption (BSFC) and the energy required by the compressor [31]. As can be seen from the figure, the fuel consumption of the compressor varies between 698.56 g/h and 1919.31 g/h. If the air conditioning system works through the cooling season, the compressor consumes fuel at the rate of 4,489 kg/year. This value is crucial because it releases emissions after the combustion process and because the consumption of fossil fuel causes vitally important environmental problems. The hourly variation of CO_2 emission caused by the VCR system is shown In Figure 4(g). It varies between 2,242.08 g/h and 6,160.16 g/h during the cooling season. If the air conditioning system works through the entire cooling season, the total of released CO_2 emission is calculated as 14,407.05 kg/year.

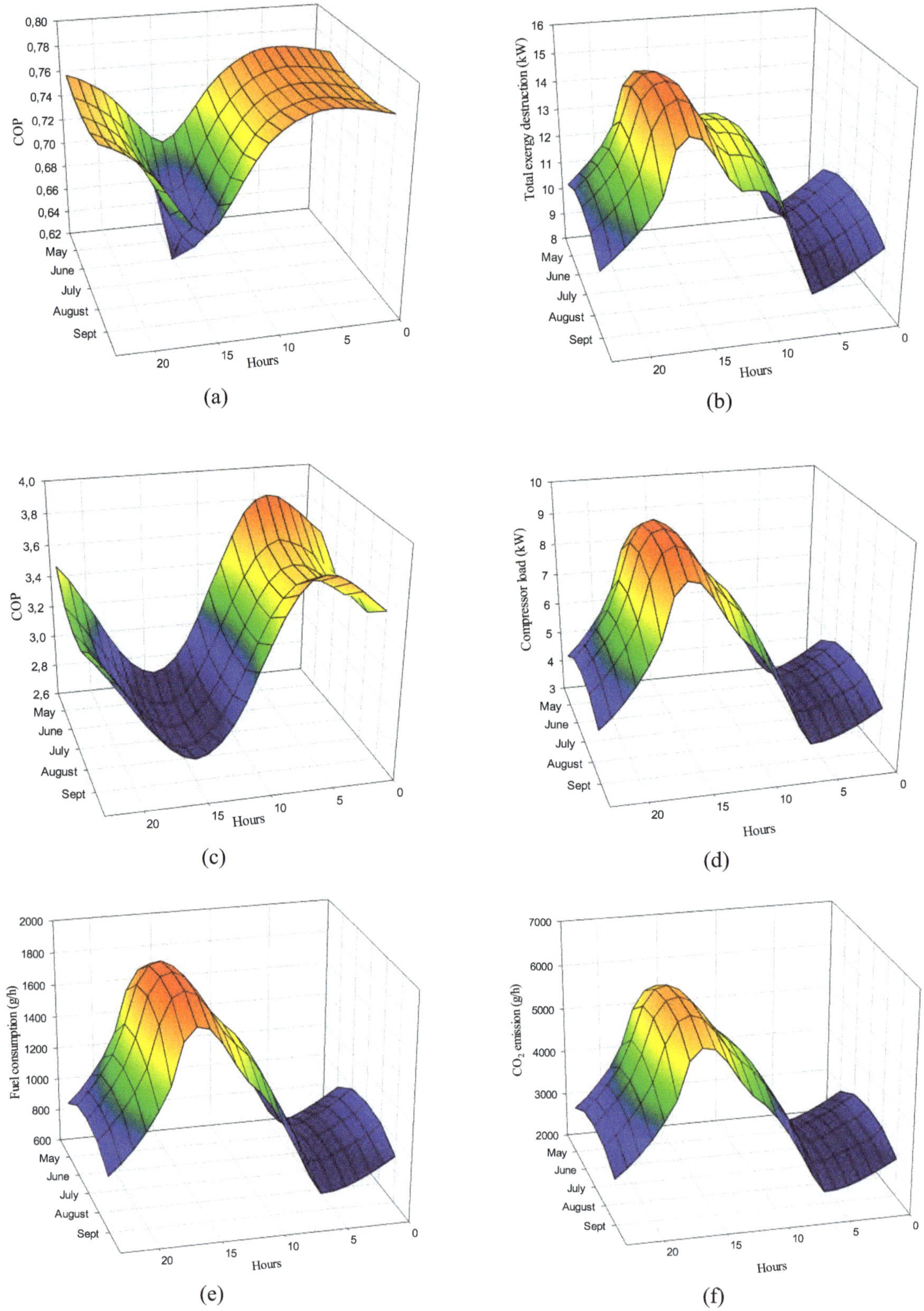

Figure 4. (a) Hourly variation of COP of the VAR system, (b) hourly variation of total exergy destruction of the VAR system, (c) hourly variation of COP of the VCR system, (d) hourly variation of compressor load of the VCR system, (e) hourly variation of fuel consumption of the VCR system, (f) hourly variation of CO_2 emission of the VCR system

The fuel consumption values and CO_2 emission quantities are calculated for the engine under 50% load. The weight issue should also be considered when the use of a VAR system driven by exhaust gas is investigated for an intercity bus. So, the VAR system components are designed and constructed by using heat and mass transfer calculations. The heat transfer capacity of the main components of the VAR system is calculated as follows;

$$\dot{Q} = U \cdot A \cdot \Delta T_m \tag{25}$$

The overall heat transfer coefficient, U can be determined as shown below;

$$\frac{1}{U \cdot A} = \sum R_{conv} + \sum R_f + \sum R_{cond} \tag{26}$$

where the thermal resistance is represented by R.

$$\Delta T_m = \frac{\Delta T_1 - \Delta T_2}{ln(\frac{\Delta T_1}{\Delta T_2})} \tag{27}$$

Here, ΔT_1 and ΔT_2 are presented as $\Delta T_1 = T_{ho} - T_{ci}$ and $\Delta T_2 = T_{hi} - T_{co}$. All of the components, except solution heat exchanger, are designed to be circular finned tube heat exchanger. Circular fins are suited at the outside of the tubes. While air and exhaust gas passes the outside of the finned tubes, the working fluid passes inside of the smooth tubes. For the calculation of the outside heat transfer coefficient, h_o of heat exchangers, the Young and Brigs [32] equation is used which is stated below:

$$Nu = 0.134Re^{0.681}Pr^{\frac{1}{3}}(\frac{l}{b})^{0.2}(\frac{l}{t})^{0.1134} \tag{28}$$

The inner side heat transfer coefficient of heat exchangers such as condensing, evaporating, film convection and mass transfer coefficient is calculated by guidelines of Florides [28] and Garousi Farshi et al. [33]. The solution heat exchanger is designed as a double pipe heat exchanger and the equation stated below was used [33]:

$$Nu = 0.023Re^{0.8}Pr^{n} \tag{29}$$

Assumptions and calculated results to be applied for the present system are summarized in Table 3. When taking the assumptions at Table 4, the point that highest exergy destruction occurs is considered, see table 5. As seen in Table 4, the total mass of the components of VAR system is 349.14kg. Currently, the Temsa TS45 model intercity bus uses ES348 model VCR system belonging SAFKAR Company. The unit provides 32 kW cooling power capacity, and its mass with the piping system and compressor is approximately 200 kg. It is estimated that a 10% increase in the mass of the vehicle decreases the fuel economy by about 7% [34]. It can be seen that, considering TS45 intercity bus's gross vehicle mass rating of 23,360 kg [35]. And the calculated maximum capacity of the evaporator shown in Figure 3a, the mass increment of the present VAR system driven by exhaust gas, compared to that of the conventional VCR system for air conditioning purposes, may be ignored in terms of fuel consumption.

After examination of the parameters related to the conventional VCR system, computation of those parameters that affect the VAR system performance and exergy destruction at specified hours throughout the cooling season becomes significant. For example, the highest total exergy destruction for the present system occurs at 4 p.m. in July when the temperature and the relative humidity are 37.6 °C and 48%, respectively. Figures 5-9 are prepared using the parameters given in Table 3, and the effectiveness of the solution heat exchanger, ϵ, is taken as 0.7 in the analysis.

Table 3. Assumptions and calculated values of the VAR system

States	Substance	T (°C)	P (kPa)	X (%)	m (kg/s)	h (kj/kg)	s (kj/kgK)
1	Superheated vapor	110	13.36	-	0.0112	2705.65	8.363
2	Saturated water	51.6	13.36	-	0.0112	216.02	0.7244
3	Water-vapor	10	1.23	-	0.0112	216.02	0.7656
4	Saturated vapor	10	1.23	-	0.0112	2518.89	8.899
5	Water-LiBr	51.6	1.23	61.2	0.1914	144.31	0.2794
6	Water-LiBr	51.6	13.36	61.2	0.1914	144.31	0.2794
7	Water-LiBr	88.14	13.36	61.2	0.1914	213.57	0.4798
8	Water-LiBr	110	13.36	65	0.1802	271.15	0.556
9	Water-LiBr	68.93	13.36	65	0.1802	197.59	0.3554
10	Water-LiBr	68.93	1.23	65	0.1802	197.59	0.3554
11	Exhaust gas	420	101.3	-	0.1697	-1323.93	7.6425
12	Exhaust gas	217.5	101.3		0.1697	-1549.52	7.2582
13	Outdoor air	37.6	101.3	-	3.823	88.54	5.9165
14	Outdoor air	44.6	101.3		3.823	95.84	5.9398
15	Indoor air	24	101.3		3.6	47.82	5.779
16	Indoor air	17	101.3		3.6	40.66	5.7546
17	Outdoor air	37.6	101.3		4.963	88.54	5.9165
18	Outdoor air	44.6	101.3		4.963	95.84	5.9398

Table4. The parameters of components of VAR system

	Absorber	Generator	Condenser	Evaporator	Sol. heat ex.
$\dot{Q}$ (kW)	35.56	37.6	28	25.95	11.19
d_i (mm)	23.4	23.4	23.4	23.4	Inner tube: 18 Outer tube:23.6
d_o (mm)	25.4	25.4	25.4	25.4	Inner tube: 21.3 Outer tube:26.9
b (mm)	10	10	10	10	-
l (mm)	2.75	2.75	2.75	2.75	-
τ (mm)	0.5	0.5	0.5	0.5	-
Fin area(m^2)	87.09	2.05	39.45	55.96	-
Tube array	3X28	2X4	2X24	3X23	-
L (mm)	1515	374	1201	1185	Total length: 21700
P_t (mm)	44	40	44	40	-
U ($W/m^{2o}C$)	36.21	84.76	62.97	41.11	380.85
ΔT_m	10.1	194.05	10.1	10.1	20.25
Components mass (kg)	145.79	3.38	66.04	93.76	40.17

Table 5 shows that the generator has the highest heat capacity, $\dot{Q}$, and exergy destruction, $\dot{E}x_{dest}$, values. The smallest amount of heat capacity and exergy destruction occurs in the solution pump.

Table 5. Heat capacity and exergy destruction of each unit of the VAR system, COP = 0.67

Parameter	Value (kW)
Evaporator heat capacity	25.79
Condenser heat capacity	27.89
Generator heat capacity	38.29
Absorber heat capacity	36.20
Solution heat exchanger	13.25
Pump power	0.0015
Exergy destruction of evaporator	0.957
Exergy destruction of condenser	0.996
Exergy destruction of generator	10.97
Exergy destruction of absorber	1.54
Exergy destruction of sol. heat ex.	0.699
Exergy destruction of ref. tho. valve	0.1376
Exergy destruction of pump	0.00136
Total exergy destruction	15.25

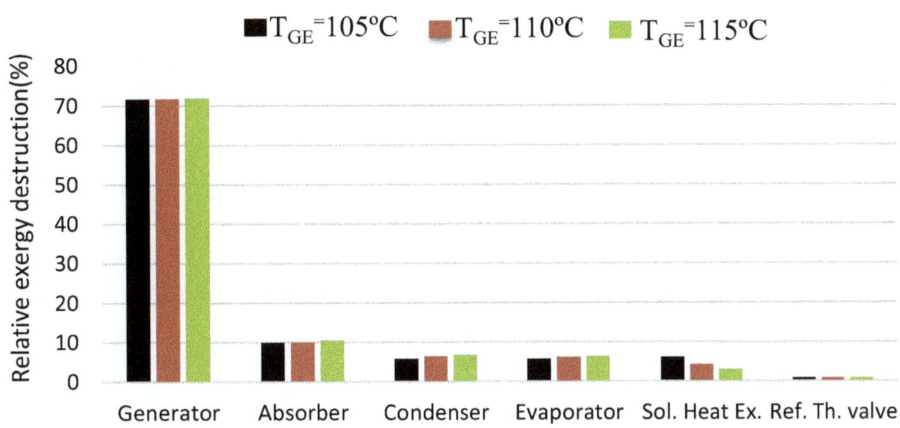

Figure 5. Relative exergy destruction of the VAR components (%)

Figure 5 presents the rate of relative exergy destruction of the VAR system components. To show the effects of generator temperature, T_{GE}, on the relative exergy destruction of the VAR system components, three different generator temperatures are examined. Because of the large heat capacity and temperature differences between streams, the highest relative exergy destruction occurs through the generator. Increasing the generator temperature upgrades the relative exergy destruction of the generator, absorber, condenser, evaporator, and refrigerant throttle valves because it decreases the total exergy destruction of the VAR system. The highest rate of relative exergy destruction of the generator is 72.04% at T_{GE} = 115 °C, and the lowest rate of relative exergy destruction of the generator is 71.67% at T_{GE} = 105 °C. The absorber has the second highest rate of relative exergy destruction, which is 10.6% at T_{GE} = 115 °C, but this rate decreases to a value of 9.93% at T_{GE} = 105 °C. The condenser and evaporator have approximately equal exergy destruction, which corresponds to 6.5% at T_{GE} = 115 °C. The smallest rate of relative exergy destruction occurs at the refrigerant throttle valve, and that is 0.93% at T_{GE} = 115 °C. Here the exergy destruction of the solution pump is ignored.

Figure 6 shows the change of total exergy destruction, $\dot{Ex}_{dest,TOT}$, and exergy destruction of generator, $\dot{Ex}_{dest,GE}$, with generator temperature, T_{GE}. The variation of parameters is investigated for three different (source) temperatures of waste heat of exhaust gas, T_S, including 410 °C, 420 °C, and 430 °C. The exergy destruction of the generator, $\dot{E}x_{dest,GE}$, tends to decrease with increasing generator temperature, T_{GE}, and decreasing source temperature, T_S. When the generator temperature increases, although the temperature of the weak solution at the generator inlet increases, the exergy of it decreases because of decreasing flow ratio, f. Since the change rate of exergy at the generator inlet is higher than that of the other streams, the exergy destruction of the generator, $\dot{E}x_{dest,GE}$, decreases. Also, an increase in the exhaust gas outlet temperature, $T_{o,exgas}$, makes a contribution to decrease the exergy destruction value of the generator. While the maximum exergy destruction of generator, $\dot{E}x_{dest,GE}$ is 13 kW at T_S= 430 °C and T_{GE}=104 °C, which corresponds to the rate of 21% reduction which represents a minimum value that is 10.18 kW at T_S = 410 °C and T_{GE} = 118 °C. On the other hand, the total exergy destruction, $\dot{E}x_{dest,TOT}$, of the VAR system has a maximum value of 18.08 kW, which occurs at T_S = 430 °C and T_{GE} = 104 °C. However, the total exergy destruction, $\dot{E}x_{dest,TOT}$, decreases with increasing generator temperature, T_{GE}, and with decreasing source temperature, T_S. Finally, it can be stated that the minimum total exergy destruction, $\dot{E}x_{dest,TOT}$, occurs with a minimum value of 14.3 kW at T_S = 410 °C and T_{GE} = 118 °C.

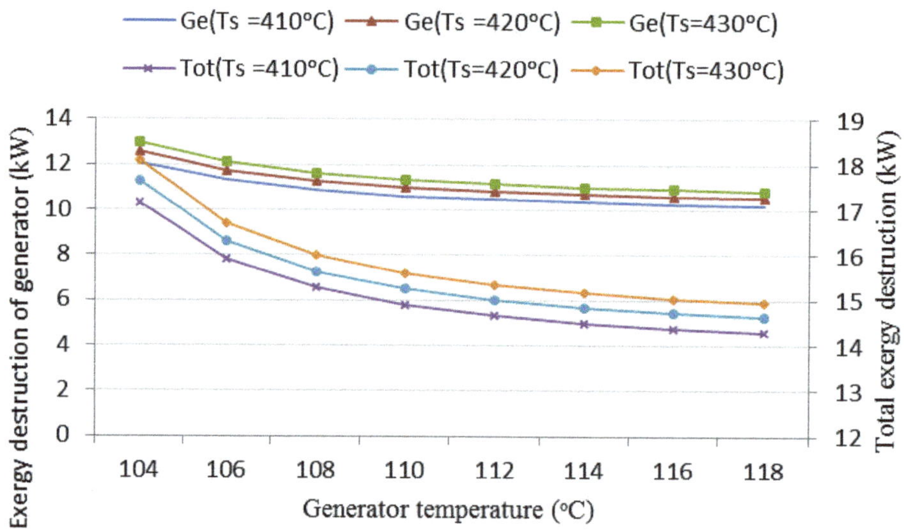

Figure 6. Variation of total exergy destruction and exergy destruction of the generator with increasing generator temperature

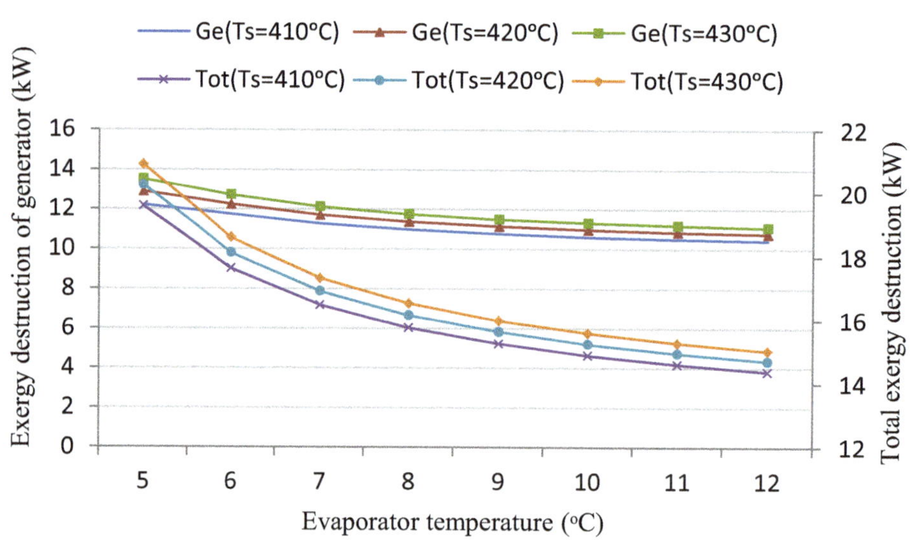

Figure 7. Variations of total exergy destruction of the VAR system and exergy destruction of the generator based on variation of evaporator temperatures

The effect of evaporator temperature, T_{EV}, on the variation of total exergy destruction, $\dot{E}x_{dest,TOT}$, of the system and exergy destruction of the generator, $\dot{E}x_{dest,GE}$, is presented in Figure 7. As illustrated in the figure, the exergy destruction of the generator, $\dot{E}x_{dest,GE}$, declines about 18% with increasing the evaporator temperature, T_{EV}, but with decreasing the source temperature, T_S. Upgrading the evaporator temperature provides a lower generator heat load, and exhaust gas outlet temperature rises. That increases the exergy value of exhaust gas at the generator outlet, leading the exergy destruction of generator, $\dot{E}x_{dest,GE}$, to decrease. The total exergy destruction, $\dot{E}x_{dest,TOT}$, of the VAR system decays about 31% when the evaporator temperature, T_{EV}, is increased but the source temperature, T_S, is decreased.

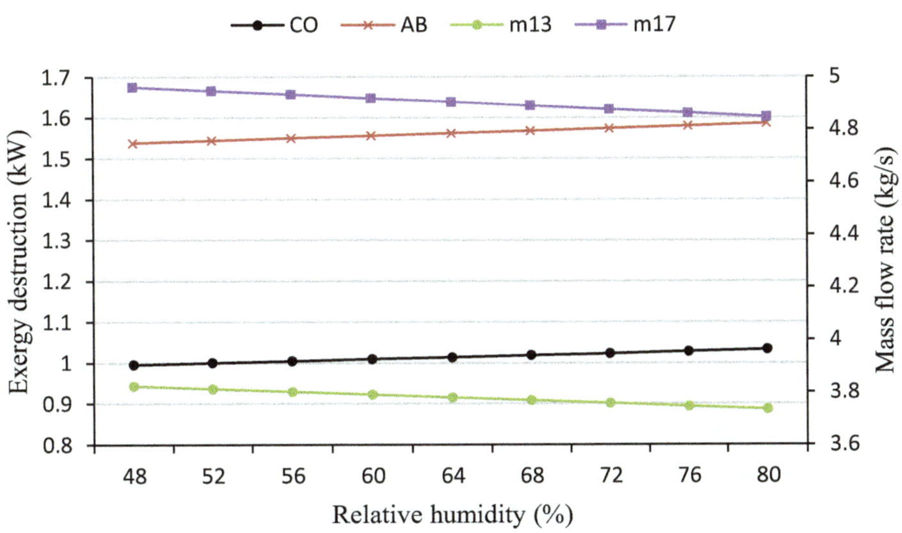

Figure 8. The effect of relative humidity on the absorber and condenser parameters

Figure 8 demonstrates changes of exergy destruction of the absorber and condenser and required mass flow rate to release heat to the ambient with changes in relative humidity, RH. Because the absorber and condenser are cooled by ambient air, changes of relative humidity, RH, of the ambient air influences only these components of the VAR system. In summary, variations of exergy destruction and required mass flow rate of condenser, $\dot{m}_{13}$, and absorber, $\dot{m}_{17}$, against relative humidity, RH, are not rapid.

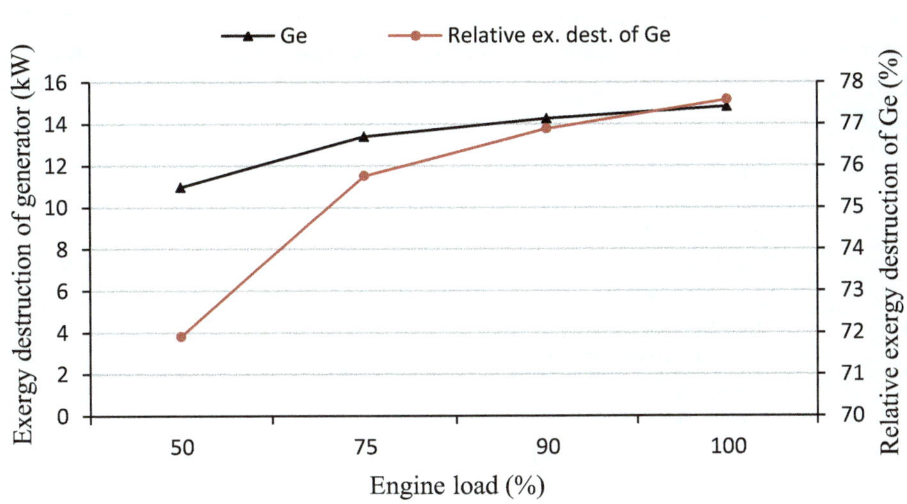

Figure 9. The effect of engine load variation on the parameters of the generator

The variations of the exergy destruction of the generator, $\dot{Ex}_{dest,GE}$, and the relative exergy destruction of the generator, which are valuable parameters, are shown in Figure 9 on the basis of the data given in Tables 1,4 and 5. As seen in the figure, exergy destruction of the generator, $\dot{Ex}_{dest,GE}$, as well as the relative exergy destruction of the generator, increases with increasing engine load. The lowest exergy destruction of the generator, $\dot{Ex}_{dest,GE}$, is 10.97 kW at an engine load rate of 50%. On the other hand, it is clearly seen that the maximum exergy destruction occurs with a value of 14.82 kW when the engine is run at a 100% load rate. The slope of the relative exergy destruction curve of the generator decreases sharply when the engine is run at a 75% load rate. In conclusion, the lowest relative exergy destruction rate of the generator is 71.91% at the engine load rate of 50%, but it increases to a maximum value of 77.57% at the engine load of 100%.

CONCLUSION

The Vapor Absorption Refrigeration system driven by waste heat of the exhaust gas received from the internal combustion engine is investigated for cabin air-conditioning purposes on an intercity bus in cooling season. To perform the energy and exergy analyses, a single-effect VAR system using water-LiBr solution is selected as working solution pair. Effects of environmental conditions and operating parameters on the system performance, exergy destruction, and heat capacities of various components of the VAR system are presented and discussed. The following conclusions can be drawn from the present research:

- During cooling season, the lowest heat capacity of the evaporator, which is 13.17 kW, is obtained at 6 a.m. in September, and the highest heat capacity of the evaporator, 25.96 kW, is obtained at 5 p.m. in July.
- The maximum COP of the VAR system is calculated as 0.778 at 5 a.m. in May, and the minimum COP is calculated as 0.6628 at 3 p.m. in July and August.
- Air-conditioning an intercity bus in cooling season by a VAR system driven by exhaust gas waste heat is possible, even at an engine load rate of 50%.
- Approximately 4,489 kg/year of fuel can be saved by using the VAR system driven by an exhaust gas waste heat in an intercity bus during the cooling season. Thus, it prevents 14,407 kg/year of CO_2 emissions released to the environment.
- The minimum total exergy destruction is obtained as 8.69 kW at 6 a.m. in September; conversely, the maximum total exergy destruction is determined to be 15.25 kW at 4 p.m. in July.
- The relative exergy destruction of each components of the VAR system, except for the solution heat exchanger, tends to rise with rising generator temperature. The generator has the highest relative exergy destruction, which correspond to the rate of 71.93% at T_{GE} = 110 °C.
- Finally, it is found that the total exergy destruction may be reduced by about 30% by increasing generator and evaporator temperatures and decreasing exhaust gas inlet temperature.

NOMENCLATURE:

A	Area (m^2)
b	fin height
BSFC	Brake specific fuel consumption (g/kWh)
COP	Coefficient of Performance
c_p	Specific heat (J/kgK)
d	tube diameter (m)
d_h	hydraulic diameter (m)
$\dot{E}$	Energy (kW)
$\dot{Ex}$	Exergy (kW)
ex	Specific exergy (kj/kg)
f	flow ratio
h	enthalpy (kj/kg), convection coefficient (W/m^2K)
k	conduction coefficient (W/m^2K)
l	fin spacing
L	length of one tube

Libr	lithium bromide
$\dot{m}$	Mass flow rate (kg/s)
n	engine efficiency
N	total number of components in system
Nu	Nusselt number, ($h*d_h/k$)
P	Pressure (kPa), pitch
Pr	Prandlt number, ($\mu*c_p/k$)
RH	relative humidity (%)
Re	Reynolds number ($u*d_h/v$)
$\dot{Q}$	heat capacity (kW)
q	specific heat capacity (kj/kg)
s	entropy (kj/kg/K)
Sept	September
T	Temperature
ΔT	Logarithmic temperature difference (°C)
U	Overall heat transfer coefficient (W/m^2 °C), velocity (m/s)
v	kinematic viscosity (m^2/s)
VAR	Vapor Absorption Refrigeration
VCR	Vapor Compression Refrigeration
W	Power (kW)
X	mass fraction
A	Ambient
AB	Absorber
c	cold fluid
CO	Condenser
cond	conduction
conv	convection
dest	destruction
e	exit
f	fouling
EV	Evaporator
exgas	Exhaust gas
GE	Generator
h	hot fluid
i	In, inlet, inner
j	number of components
m	mean
o	Out, outlet, outer
p	Pump
r	refrigerant
rtv	Refrigerant throttle valve
s	solution, source
she	Solution heat exchanger
t	tube
Tot	total
1.2…	state points
0	reference state
ϵ	effectiveness
η	exergy efficiency; efficiency
μ	dynamic viscosity (kg/ms)
ϑ	specific volume (m^3/kg)
τ	fin thickness

Acknowledgements
The authors would like to thank to TEMSA Global for its contribution.

REFERENCES

[1] Hwang, Y. (2004). Potential energy benefits of integrated refrigeration system with microturbine and absorption chiller. International Journal of Refrigeration, 27(8), 816-829.
[2] www.johnsoncontrols.com. Application opportunities for absorption chillers. (accessed 17 January 2017)

[3] http://thermaxglobal.com. (accessed 17 January 2017)
[4] Horuz, I. (1998). A comparison between ammonia-water and water-lithium bromide solutions in vapor absorption refrigeration systems. International communications in heat and mass transfer, 25(5), 711-721.
[5] Little, A. B., & Garimella, S. (2011). Comparative assessment of alternative cycles for waste heat recovery and upgrade. Energy, 36(7), 4492-4504.
[6] Saidur, R., Rezaei, M., Muzammil, W. K., Hassan, M. H., Paria, S., & Hasanuzzaman, M. (2012). Technologies to recover exhaust heat from internal combustion engines. Renewable and sustainable energy reviews, 16(8), 5649-5659.
[7] Boatto, P., Boccaletti, C., Cerri, G., & Malvicino, C. (2000). Internal combustion engine waste heat potential for an automotive absorption system of air conditioning part 1: tests on the exhaust system of a spark-ignition engine. Proceedings of the Institution of Mechanical Engineers, Part D: Journal of Automobile Engineering, 214(8), 979-982.
[8] Manzela, A. A., Hanriot, S. M., Cabezas-Gómez, L., & Sodré, J. R. (2010). Using engine exhaust gas as energy source for an absorption refrigeration system. Applied energy, 87(4), 1141-1148.
[9] Koehler, J., Tegethoff, W. J., Westphalen, D., & Sonnekalb, M. (1997). Absorption refrigeration system for mobile applications utilizing exhaust gases. Heat and Mass Transfer, 32(5), 333-340.
[10] Horuz, I. (1999). Vapor absorption refrigeration in road transport vehicles. Journal of Energy Engineering, 125(2), 48-58.
[11] Boatto, P., Boccaletti, C., Cerri, G., & Malvicino, C. (2000). Internal combustion engine waste heat potential for an automotive absorption system of air conditioning part 1: tests on the exhaust system of a spark-ignition engine. Proceedings of the Institution of Mechanical Engineers, Part D: Journal of Automobile Engineering, 214(8), 979-982.
[12] Lambert, M. A., & Jones, B. J. (2006). Automotive adsorption air conditioner powered by exhaust heat. Part 1: conceptual and embodiment design. Proceedings of the Institution of Mechanical Engineers, Part D: Journal of Automobile Engineering, 220(7), 959-972.
[13] Kilic, M., & Kaynakli, O. (2007). Second law-based thermodynamic analysis of water-lithium bromide absorption refrigeration system. Energy, 32(8), 1505-1512.
[14] Şencan, A., Yakut, K. A., & Kalogirou, S. A. (2005). Exergy analysis of lithium bromide/water absorption systems. Renewable energy, 30(5), 645-657.
[15] Talbi, M. M., & Agnew, B. (2000). Exergy analysis: an absorption refrigerator using lithium bromide and water as the working fluids. Applied Thermal Engineering, 20(7), 619-630.
[16] Gomri, R. (2009). Second law comparison of single effect and double effect vapour absorption refrigeration systems. Energy Conversion and Management, 50(5), 1279-1287.
[17] Kaynakli, O., Saka, K., & Kaynakli, F. (2015). Energy and exergy analysis of a double effect absorption refrigeration system based on different heat sources. Energy Conversion and Management, 106, 21-30.
[18] Arora, A., Dixit, M., & Kaushik, S. C. (2016). Energy and exergy analysis of a double effect parallel flow LiBr/H2O absorption refrigeration system. Journal of Thermal Engineering, 2(1), 541-549.
[19] Onan, C., Ozkan, D. B., & Erdem, S. (2010). Exergy analysis of a solar assisted absorption cooling system on an hourly basis in villa applications. Energy, 35(12), 5277-5285.
[20] Yılmaz, A. (2015). Transcritical organic Rankine vapor compression refrigeration system for intercity bus air-conditioning using engine exhaust heat. Energy, 82, 1047-1056.
[21] Shu, G., Liu, L., Tian, H., Wei, H., & Xu, X. (2013). Performance comparison and working fluid analysis of subcritical and transcritical dual-loop organic Rankine cycle (DORC) used in engine waste heat recovery. Energy Conversion and Management, 74, 35-43.
[22] Yamankaradeniz, R., Horuz, I., Kaynakli, O., Coskun, S., & Yamankaradeniz, N. (2009). Refrigeration techniques and heat pump applications.
[23] Dincer, I., & Rosen, M. A. (2012). Exergy: energy, environment and sustainable development. Newnes.
[24] Klein, S. A. (2003). EES-engineering equation solver, professional version. F-chart software, Middleton, WI.
[25] Florides, G. A., Kalogirou, S. A., Tassou, S. A., & Wrobel, L. C. (2003). Design and construction of a LiBr–water absorption machine. Energy conversion and management, 44(15), 2483-2508.
[26] Chua, H. T., Toh, H. K., Malek, A., Ng, K. C., & Srinivasan, K. (2000). Improved thermodynamic property fields of LiBr–H2O solution. International Journal of Refrigeration, 23(6), 412-429.
[27] Y.A. Çengel, M. Boles, Thermodynamics: An Engineering Approach. New York: McGraw-Hill Series, 2011.
[28] Anand, D. K., & Kumar, B. (1987). Absorption machine irreversibility using new entropy calculations. Solar Energy, 39(3), 243-256.
[29] S.C. Kaushik, A. Arora, "Energy and exergy analysis of single effect and series flow double effect water–lithium bromide absorption refrigeration systems", International Journal of Refrigeration, 2009.

[30] Wikipedia, The free encyclopedia, https://en.wikipedia.org/wiki/Brake_specific_fuel_consumption. (accessed 17 January 2017)
[31] Hegar, M., Kolda, M., Kopecka, M., Rajtmajer, V., & Ryska, A. (2013). Bus HVAC energy consumption test method based on HVAC unit behavior. International Journal of Refrigeration, 36(4), 1254-1262.
[32] Robinson, K. K., & Briggs, D. E. (1966). Pressure drop of air flowing across triangular pitch banks of finned tubes. In Chem. Eng. Prog. Symp. Ser (Vol. 62, No. 64, pp. 177-184).
[33] Farshi, L. G., Mahmoudi, S. S., Rosen, M. A., Yari, M., & Amidpour, M. (2013). Exergoeconomic analysis of double effect absorption refrigeration systems. Energy Conversion and Management, 65, 13-25.
[34] Ghassemieh, E. (2011). Materials in automotive application, state of the art and prospects. In New trends and developments in automotive industry. InTechOpen..

EXERGY ANALYSIS OF THE ORGANIC RANKINE CYCLE BASED ON THE PINCH POINT TEMPERATURE DIFFERENCE

A.H. Bademlioglu[1,*], R. Yamankaradeniz[2], O. Kaynakli[2]

ABSTRACT

Organic Rankine Cycle (ORC) is a system that uses working fluids with hydrocarbon components instead of water and generates power from the heat recovery of different heat sources. In this study, the exergy analysis of a simple ORC, which produces electrical energy with the help a geothermal source (125°C), was performed. R123, R152a, R245fa and R600a were determined as the fluids to be used in the Cycle. In this analysis, which was carried out according to the pinch point temperature differences (5-20°C) in the evaporator, the exergy performance of the cycle components was evaluated for the geothermal resource unit flow rate and the variation of the exergy efficiency of the system was calculated. With the increase of the pinch point temperature difference in the evaporator, the decrease of the system's exergy efficiency became maximal (11.7%) with the use of R152a as a refrigerant and the loss in the system's exergy efficiency became minimal (9.03%) with the use of R123 as a refrigerant.

Keywords: *Organic Rankine Cycle, Exergy Analysis, Pinch Point Temperature, Refrigerant, Geothermal Source*

INTRODUCTION

The Rankine cycle is a thermodynamic cycle that converts heat energy into work and water is used as the working fluid in this cycle. The water used in the cycle is preferred to generate electricity in medium and large power plants. In the recent years, hydrocarbon-based fluids with lower critical temperature and pressure, higher molecular mass and less corrosion hazard have started to be used in the Rankine cycle instead of water. With the use of these fluids, these systems have taken the name of the Organic Rankine cycle (ORC). Being used in applications such as recycling of waste heat in factories, solar energy and especially in geothermal applications, the importance of this cycle is increasing day by day in terms of energy efficiency.

Thermodynamic properties of certain fluids used in the Rankine cycle are presented in Table 1. Many studies have shown that; among the hydrocarbon compound organic fluids used in the ORC systems, the ones that have higher molecular mass, low critical temperature and pressure and that are dry and isentropic in the meantime are more appropriate [1-5]. Wang et al. [6] investigated the performance of the ORC systems, which work around the critical range, in terms of thermodynamics. Ergun et al. [7] conducted extensive research on the application areas of ORC systems, and made suggestions for ORC systems that could be used in Turkey. Kaynakli et al. [8] studied on the thermodynamic analysis of a basic and simple ORC for some determined operation conditions in which auxiliary heat exchanger does not exist. Akkaya [9] examined an Organic Rankine Cycle based power generation system, which uses the thermal energy of exhausted gases from an industrial plant. Cihan [10] modeled a system in which the organic Rankine cycle that operates with low-temperature waste heat is combined with the traditional vapor-compression refrigeration cycle. Wang et al. [11] studied a regenerative organic Rankine cycle to utilize the solar energy over a low temperature range using flat plate solar collectors. Kai et al. [12] optimized the ORC system parameters and examined the effects of these optimal parameters for six working fluids (Butane, R236fa, R227ea, R236ea, R245fa and R245ca). Gao et al. [13] examined the performance of solar based organic Rankine cycle through the various working conditions like inlet pressure and temperature of the turbine. Mago et al. [14] analysed the performance of some working fluids under different heat source temperatures and indicated that the boiling point of the working fluids has a strong influence on the system thermal efficiency.

Roy et al. [15] studied the outlet power, the system and second law efficiency, exergy destruction of the system and so on in the case of two different heat source temperatures Dai et al. [16] conducted parametric

This paper was recommended for publication in revised form by Ozgen Acikgoz
[1]*Department of Energy Systems Engineering, Bursa Technical University, Bursa, TURKEY*
[2]*Department of Mechanical Engineering, Uludag University, Bursa, TURKEY*
**E-mail address: husnu.bademlioglu@btu.edu.tr*

optimization of ORC with exergy efficiency. Kerme et al. [17] analyzed the energetic and exergetic performance of organic Rankine cycle driven by solar energy received by parabolic trough solar collectors. Li et al. [18] explored the effect of the evaporating temperature on the system thermal and exergy efficiencies and the net power outlet of an ORC system. Yaglı et al. [19] compared the exergetic and thermal performance of a R245fa based subcritical and supercritical ORC for the recovery of exhaust waste gas heat of the combined heat and power engine, which is fuelled by biogas. Guo et al. [20] analyzed and compared the performance of an ORC with internal heat exchanger to that of a basic ORC, using R600a, R245fa and R290 as working fluids. This analysis is based on the locations of heat transfer pinch point.

Table 1. Thermodynamic properties of certain organic fluids [8]

Organic Fluid	Type of Fluid	Molecular Mass (g/mol)	Critical Temperature (K)	Critical Pressure (Mpa)	ODV	GWP
R123 ($CHCl_2CF_3$)	isentropic	152.93	456.8	3.66	0.060	77 (low)
R134a (CF_2CH_2F)	wet	102.03	374.2	4.06	0	1430 (medium)
R152a ($C_2H_4F_2$)	wet	66.05	386.2	4.52	0	124 (low)
R236fa ($CF_3CH_2CF_3$)	isentropic	152.04	398.1	3.2	0	9810 (high)
R245fa ($C_3H_3F_5$)	isentropic	134.05	427.2	3.64	0	1030 (medium)
R600a (C_4H_{10})	dry	58.1	408.1	3.65	0	3 (low)

In this study, exergy analysis of an Organic Rankine cycle sample that produces electrical energy was made using a geothermal source with a temperature of 125°C. R123, R152a, R245fa and R600a were identified as fluids to be used in the cycle. In this analysis which is based on the pinch point temperature differences, pptd (5-20°C) in the evaporator, the exergy performance of the cycle elements for the geothermal source unit flow rate was theoretically investigated and the change of the exergy efficiency of the system was calculated. The obtained results are shown in the figures. It is important to consider the pinch point temperature difference in order to determine the optimum working conditions and to optimal design the system. For this reason, one of the main contributions to this study is to perform detailed exergy analysis for the different refrigerants based on the varying pinch point temperature difference in the evaporator.

MATERIAL AND METHODS

The schematic diagram of Organic Rankine cycle in which exergy analysis is performed in this study is given in Figure 1 (a) and T-s diagram in Figure 1 (b).

Many studies [4,6,8,14,19,20] in the literature show that the temperature difference between the refrigerant and the heat source has a significant effect on system performance. The minimum temperature difference between the evaporation temperature in the evaporator of refrigerant used in the cycle and the geothermal fluid temperature is defined as the pinch point temperature difference.

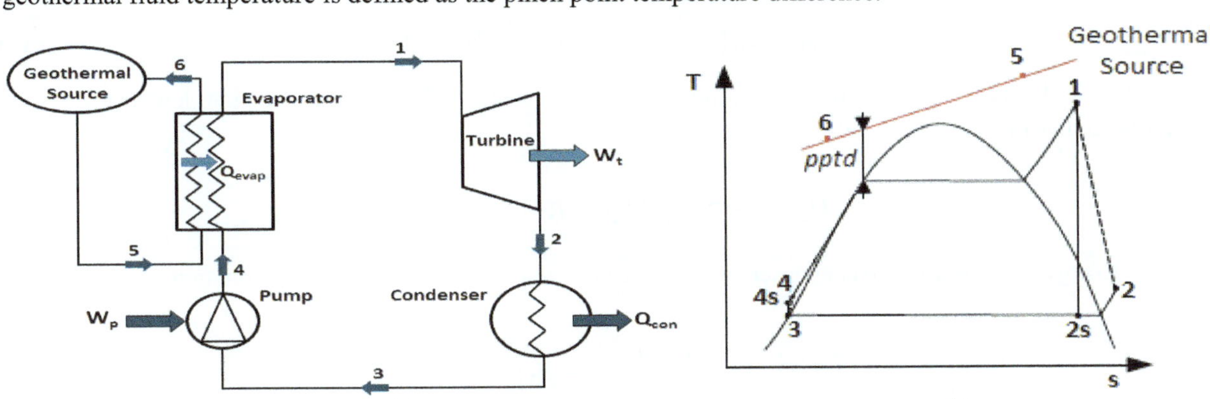

(a) (b)

Figure 1. The schematic (a) and T-s (b) diagrams of Organic Rankine cycle

Exergy Analysis of Organic Rankine Cycle

Within the scope of this study, it is aimed to make the exergy analysis of the cycle elements depending on the pinch point temperature difference in the evaporator by using the second law equations of thermodynamics and to calculate the exergy efficiencies depending on the exergy destruction [14,17,20,21].

For each point in the cycle, specific exergy is shown as e_i,

$$e_i = h_i - h_0 - [T_0(s_i - s_0)] \tag{1}$$

Here, T_0 represents the dead-state temperature for exergy calculations, h_0 and s_0 stand for the enthalpy and entropy of the operating fluid in the dead-state conditions (at the pressure and temperature), respectively. For each point in the cycle, exergy E_i is calculated with the equation,

$$E_i = \dot{m}_i e_i \tag{2}$$

Here, $\dot{m}_i$, the refrigerant flow rate for the points on the cycle (points 1, 2, 3, 4) is used as $\dot{m}_{ref}$ and geothermal resource flow are for the geothermal resource inlet and outlet points (5, 6 points) is used as $\dot{m}_{geo}$.

Exergy balance for each process in the cycle is shown as,

$$\sum E_{input} - \sum E_{output} = I \tag{3}$$

Exergy Analysis for the Steam Turbine

Exergy destruction in the turbine between 1 and 2 is shown as,

$$I_t = E_1 - (E_2 + W_t) \tag{4}$$

Here, W_T is the work obtained from the turbine. It is calculated by using the first law of thermodynamics equation as W_T,

$$W_t = \dot{m}_{ref}(h_1 - h_2) \tag{5}$$

Here, h_1 and h_2 are the enthalpy values at the inlet and outlet of the turbine, respectively. Exergy efficiency of the turbine, $\eta_{ex,T}$ is calculated by using the following equation,

$$\eta_{ex,t} = {W_T}/{(E_1 - E_2)} \tag{6}$$

Exergy Analysis for the Condenser

Exergy destruction in the condenser between 2 and 3 is shown as,

$$I_{con} = T_0\left(\dot{m}_{ref}(s_2 - s_3) + {Q_{con}}/{T_{con}}\right) \tag{7}$$

Here Q_{con} is the heat rejected from the condenser and is calculated by using the first law equations of thermodynamics as Q_{con},

$$Q_{con} = m_{ref}\ (h_2 - h_3) \tag{8}$$

Exergy efficiency of the condenser, $\eta_{ex,con}$ is calculated by using the following equation,

$$\eta_{ex,con} = 1 - \left({I_{con}}/{(E_2 - E_3)}\right) \tag{9}$$

Exergy Analysis for the Pump

Exergy destruction in the pump between 3 and 4 is shown as,

$$I_P = (E_3 + W_P) - E_4 \tag{10}$$

Here, W_P, is the required work to compress the operating refrigerant in the pump. It is calculated by using the first law equations of thermodynamics as W_P,

$$W_P = \dot{m}_{ref}(h_4 - h_3) \tag{11}$$

Here h_3 and h_4 are the enthalpy values at the inlet and outlet of the pump. Exergy efficiency at the pump, $\eta_{ex,P}$ is calculated by using the following equation,

$$\eta_{ex,P} = {(E_4 - E_3)}/{W_P} \tag{12}$$

Exergy Analysis for the Evaporator

Exergy destruction in the evaporator between 4 and 1 is shown as,

$$I_{evap} = (E_4 + E_5) - (E_1 + E_6) \tag{13}$$

Here, points 5 and 6 are specified for the inlet and outlet points of the geothermal resource, respectively. E_5 and E_6 are the exergies at these points. Exergy efficiency in the evaporator, $\eta_{ex,evap}$ is calculated by using the equation,

$$\eta_{ex,evap} = {(E_1 - E_4)}/{(E_5 - E_6)} \tag{14}$$

Exergy Efficiency of the Cycle

By using the second law equations of thermodynamics, exergy efficiency of the cycle is shown as $\eta_{ex,cycle}$,

$$\eta_{ex,cycle} = {(W_T - |W_P|)}/{(E_5 - E_6)} \tag{15}$$

In this study, the exergy analysis of the system is performed for the pinch point temperature difference in the evaporator by using the equations given above. Firstly, the flow rates of refrigerant have been determined for the unit flow rate of the geothermal source. Then, the thermophysical properties of each point in the system have been determined and the exergy changes at these points have been calculated. The exergy destruction in each component (evaporator, turbine, heat exchanger, condenser and pump) has been determined with the equations given above. Finally, the exergy efficiency of the cycle has been calculated by using Eq. (15).

Assumptions

In order to simplify the exergy analysis following assumptions are made;

- The system runs under steady-state conditions.
- Friction and heat losses as well as kinetic and potential energies are neglected.
- The evaporator capacity is fixed.
- The water at the outlet of condenser is saturated liquid.
- The specific volume of the working fluid remains constant during pumping.
- The efficiency of the turbine and pump is assumed to be constant for all working fluids.
- Chemical exergies of the substances are neglected.

RESULTS AND DISCUSSION

Design parameters and working conditions of system elements to be used for exergy analysis are given in Table 2.

In the exergy analysis made with the help of parameters given in Table 1, point 1 is taken 8°C higher than the evaporation temperature in the evaporator so that refrigerant could be superheated steam in the turbine inlet.

Table 2. Design parameters and operating conditions [6,13]

Parameters	**Value**
Evaporation temperature	100°C
Turbine inlet temperature	108°C
Geothermal water inlet temperature	125°C
Geothermal water pressure	500 kPa
Pinch point temperature difference (*pptd*) in the evaporator	5-20°C
Condensing temperature	30°C
Turbine isentropic efficiency	85%
Pump isentropic efficiency	80%
Ambient temperature	20°C

For the geothermal resource unit flow rate, the change of the flow rates of different refrigerants is shown in Figure 2, which are used in the cycle depending on the pinch point temperature difference in the evaporator. As can be seen in the figure, refrigerant flow rates decrease because of the increase in pinch point temperature difference. With the increase of pinch point temperature difference from 5°C to 20°C, the change of the flow rate of R123 becomes maximum and the flow rate drops from 0.60 kg/s to 0.15 kg/s. However, the change of the flow rate of R600a becomes minimum and the flow rate drops from 0.36 kg/s to 0.09 kg/s.

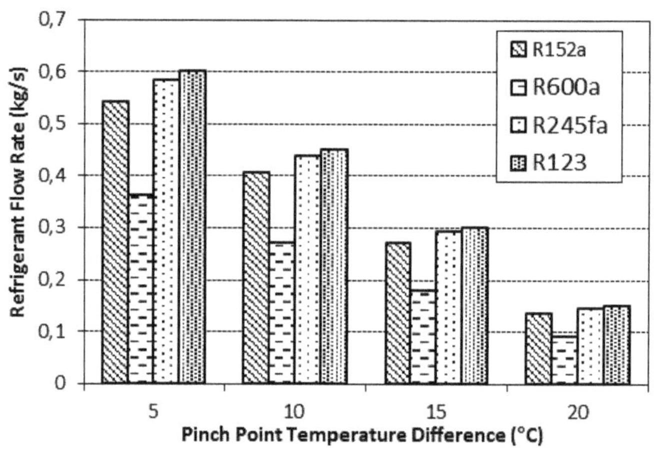

Figure 2. The change of the refrigerant flow rates depending on the pinch point temperature difference in the evaporator

For different refrigerants, the change of exergy destruction in the turbine is shown in Figure 3 depending on the pinch point temperature difference in the evaporator. As can be seen in the figure, the exergy destruction in the turbine is decreasing with the effect of increasing pinch point temperature difference. For the geothermal resource unit flow rate, while the decrease in exergy destruction in the turbine is maximum (2.893kW) when R152a is used in cycle, the decrease in exergy destruction in the turbine is minimum (2.094kW) when R123 is used depending on the pinch point temperature difference. Depending on the refrigerants and operating conditions, the exergy destruction in the turbine is decreased by a maximum of about 74.93%.

For different refrigerants, the change of exergy destruction in the pump is shown in Figure 4 depending on the pinch point temperature difference in the evaporator. As can be seen in the figure, the exergy destruction in the pump is decreasing with the effect of increasing pinch point temperature difference. For the geothermal resource unit flow rate, while the decrease in exergy destruction in the pump is maximum (0.316kW) when

R152a is used in cycle, the decrease in exergy destruction in the pump is minimum (0.05kW) when R123 is used depending on the pinch point temperature difference.

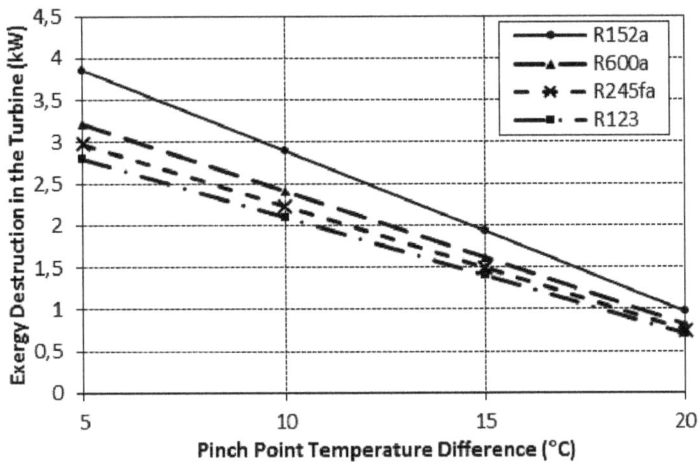

Figure 3. For different refrigerants, the change of exergy destruction in the turbine depending on the pinch point temperature difference in the evaporator

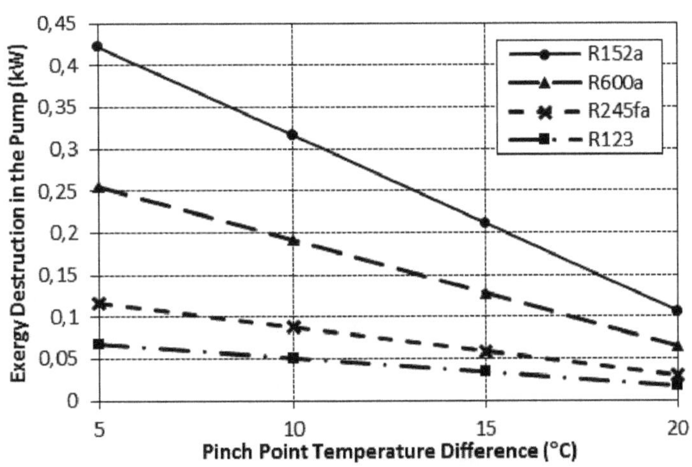

Figure 4. The change of exergy destruction in the pump depending on the pinch point temperature difference in the evaporator

For different refrigerants, the change of exergy destruction in the evaporator is shown in Figure 5 depending on the pinch point temperature difference in the evaporator. As can be seen in the figure, the exergy destruction in the evaporator is decreasing with the effect of increasing pinch point temperature difference. For the geothermal resource unit flow rate, while the decrease in exergy destruction in the evaporator is maximum (4.492kW) when R600a is used in cycle, the decrease in exergy destruction in the evaporator is minimum (3.752kW) when R123 is used depending on the pinch point temperature difference. Depending on the refrigerants and operating conditions, the exergy destruction in the evaporator is decreased by a maximum of about 62.32%.

The change of the outlet temperature of the geothermal fluid from the evaporator depending on the pinch point temperature difference in the evaporator is examined for different refrigerants in Figure 6. As can be seen in the figure, the outlet temperature of geothermal fluid from evaporator is increasing with the effect of increasing pinch point temperature difference. Outlet temperature of geothermal fluid from the evaporator increases by maximum about 34% when R152a is used and temperature rises from 86.1°C to 115.3°C. Additionally, if R123 is used in the cycle, the maximum outlet temperature is obtained for different pinch point temperature differences.

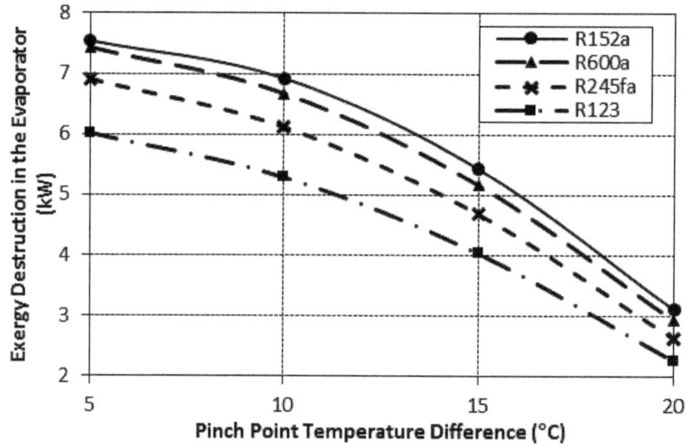

Figure 5. The change of exergy destruction in the evaporator depending on the pinch point temperature difference in the evaporator

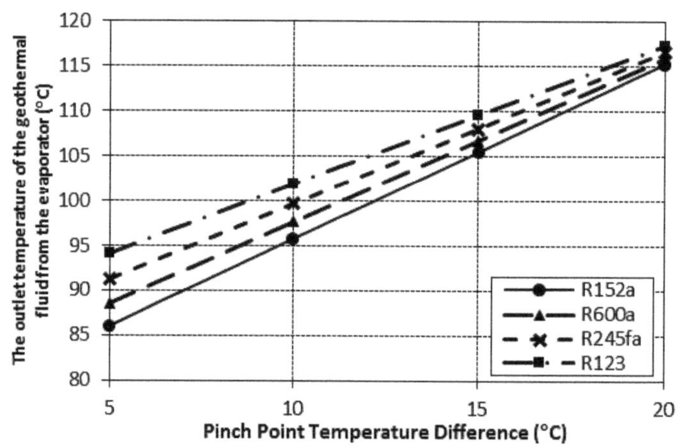

Figure 6. The change of outlet temperature of geothermal fluid from the evaporator depending on the pinch point temperature difference in the evaporator

For different refrigerants, the change of the exergy efficiency of cycle is shown in Figure 7 depending on the pinch point temperature difference in the evaporator. As can be seen in the figure, exergy efficiency of the cycle is decreasing with the effect of increasing pinch point temperature difference. When R152a is used as refrigerant in the cycle, the decrease in exergy efficiency of cycle is maximum and about 11.7% depending on the pinch point temperature difference. However, in case that R123 is used, the decrease in exergy efficiency of cycle is minimum and about 9.03%.

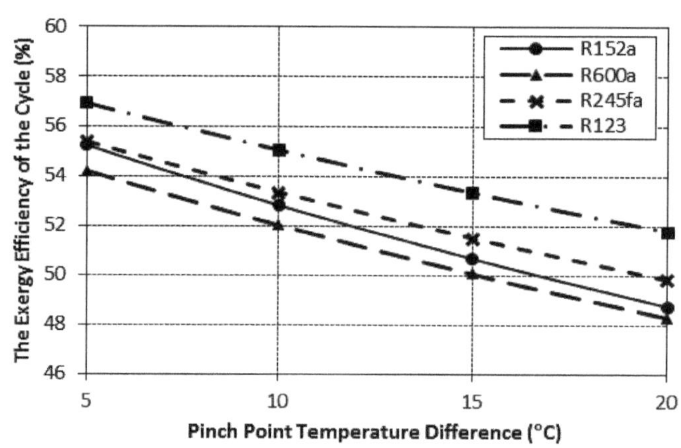

Figure 7. For different refrigerants, the change of exergy efficiency of the cycle depending on the pinch point temperature difference in the evaporator

CONCLUSION

For a comprehensive thermodynamic evaluation in Organic Rankine cycles, exergy analysis should be done alongside energy analysis, as well. Within the scope of this study, the impact of pinch point temperature difference in the evaporator on the exergy performance is studied for different refrigerants. The results obtained:

- In general, the exergy efficiency of cycle decreases with the increase of pinch point temperature difference in the evaporator. For this reason it is recommended that the pinch point temperature difference be lower.
- Depending on the pinch point temperature difference, while the exergy loss in the cycle is maximum and about 11.7% when R152a is used, the exergy loss in the cycle is minimum and about 9.03% when R123 is used. Therefore, it is seen that R123 is more preferable as the refrigerant in the cycle.
- Depending on the refrigerant and the pinch point temperature difference, exergy destruction in the evaporator decreases about maximum of 62.32%. Similarly, the exergy destruction in the turbine decreases about maximum of 74.93%.
- The outlet temperature of geothermal fluid from the evaporator increases about maximum of 34% depending on the pinch point temperature difference when R152a is used as the working fluid in the cycle.

NOMENCLATURE

e	Specific exergy (kJ/kg)
E	Exergy (kW)
I	Exergy destruction (kW)
$\dot{m}$	Refrigerant flow rate (kg/s)
T	Temperature (°C)
h	Enthalpy (kJ/kg)
s	Entropy (kJ/kg K)
η_{ex}	Exergy efficiency (%)
W	Work (kW)
Q	Heat transfer rate (kW)
P	Pressure (kPa)
pptd	Pinch temperature difference in the evaporator (°C)
t	Turbine
con	Condenser
p	Pump
evap	Evaporator
geo	Geothermal
ref	Refrigerant
0	Dead states
i	Point

REFERENCES

[1] Bertrand, F.T., George, P., Gregory, L., Antonios, F. (2009). Fluid selection for a low-temperature solar Rankine cycle. Applied Thermal Engineering, 29, 2468-2476.

[2] Drescher, U., Bruggemann, D. (2007). Fluid selection for the Organic Rankine Cycle (ORC) in biomass power and heat plants. Applied Thermal Engineering, 27, 223-228.

[3] Rayegan, R.,Tao, X. (2011). A procedure to select working fluids for solar Rankine cycle (ORCs). Renewable Energy, 36, 659-670.

[4] He, C., Liu, C., Gao, H., Xie, H., Li, Y., Wu, S., Xu, J. (2012). The optimal evaporation temperature and working fluids for subcritical organic Rankine cycle. Energy, 38, 136-143.

[5] Quoilin, S., Van den Broeck, M., Declaye, S., Dewallef, P., Lemort, V. (2013). Techno-economic survey of Organic Rankine Cycle (ORC) systems. Renewable & Sustainable Energy Reviews, 22, 168-186.

[6] Wang, Y., Ding, X., Tang, L., Weng, Y. (2016). Effect of evaporation temperature on the performance of organic Rankine Cycle in near-critical condition. ASME Journal of Energy Resources Technology, 138, 032001-032008.

[7] Ergun, A., Ozkaymak, M., Kılıcaslan, E. (2016). Power generation applications with Organic Rankine Cycle from low temperature heat sources. Duzce University Journal of Science & Technology, 4, 686-696.

[8] Kaynakli, O., Bademlioglu, A.H., Yamankaradeniz, N., Yamankaradeniz, R. (2017). Thermodynamic analysis of the Organic Rankine Cycle and the effect of refrigerant selection on cycle performance. International Journal of Energy Applications and Technologies, 4, 101-108.

[9] Akkaya, A.V. (2017). Performance analyzing of an organic Rankine cycle under different ambient conditions. Journal of Thermal Engineering, 3, 1498-1504.

[10] Cihan, E. (2014). Cooling performance investigation of a system with an organic Rankine cycle using waste heat sources. Journal of Thermal Science and Technology, 34, 101-109.

[11] Wang, M., Wang, J., Zhao, Y., Zhao, P., Dai, Y. (2013). Thermodynamic analysis and optimization of a solar-driven regenerative organic Rankine cycle (ORC) based on flat-plate solar collectors. Applied Thermal Engineering, 50, 816-825.

[12] Kai, Z., Mi, Z., Yabo, W., Zhili, S., Shengchun, L., Jinghong, N. (2015). Parametric optimization of low temperature ORC system. Energy Procedia, 75, 1596-1602.

[13] Gao, W., Li, H., Xu, G., Quan, Y. (2014). Working fluid selection and preliminary design of a solar organic Rankine cycle system. Environmental Progress & Sustainable Energy, 34, 619–626.

[14] Mago, P.J., Chamra, L.M., Somayaji, C. (2007). Performance analysis of different working fluids for use in Organic Rankine Cycles. Proceedings of the Institution of Mechanical Engineers, Part A: Journal of Power and Energy, 221, 255-264.

[15] Roy, J.P., Mishra, M.K., Mishra, A. (2011). Performance analysis of an Organic Rankine Cycle with superheating under different heat source temperature conditions. Applied Energy, 88, 2995-3004.

[16] Dai, Y.P., Wang, J.F., Lin, G. (2009). Parametric optimization and comparative study of organic Rankine cycle (ORC) for low grade waste heat recovery. Energy Conversion and Management, 50, 576-582.

[17] Kerme, E.D., Orfi, J. (2015). Exergy-based thermodynamic analysis of solar driven organic Rankine cycle. Journal of Thermal Engineering, 1, 192-202.

[18] Li, W., Feng, X., Yu, L.J., Xu, J. (2011). Effects of evaporating temperature and internal heat exchanger on organic Rankine cycle. Applied Thermal Engineering, 31, 4014-4023.

[19] Yaglı, H., Koc, Y., Koc, A., Gorgulu, A., Tandiroglu, A. (2016). Parametric optimization and exergetic analysis comparison of subcritical and supercritical organic Rankine cycle (ORC) for biogas fuelled combined heat and power (CHP) engine exhaust gas waste heat. Energy, 111, 923–932.

[20] Guo, C., Du, X., Yang, L., Yang, Y. (2014). Performance analysis of organic Rankine cycle based on location of heat transfer pinch point in evaporator. Applied Thermal Engineering, 62, 176–186.

[21] Safarian, S., Aramoun, F. (2015). Energy and exergy assessments of modified Organic Rankine Cycles (ORCs). Energy Reports, 1, 1-7.

3

THERMOECONOMIC ANALYSIS OF BIOGAS ENGINES POWERED COGENERATION SYSTEM

N. Doseva[1,*], D. Chakyrova

ABSTRACT

This study presents an analysis of an existing biogas engines-powered cogeneration system installed in Varna Wastewater Treatment Plant, Bulgaria using thermoeconomic method. The thermoeconomic analysis is conducted using three different groups of environmental conditions. The SPECO (specific exergy cost) method is applied to the system and cost balance equations are formulated for each component. The fuel F and the product P rules are used to obtain auxiliary equations. Moreover, in this paper, various thermoeconomic performance parameters such as the cost of exergy destruction, the relative cost difference and the exergoeconomic factor are determined. The obtained results show that the specific unit exergetic cost of the electrical power produced by the cogeneration system are found to be 30.0€/GJ = 0.11€/kWh, while the conducted thermoeconomic analysis based on energy delivers a result of 0.19€/kWh for the electrical work produced by biogas engines. In addition, the obtained results are compared to those seen in similar studies.

Keywords: *Biogas Engine, Cogeneration, Exergy, Thermoeconomics*

INTRODUCTION

Growing interest about energy efficiency of thermal systems has led to the presence of many studies dealing with development of methodologies for cost allocation and thermoeconomic optimization of energy conversion systems. Some authors [1-4] introduce energy as a base for assessment and monetary cost calculations. Nevertheless, analyses based on firs law of thermodynamics are inadequate for evaluating some features of energy resource utilization [5, 6]. Therefore, this paper claims that if the objective is optimization, specifying the saving opportunities (i.e. energy audit) or determining of unit product cost of an installation, a detailed study of thermodynamic irreversibilities occurring within system components should be carried out.

Some researches including [7], [8] and [9] conducted thermoeconomic analysis of diesel engine powered cogeneration (CHP) systems. Surveys such as that performed by Cavalcanti and Motta [10] and Tempesti et al. [11] examined the relationship between monetary unit costs and exergy destruction of CHP systems with integrated renewable energy resources. Trigeneration systems are investigated from thermoeconomic point of view by Palomino and Nebra [12], Temir and Bilge [13], and Balli and Aras [14]. A large and growing body of literature has investigated gas and steam turbine based cogeneration systems [15-17]. In recent years, an increasingly studies reported results of the thermoeconomic performance parameters of biomass burned CHP systems [18-22].

Although a recent study by Abusoglu et al.[23] investigated a similar to this cogeneration system, it can be argued that there is an area required applying of the thermoeconomic methodologies to the combined heat and power system fueled biofuels. Furthermore, in reviewing the literature, no surveys was found on thermoeconomic performance parameters of CHP systems located in Bulgaria, and an undeniable fact is that the thermoeconomic analysis, especially the optimum between thermodynamic and economic parameters, immediately depends on the system location.

Therefore, this paper seeks to remedy the highlighted gap of knowledge in the field of the study by conducting the thermoeconomic evaluation based on exergy analysis with economic principles of the biogas engines powered CHP system. Thus, it is possible to explain the process of cost formation of products and, then, to formulate the theoretical framework of the procedures for exergy saving and optimization.

This paper was recommended for publication in revised form by Regional Editor Jovana Radulovic
[1]Department of Thermal Engineering, Varna Technical University, Varna, BULGARIA
**E-mail address: chakyrova_d@abv.bg; nadezhda.doseva@abv.bg*

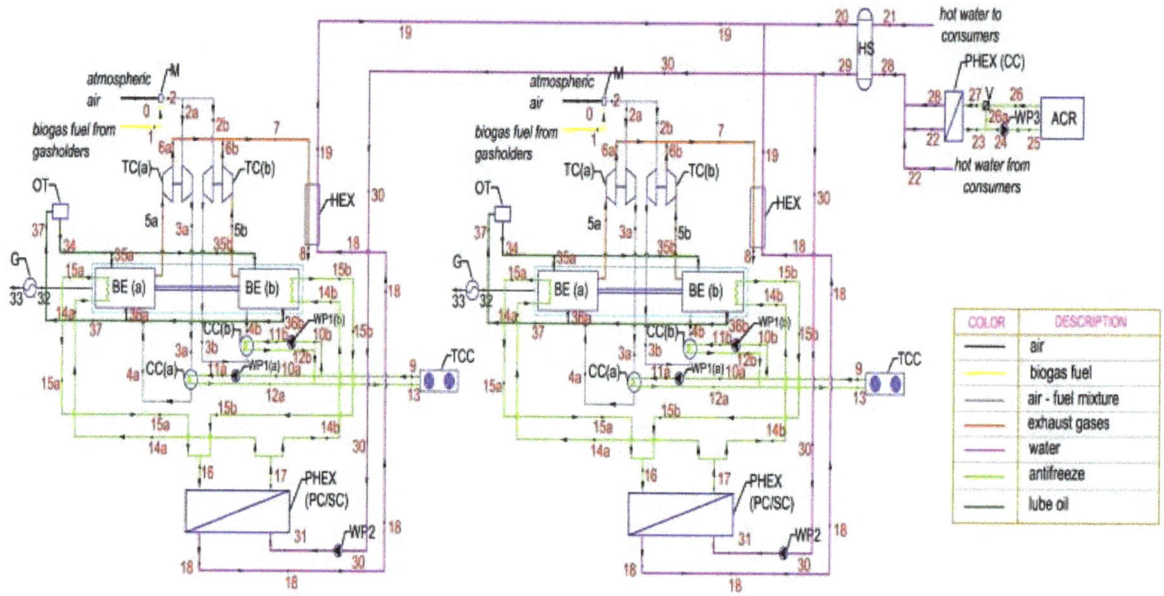

Figure 1. A schematic representation of the analyzed cogeneration system
M – mixer; TC(a), TC(b) – turbochargers; BE(a), BE(b) – internal combustion engines; CC(a), CC(b) – charge coolers; WP1(a), WP1(b) – technological circuit water pumps; TCC – technological circuit cooler; PHEX PC/SC – plate heat exchanger from secondary circuit; WP2 – secondary circuit water pump; HEX - heat exchanger; G – generator, OT – oil tank; HS – hydraulic separator; PHEX (CC) – plate heat exchanger from cooling circuit; V – 3 –way valve; WP3 - cooling circuit water pump; ACR – air – cooled radiator.

For this purpose, the paper begins by formulating of the cost rate balances for each component of the cogeneration system separately by applying of the SPECO method. It will then go on to determining of various thermoeconomic performance parameters, such as exergoeconomic factor, relative cost difference and cost rate of exergy destruction. Finally, the results obtained from thermoeconomic analysis are compared with those from traditional economic cost analysis.

Conducting the introduced methodology for thermoeconomic analysis of a cogeneration system driven by biogas engines, it is possible to get a notion about the real picture of the products costs of the system, that are observed on the Bulgarian market. The results of the current investigation, as well as the availability of the similar systems operating in various Bulgarian wastewater treatment plants creates the possibility to launching the work out strategies for the sustainable development of biogas technology in Bulgaria, based on real data.

DESCRIPTION OF THE SYSTEM

This study is made for a cogeneration plant driven by biogas internal combustion engines (ICE) and its purpose is to utilize the methane produced as a by-product of the sludge stabilization process at Varna Wastewater Treatment Plant (Varna WWTP) (Figure 2). The plant consists of two CHP modules, each of which is driven by ICE burned biogas - model Cento T300 SP BIO+ZP (Tedom) (Figure 1). The plant produced electrical energy and hot water. The electricity is generated by two, biogas engine actuated generator set. Each of the biogas engines – generators sets produce 320kW electricity at 100% load. The current thermoeconomic model of the cogeneration plant was made for typical operation conditions of the system, namely 75% load. For these conditions the system produces 2x240kW electricity.

In the heat exchanger (HEX) of the plant, high temperature exhaust gas energy is used to heat water. Thus, the produced hot water has mass flow rate 7.6 kg/s. The heat energy consumption depends on thermal needs of mesophilic fermentation process, occurring within the digesters. Therefore, it seems that the thermoeconomic analysis should be conducted using different groups of ambient reference conditions. In current investigation, the temperature, pressure and relative humidity of the different dead (environmental) states are taken as follows:

Figure 2. Biogas treatment and utilization system at Varna WWTP

- Summer operation condition of the system: $T_0 = 298.15K$, $p_0 = 1.013bar$ and $\varphi_0 = 55\%$;
- Spring - autumn operation condition of the system: $T_0 = 288.15K$, $p_0 = 1.013bar$ and $\varphi_0 = 60\%$;
- Winter operation condition of the system: $T_0 = 276.15K$, $p_0 = 1.013bar$ and $\varphi_0 = 80\%$.

THERMOECONOMIC ANALYSIS OF THE BIOGAS POWERED COGENERATION SYSTEM

In this study, the thermoeconomic analysis consists of the conventional exergy analysis and economic analysis. An aim of the performed here thermoeconomic analysis is to reveal the process of cost formation and assess unit exergy cost of product streams. This calls the use of cost rate balances formulated for each component of the cogeneration system separately.

For a unit operating in steady state, the cost rate balance is defined as given below [24, 26]:

$$\sum_e \dot{C}_{e,k} + \dot{C}_{W,k} = \dot{C}_{q,k} + \sum_i \dot{C}_{i,k} + \dot{Z}_k , \tag{1}$$

where:

$$\dot{C}_e = c_e \dot{E}x_e \tag{2}$$

$$\dot{C}_i = c_i \dot{E}x_i \tag{3}$$

$$\dot{C}_W = c_W \dot{W} \tag{4}$$

$$\dot{C}_q = c_q \dot{E}x_q \tag{5}$$

Therefore, the cost rate associated with exergy of the product, $\dot{C}_P$, is the sum of the cost rate associated with exergy of the fuel, $\dot{C}_F$, capital investment (CI) and operating and maintenance costs (O&M).

The term $\dot{Z}_k$ is determined by first calculating the carrying charges (CC) and O&M costs of the CHP system. This was performed in economic analysis of the plant [25]. After that are computed the levelized values of these expenditures per unit of time (hour) of system operation and associated with the kth component using the following expression [24]:

$$\dot{Z}_k = \dot{Z}_k^{CI} + Z_k^{O\&M} = \frac{CC_L + O\&M_L}{\tau} \cdot \frac{PEC_k}{\sum_{cogen} PEC} \tag{6}$$

Applying Eq. (1) to the components of the system yields many unknowns and requires auxiliary equations to determine all cost per unit exergy entering and leaving control volumes of the system. These auxiliary equations are written according to the *F* (Fuel) rule and *P* (Product) rule [26] and along with the cost rate balances are presented in Table 1.

THERMOECONOMIC PERFORMANCE PARAMETERS FOR COMPONENT EVALUATION

The thermoeconomic performance parameters include the cost of exergy destruction, the relative cost difference and the exergoeconomic factor. The approach used for determining of the thermoeconomic performance parameters is described in the following paragraphs.

Table 1. Cost rate balances and axially equation for the cogeneration system components

Component	Control volume	Cost rate balances and auxiliary equations
M	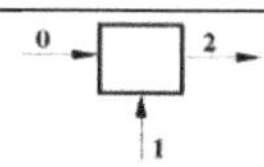	$\dot{C}_2 = \dot{C}_0 + \dot{C}_1 + \dot{Z}_M$ $c_0 = 0$ *(Assumption)*
TCa,b	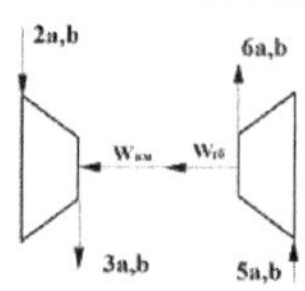	$\dot{C}_{3a,b} - \dot{C}_{2a,b} = \dot{C}_{5a,b} - \dot{C}_{6a,b} + \dot{Z}_{TCa,b}$ $\frac{\dot{C}_{2a,b}}{\dot{Ex}_{2a,b}} = \frac{\dot{C}_{3a,b}}{\dot{Ex}_{3a,b}}$ *(P-rule)*
CCa,b	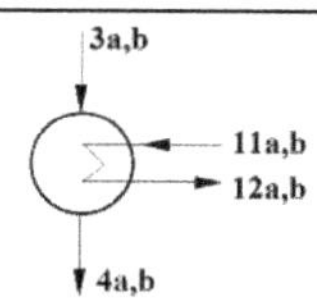	$\dot{C}_{4a,b} - \dot{C}_{3a,b} = \dot{C}_{11a,b} - \dot{C}_{12a,b} + \dot{Z}_{CCa,b}$ $\frac{\dot{C}_{3a,b}}{\dot{Ex}_{3a,b}} = \frac{\dot{C}_{4a,b}}{\dot{Ex}_{4a,b}}$ *(F-rule)*
WP1a,b	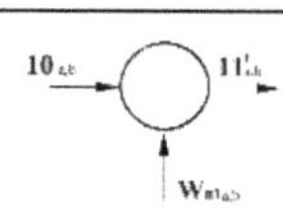	$\dot{C}_{W,WP1} + \dot{Z}_{WP1} = \dot{C}_{11a,b} - \dot{C}_{10a,b}$ $c_{W,WP1} = 0.19€/kWh$ *(Assumption)*
TCC	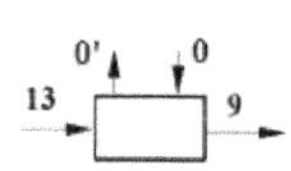	$\dot{C}_9 + \dot{C}_{0'} = \dot{C}_{13} + \dot{C}_0 + \dot{Z}_{TCC}$ $\frac{\dot{C}_{13}}{\dot{Ex}_{13}} = \frac{\dot{C}_9}{\dot{Ex}_9}$ *(F-rule)*
BEa,b	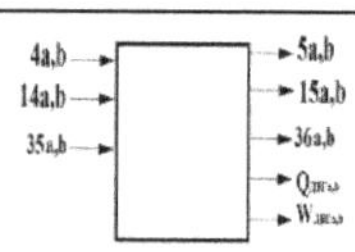	$\left(\dot{C}_{4a,b} - \dot{C}_{5a,b}\right) + \dot{Z}_{BEa,b} = \dot{C}_{W,32} + \left(\dot{C}_{15a,b} - \dot{C}_{14a,b}\right) + \left(\dot{C}_{36a,b} - \dot{C}_{35a,b}\right) +$ $+\dot{C}_{q,BEa,b}$ $c_{5a,b} = c_{6a,b} = c_{15a,b} = c_{36a,b}$ *(F-rule)*
OT	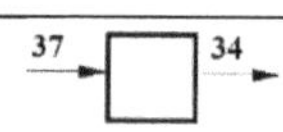	$\dot{C}_{34} = \dot{C}_{37} + \dot{Z}_{OT}$
PHEX PS/SC	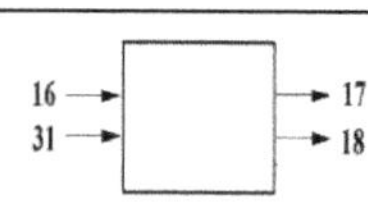	$\dot{C}_{18} - \dot{C}_{31} = \dot{C}_{16} - \dot{C}_{17} + \dot{Z}_{PHEX(SC/PC)}$ $\frac{\dot{C}_{16}}{\dot{Ex}_{16}} = \frac{\dot{C}_{17}}{\dot{Ex}_{17}}$ *(F-rule)*
HEX	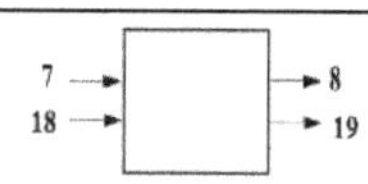	$\dot{C}_{19} - \dot{C}_{18} = \left(\dot{C}_7 - \dot{C}_8\right) + \dot{Z}_{HEX}$ $\frac{\dot{C}_7}{\dot{Ex}_7} = \frac{\dot{C}_8}{\dot{Ex}_8}$ *(F-rule)*
WP2	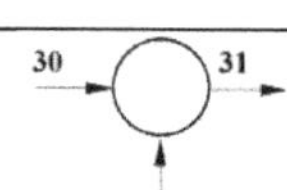	$\dot{C}_{W,WP2} + \dot{Z}_{WP2} = \dot{C}_{31} - \dot{C}_{30}$ $c_{W,WP2} = 0.19€/kWh$ *(Assumption)*
HS	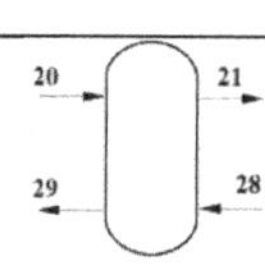	$\dot{C}_{21} + \dot{C}_{29} = \dot{C}_{20} + \dot{C}_{28} + \dot{Z}_{HS}$ $\frac{\dot{C}_{29}}{\dot{Ex}_{29}} = \frac{\dot{C}_{21}}{\dot{Ex}_{21}}$ *(F-rule)*
WP3	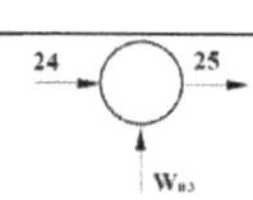	$\dot{C}_{W,WP3} + \dot{Z}_{WP3} = \dot{C}_{25} - \dot{C}_{24}$ $c_{W,WP2} = 0.19€/kWh$ *(Assumption)*

Table 1. (Cont.) Cost rate balances and axially equation for the cogeneration system components

ACR	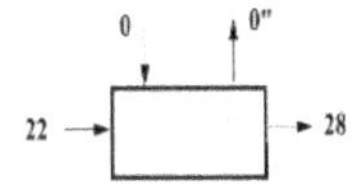	$\dot{C}_{26}+\dot{C}_{0''}=\dot{C}_{25}+\dot{C}_{0}+\dot{Z}_{ACR}$ $\frac{\dot{C}_{25}}{\dot{E}x_{25}}=\frac{\dot{C}_{26}}{\dot{E}x_{26}}$ *(F-rule)*
PHEX CC	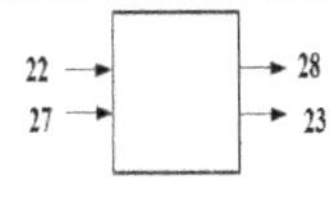	$\dot{C}_{28}+\dot{C}_{23}=\dot{C}_{22}+\dot{C}_{27}+\dot{Z}_{PHEX(CC)}$ $\frac{\dot{C}_{22}}{\dot{E}x_{22}}=\frac{\dot{C}_{28}}{\dot{E}x_{28}}$ *(F-rule)*
V	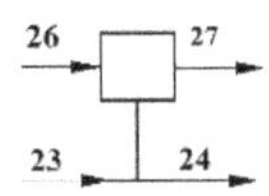	$\dot{C}_{26}+\dot{C}_{23}+\dot{Z}_{V}=\dot{C}_{27}+\dot{C}_{24}$ $\frac{\dot{C}_{26}}{\dot{E}x_{26}}=\frac{\dot{C}_{23}}{\dot{E}x_{23}}$ *(F-rule)*
G	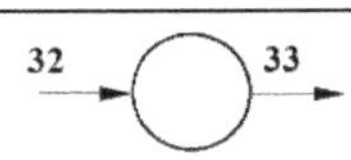	$\dot{C}_{W,32}+\dot{Z}_{G}=\dot{C}_{W,33}=c_{W}\dot{W}_{33}$

The thermoeconomic performance parameters include the cost of exergy destruction, the relative cost difference and the exergoeconomic factor. The approach used for determining of the thermoeconomic performance parameters is described in the following paragraphs.

In cost rate balances formulated for the cogeneration system components (Table 1), there is no term directly associated with exergy destruction, i.e. the cost rate of exergy destruction is hidden cost and it can be determine only by thermoeconomic evaluation on component level. According to the assumptions made in this paper, the cost rate of exergy destruction of the *k*th component can be defined by [24], [26]:

$$\dot{C}_{D,k}=c_{F,k}\dot{E}x_{D,k} \tag{7}$$

Another parameter determined in this study is the relative cost difference, *r*. This variable expresses the relative increase in the unit exergy cost between fuel and product of the component and it is defined by [24, 26]:

$$r_{k}=\frac{c_{P,k}-c_{F,k}}{c_{F,k}} \tag{8}$$

In evaluating the thermoeconomic performance of a component, it is useful to know the contribution of the non-exergy related cost (CI and O&M costs) to the total cost increase. This is provided by the exergoeconomic factor represented by the following equation [24], [26]:

$$f_{k}=\frac{\dot{Z}_{k}}{\dot{Z}_{k}+c_{F,k}\left(\dot{E}x_{D,k}+\dot{E}x_{L,k}\right)} \tag{9}$$

RESULTS AND DISCUSSION

In this study, the SPECO method is applied to an existing CHP system which uses biogas engines. The following assumptions were made during the analysis:

(i) The CHP system and its components operate at steady state;

(ii) The exergy flow rates ($\dot{E}x_{e}$, $\dot{W}_{k}$, $\dot{E}x_{q,k}$ and $\dot{E}x_{i}$) exiting and entering the *k*th component of the CHP system are calculated in another study conducted by authors [6]. It was conducted error analysis of the thermodynamic model, comparing the measured and obtained numerical data at some states of the system [25];

(iii) The annual carrying charges and operating and maintenance costs are apportioned among the system components according to the contribution of each component to the sum of purchased-equipment costs [24];

(iv) A zero unit cost is assumed for air entering mixer (M);

(v) The average unit cost of the electrical energy entering water pumps (WP1a,b; WP2 and WP3) and the water entering the CHP system is assumed according to the current prices of electricity distribution company and municipal water and sewage department, respectively;

(vi) The exergy of product ($\dot{E}x_{P,k}$) is fixed and the unit cost of fuel ($c_{F,k}$) of the *k*th component is independent of the exergy destruction.

According to assumption (*ii*) and Eq.(6), the term $\dot{Z}_k$ is calculated with the assistance of the annual levelized carrying charges (CC), O&M costs and total purchased equipment cost (PEC) of the CHP system and its components. Using the data published in another investigation [25], the hourly levelized CC and O&M costs are 247610.44€ and 30181€, respectively. Furthermore, the total annual number hours of system operation at 75% load (τ), the life time of the system (n) and capital recovery factor (CRF) are taken as 5256 h, 20 yr and 0.07708, respectively [25]. The hourly levelized total costs of the *k*th system component are presented in Figure 3 and Table 2.

Annual fuel cost rate of biogas, FC_L, is calculated by Doseva, N. [25] to be 151982.90 €/yr. On exergetic base, the hourly levelized fuel cost rates is 28.82€/h. The average unit exergy cost of total fuel entering a CHP module,c_1, is calculated as 0.022 €/kWh.

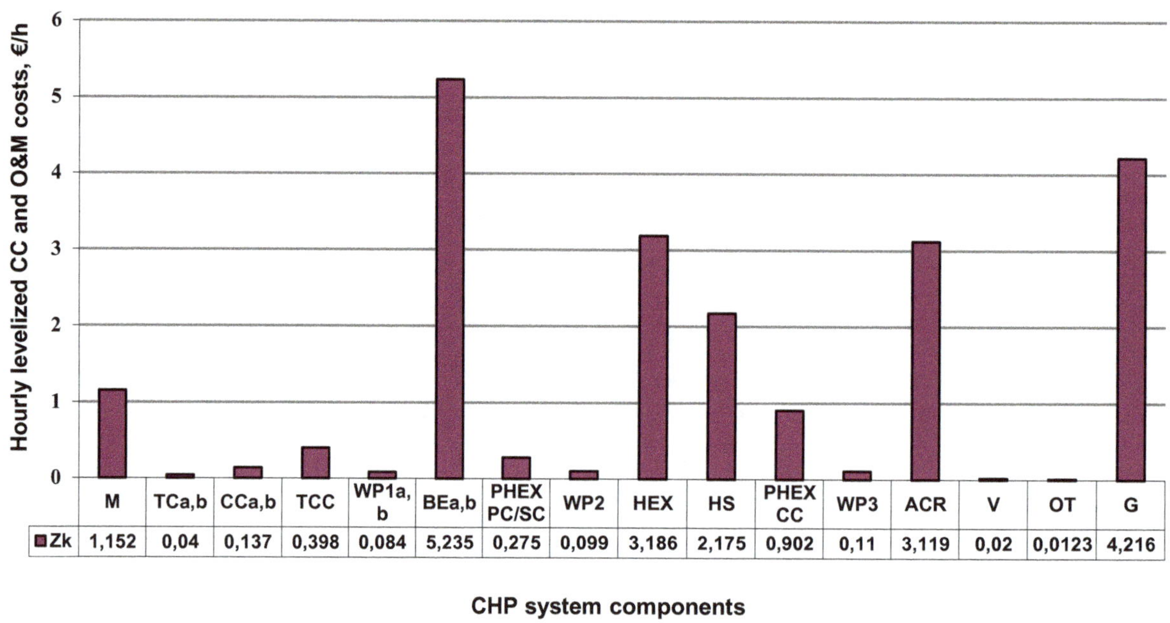

Figure 3. Cost rate associated with the sum of carrying charges and O&M of the cogeneration system components

From Figure 3 and Table 2 is clear that the biogas engines, the generator and the air-cooled radiator are components with the higher value of the parameter $\dot{Z}_k$ due to high values of the hourly levelized carrying charges and O&M costs of the components.

Using Eq. (1) and *P*- and *F*-rule [26] to each system component lead to system of linear equations and unknowns indicated by equations in Table 1. This system of linear equations is solved to obtain the cost per unit exergy associated with each stream of the cogeneration system. The obtained results at the summer operating condition of the CHP system are presented in Table 2 and a cost flow diagram of the part of the system is shown in Figure 5.

Table 2. Cost per unit of exergy and cost rates of the system flows at the summer operation condition of the system ($T_0 = 298.15K$, $p_0 = 1.013bar$ and $\varphi_0 = 55\%$)

Mixer (M) $\dot{Z}_M = 1.152$ **€/h**				
№	Stream	$\dot{E}x$, GJ/h	c, €/GJ	$\dot{C}$, €/h
0	Air	0	0	0
1	Biogas	2.395	6.11	14.63
2	Air-fuel mixture	2.382	6.67	15.88
Turbochargers (TCa,b) $\dot{Z}_{TCa,b} = 0.05$ **€/h**				
2a,b	Air-fuel mixture	1.191	6.67	7.94
3a,b	Air-fuel mixture	1.257	6.67	8.38
5a,b	Exhaust gases	0.458	4.44	2.04
6a,b	Exhaust gases	0.370	4.44	1.65
Charge coolers (CCa,b) $\dot{Z}_{CCa,b} = 0.137$ **€/h**				
3a,b	Air-fuel mixture	1.257	6.67	8.38
4a,b	Air-fuel mixture	1.242	6.67	8.28
11a,b	Antifreeze	38.207	44.53	1701.18
12a,b	Antifreeze	38.215	44.53	1701.62
Technological circuit cooler (TCC) $\dot{Z}_{TCCa,b} = 0.398$ €/h				
9	Antifreeze	76.413	44.53	509.42
13	Antifreeze	76.430	44.53	509.54
0	Air	0.026	8.33	1.16
0′	Air	0.033	45.28	1.46
Water pumps (WP1a,b) $\dot{Z}_{WP1a,b} = 0.084$ **€/h**				
10a,b	Antifreeze	38.207	44.52	1701.12
11a,b	Antifreeze	38.207	44.53	1701.18
	Mechanical work	0.002	53.06	0.11
Biogas engines (BEa,b) $\dot{Z}_{BEa,b} = 5.235$ **€/h**				
4a,b	Air-fuel mixture	1.242	6.67	8.28
14a,b	Antifreeze	19.853	4.47	88.79
35a,b	Lube oil	0.004	13.89	0.05
36a,b	Lube oil	0.009	4.47	0.04
15a,b	Antifreeze	19.887	4.47	88.94
5a,b	Exhaust gases	0.458	4.47	2.05
32	Mechanical work	0.450	24.17	10.88
	Loss	0.015	6.72	0.10
Plate heat exchanger PC/SC (PHEX PC/SC) $\dot{Z}_{PHEX_{PC/SC}} = 0.275$ **€/h**				
16	Antifreeze	39.773	4.47	177.87
17	Antifreeze	39.706	4.47	177.57
18	Water	2.548	0.69	1.77
31	Water	2.505	0.47	1.18
Water pump (WP2) $\dot{Z}_{WP2} = 0.099$ **€/h**				
30	Water	2.505	0.42	1.04
31	Water	2.505	0.47	1.18
	Mechanical work	0.002	53.06	0.08
Heat exchanger (HEX) $\dot{Z}_{HEX} = 3.186$ **€/h**				
7	Exhaust gases	0.740	4.47	3.31
8	Exhaust gases	0.507	4.47	2.27
18	Water	2.548	0.69	1.77
19	Water	2.654	2.29	6.09

Table 2. (Cont.) Cost per unit of exergy and cost rates of the system flows at the summer operation condition of the system

Hydraulic separator (HS) $\dot{Z}_{HS} = 2.175$ €/h				
20	Water	5.307	2.29	12.18
28	Water	4.632	0.42	1.93
21	Water	4.926	2.85	14.03
29	Water	5.009	0.42	2.09
Plate heat exchanger CC (PHEX CC) $\dot{Z}_{PHEX_{CC}} = 0.902$ €/h				
22	Water	4.737	0.42	1.97
28	Water	4.632	0.42	1.93
27	Antifreeze	252.454	11.23	2835.97
23	Antifreeze	252.512	11.23	2836.76
Water pump (WP3) $\dot{Z}_{WP3} = 0.11$ €/h				
24	Antifreeze	534.046	11.23	5998.37
25	Antifreeze	534.046	11.23	5998.67
	Mechanical work	0.004	53.06	0.21
Air – cooled radiator (ACR) $\dot{Z}_{ACR} = 3.119$ €/h				
25	Antifreeze	534.046	11.23	5998.67
26	Antifreeze	533.996	11.23	5998.11
0	Air	0.025	8.33	0.21
0″	Air	0.029	133.06	3.89
3-way valve (V) $\dot{Z}_{V} = 0.020$ €/h				
26	Antifreeze	533.996	11.23	5998.11
27	Antifreeze	252.454	11.23	2835.97
23	Antifreeze	252.512	11.23	2836.76
24	Antifreeze	534.046	11.23	5998.37
Oil tank (OT) $\dot{Z}_{OT} = 0.023$ €/h				
37	Lube oil	0.018	4.47	0.08
34	Lube oil	0.007	13.89	0.10
Generator (G) $\dot{Z}_{G} = 4.216$ €/h				
32	Mechanical work	0.900	24.14	21.73
33	Electrical energy	0.864	30.00	25.92

Table 2 is quite revealing in several ways. First, the obtained data demonstrate that the unit exergy cost of the net electrical power produced by the CHP system at the summer operating conditions of the CHP system ($T_0 = 298.15K$, $p_0 = 1.013bar$ and $\varphi_0 = 55\%$) is 30.0€/GJ = 0.11 €/kWh, while the same cost of the by-product (thermal energy) is 0.42€/GJ. Moreover, as can be seen from the table above, the cost flow rate associated with gases leaving the heat exchanger (HEX) and exhaust to the environment is 2.27€/h.

Figure 4 compares the obtained results of the unit exergy cost of the produced net electrical energy at the considered different groups of ambient reference conditions. This figure seems to show generally closer values. On the other hand, there is a trend of decreasing of the unit exergy cost of produced electricity with declining of the ambient temperature. These results may be explained by the fact that the exergy efficiency of the CHP system slowly increase with decreasing of the reference temperature [25].

To verify the results of calculation of the flow costs of the system, some previous studies are reviewed. The data obtained in these studies is listed in Table 3.

Table 3 shown that there are similarities between the unit exergy cost of the net electrical power produced by the CHP system calculated in this study and this presented by Abusoglu *et al.* [23]. Although the cited paper also investigates the process of cost formation of the products of a biogas engine powered CHP system, the slight difference in the exergy costs is due to variation of the fuel price, the power capacity of system, the type and purchased cost of equipment and the system location

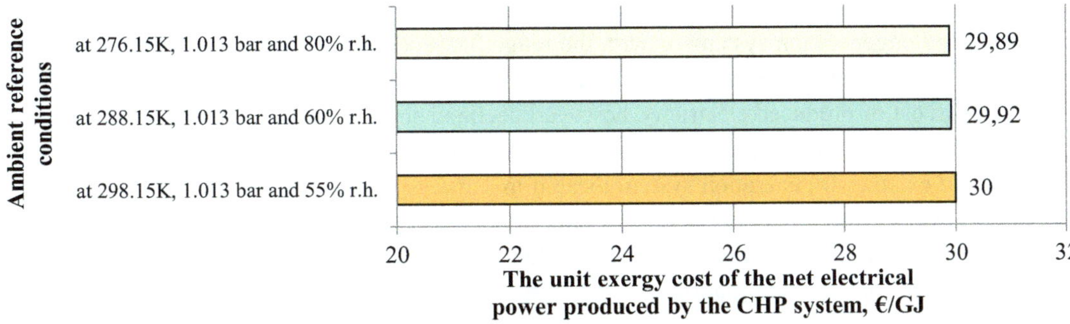

Figure 4. Comparison of results of the unit exergy cost of the produced net electrical energy at the considered different groups of ambient reference conditions

Table 3. Results obtained in similar studies

Exergy costs	Type of CHP system	Reference
	Engine driven CHP systems	
Electricity: 0.03€/kWh	Diesel engine powered CHP system installed in Turkey	[7]
Electricity: 0.13€/kWh	Micro CHP system based on an Ericsson engine and located in France	[27]
	Turbine driven CHP systems	
Electricity: 0.07€/kWh *Heat*: 0.02€/kWh	Micro GT CHP system burned natural gas	[15]
Electricity: 0.08€/kWh *Heat*: 0.02€/kWh	Micro GT-ORC CHP system burned natural gas	[15]
Electricity: 0.06€/kWh *Heat*: 0.05€/kWh	GT CHP system located in Turkey	[16]
Electricity: 0.1/kWh *Heat:* 0.03€/kWh	Micro GT CHP system situated in Turkey	[28]
Electricity: 0.07 €/kWh	GT / ST CHP system installed in Turkey	[29]
Electricity: GT: 0.05€/kWh; ST: 0.12 €/kWh; Average: 0.0622€/kWh	GT / ST CHP system installed in Turkey	[17]
Electricity: 0.09€/kWh	Rankine power cycle (ST CHP system)	[30]
Electricity: 0.05€/kWh	GT CHP at 100% load with an electricity output 1000kW	[31]
Electricity: 0.06€/kWh *Heat:* 0.09€/kWh	Base case GT CHP system (CGAM problem)	[24]
	CHP systems with integrated renewable energy resources	
Electricity: 0.03 €/kWh *Heat:* 0.10€/kWh	Solar – powered / fuel assisted Rankine cycle for power generation located in Brazil	[10]
Electricity: 0.08€/kWh	Biogas engine powered CHP system	[23]
Electricity: 0.05 €/kWh (at working fluid R245fa in July)	Micro CHP ORC system fuelled by two renewable energy resources (solar and low-temperature geothermal) situated in Italy	[11]
Electricity: 0.21€/kWh *Heat*: 0.20 €/kWh	ORC CHP system burned biomass	[19]
	Trigeneration systems	
Electricity: 0.035€/kWh	Tetra-combined trigeneration system developed by Garagatti, Arriola and Olivera (2001)	[32]
Electricity: 0.15€/kWh	Trigeneration system located in Turkey	[14]

Referring to the data in Table 3, a comparison of the unit cost of produced electricity with those of the other configurations cogeneration systems reveals the following: the unit exergy cost of electricity produced by gas turbine (GT) based cogeneration systems is with the range 0.05€/kWh - 0.1€/kWh, while the stream turbine (ST) based CHP systems are characterized by unit exergy cost of produces electrical power around 0.11€/kWh [17]. The unit exergy cost of produced electricity, however, declined sharply to 0.03€/kWh when a diesel engine based cogeneration systems is used [7] and it is scarcely smaller in case of external combustion engine powered CHP system [27]. An existing trigeneration system located in Turkey [14] has an exergy cost per unit produced electrical power amounting to 0.15€/kWh and it is somewhat bigger than the electricity cost of the analyzed system. In general, therefore, it seems that the investigated CHP system produces electrical power with price, which is competitive to others configurations CHP systems and those with integrated renewable energy resources.

The cost of exergy destruction, the relative cost difference and exergoeconomic factor of the CHP system components at the summer operating conditions of the CHP system ($T_0 = 298.15K$, $p_0 = 1.013bar$ and $\varphi_0 = 55\%$) are calculated and represented in Table 4. From this data, we can see that components such as heat exchangers (PHEX CC, HEX, ACR and TCC) and water pumps (WP3, WP2 and WP1a,b) have very high values of the relative cost difference. These results may be explained by the fact that their values of the cost per unit exergy of the product are many times greater than the cost per unit exergy of the fuel due to high value of exergy destruction rate and low values of exergy efficiency for these components determined in another study [6]. This statement is somewhat confirmed by the exergoeconomic factor values. Among all the system components, it is noticeable that the heat exchangers (PHEX CC, HEX and ACR) have high values of exergoeconomic factor. The exergoeconomic factor results of PHEX CC, HEX and ACR may be explained by the fact that the investment and O&M costs are greater than the costs rate of exergy destruction of these components, and the low values of cost rate of exergy destruction are caused by low exergy cost of the fuel. Not always, however, high values of r_k lead to high value of f_k. An example of this is a case of the mixer (M) – the relative cost difference is determined as 8.47% and it is one of the lowest values, while the exergoeconomic factor is calculated to be 95.53%.

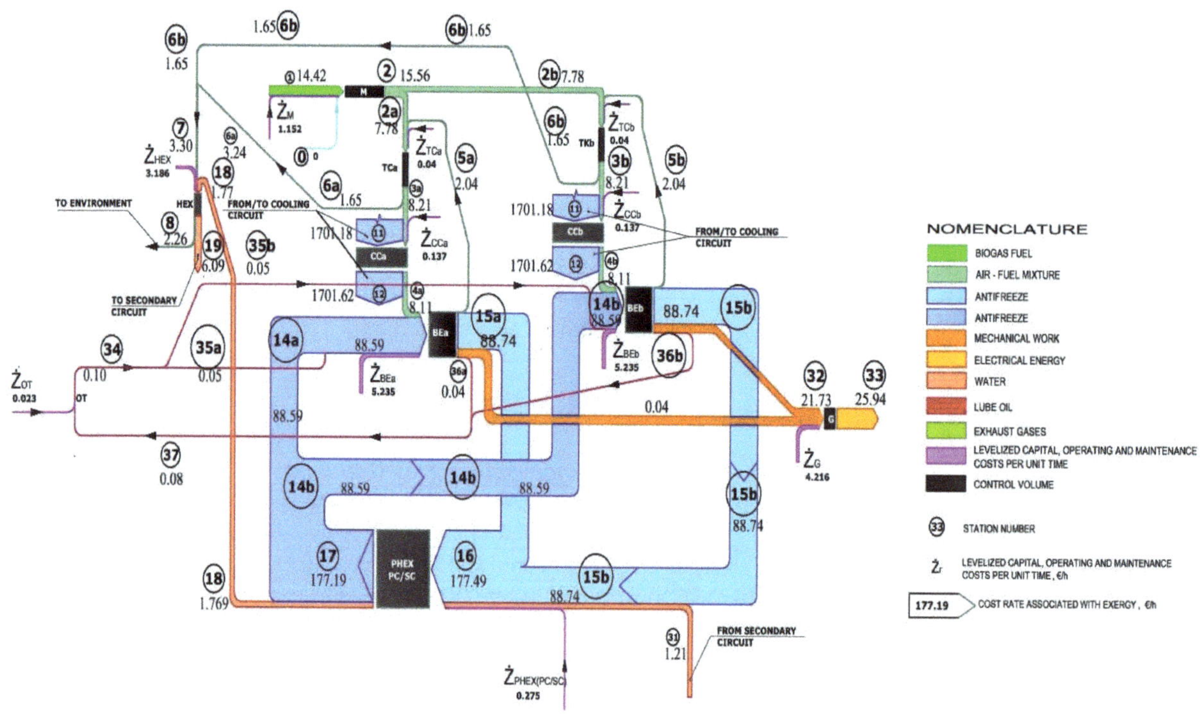

Figure 5. Cost flow diagram of the CHP module (at the summer operating conditions - $T_0 = 298.15K$, $p_0 = 1.013bar$ and $\varphi_0 = 55\%$)

According to the data from Table 4 about the sum $\dot{C}_D + \dot{Z}_k$, the biogas engines (BEa,b), the generator (G), heat exchanger (HEX) and the air-cooled radiator (ACR) can be considered as the most important system components from thermoeconomic optimization point of view and it remains unchanged at the considered different groups of ambient reference conditions (Figure 6).

Table 4. Thermoeconomic performance parameters at the summer operation condition of the system

($T_0 = 298.15K$, $p_0 = 1.013bar$ and $\varphi_0 = 55\%$)

Component	c_P, €/GJ	c_F, €/GJ	$\dot{C}_{D,k} + \dot{Z}_k$ €/h	f_k, %	r_k, %
BEa,b	22.50	7.72	7.28	69.86	190.66
G	30.00	24.08	5.08	82.77	24.65
HEX	64.44	4.47	3.93	80.89	1345.7
ACR	369.44	11.22	3.56	87.43	3193.6
HS	5.28	0.42	2.18	99.92	1186.2
M	6.67	6.03	1.20	95.53	8.47
TCC	490.83	44.53	1.06	37.34	1002.6
PHEX CC	5.56	0.42	0.92	97.92	1244.8
PHEX PC/SC	8.33	4.47	0.37	73.99	99.68
WP3	1494.44	52.83	0.21	51.0	2728.3
CCa,b	27.50	6.53	0.18	75.47	321.9
OT	13.89	4.47	0.17	14.18	209.5
WP2	1073.89	52.83	0.16	60.52	1932.2
WP1a,b	1533.33	52.83	0.15	58.20	2802.0
TCa,b	6.67	4.47	0.14	29.04	46.35
V	29.44	11.22	0.11	18.79	0.0034

The biogas engines and generator have the highest value of $\dot{C}_D + \dot{Z}_k$, but their values of the relative cost difference are not among the highest. Furthermore, according to previous studies made by author [6], [33], the exergy efficiency of the generator is very high (96%), while the exergy destruction of the BEa,b that can be avoided is a small part (18% of total exergy destruction within this component) due to the source of the thermodynamic inefficient in this unit – the chemical reaction of combustion, which is a high irreversible process. Therefore, the second highest value of the sum $\dot{C}_D + \dot{Z}_k$ is attributable to the high value of $\dot{Z}_k$ for the generator (86.7% of the sum $\dot{C}_D + \dot{Z}_k$), and not due to cost rate of exergy destruction. It is important to note that by reason of technological limitations (M, TCa,b, CCa,b, OT, BEa,b and G are mounted together to form a single piece of equipment called the CHP module), the capital investment for the generator, constituting the value of $\dot{Z}_k$ could not be decreased. It would to be appropriate to investigate, however, how a slightly increasing of the biogas engines exergy efficiency may be affect to the overall cost effectiveness.

The heat exchanger (HEX) and the air-cooled radiator (ACR) are the components having the next highest values of $\dot{C}_D + \dot{Z}_k$. The high values of f_k and r_k suggest that it would be cost effective to decrease their capital investments and to increase their exergy efficiencies.

According to the control algorithm of the CHP system, ACR and WP3 are switched on when the temperature of the returned from the digesters water, T_{22}, becomes equal to 68°C. At the winter ambient reference conditions, the thermal needs of mesophilic fermentation process, occurring within the digesters are high. Thus, the temperature of the returned from the digesters water is below 68°C and, as can be seen from Figure 6, the air-cooled radiator is not exploited at the winter operating conditions. Therefore, the heat exchanger (HEX) is the third major component from thermoeconomic optimization point of view. Consequently, if partial load of the CHP modules be changed, the optimum between the thermodynamic and economic system parameters could be achieved.

It is apparent from Table 4 that the smallest value of the sum $\dot{C}_D + \dot{Z}_k$ is observed in the water pumps (WP1a,b and WP2), the turbochargers (TCa,b) and the 3-way valve (V). Contribution of these components to the process of cost formation of the products in the CHP system is insignificant.

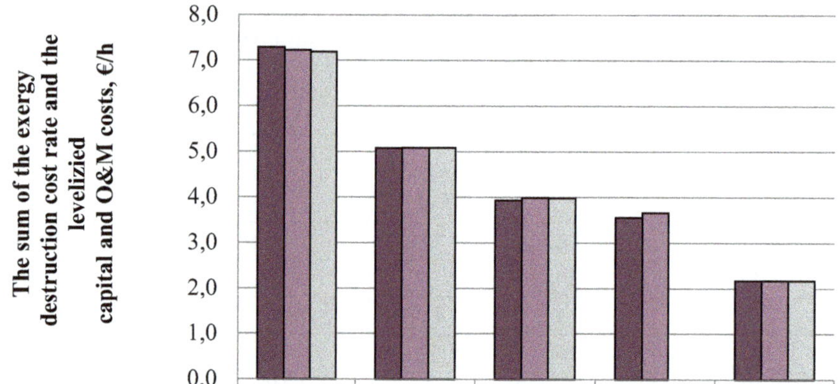

	BEa,b	G	HEX	ACR	HS
■ at the summer operating condition	7,28	5,08	3,92	3,55	2,18
■ at the spring-autumn operating condition	7,22	5,08	3,98	3,66	2,18
□ at the winter operating condition	7,18	5,08	3,98		2,18

Figure 6. The most important system components from thermoeconomic optimization point of view at different ambient reference conditions

Both thermoeconomic and traditional economic cost analysis ensure the attribution of all system expenditures (capital investment, O&M costs, fuel cost and etc.) to the costs of electricity and heat produced. Notwithstanding, from the Figure 7 we can see that there is a difference between the results obtained by the two methods.

The analysis of the results presented in Figure 7 reveals that the thermoeconomic cost of produced electrical power relative to cost of thermal energy is comparable to that of market price of electricity to thermal and this statement corroborates the idea of Bagdanaviciusa, Sansom *et al.* [15], who suggested that the thermoeconomic analysis provides a better estimate of heat value than that from the economic cost analysis. This is because of the fact that the economic analysis does not take into account the quality of the products of the system.

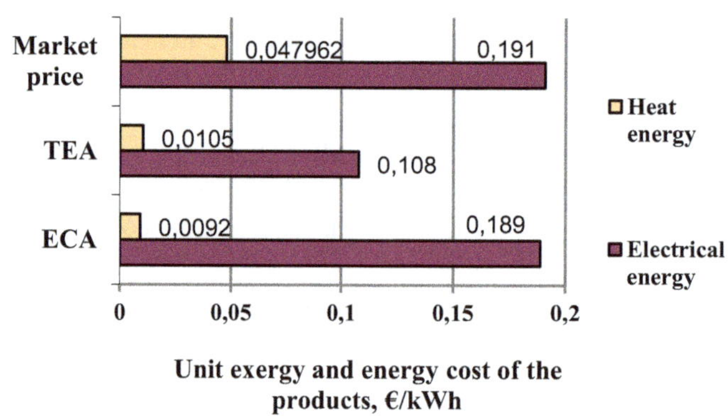

Figure 7. Comparison of results from economic and thermoeconomic analysis of the biogas CHP system

CONCLUSION

This paper has explained processes of cost formation in a cogeneration system driven by biogas engines by applying the thermoeconomic analysis based on the SPECO method. One of the more significant findings to emerge from this study is that the specific unit exergetic cost of the net electrical power and heat energy produced by the biogas engine driven cogeneration system, situated in Bulgaria are calculated to be 0.11 €/kWh and 0.01 €/kWh, respectively. An issue that was not addressed in this study was how some design parameters have an effect on the cost of the total system product.

Returning to the problems and aims posed at the beginning of this study, it is now possible to state that the thermoeconomic interconnections of the components were defined, and the correct theoretical framework for optimization can be determined. It is clear that, if the optimization objective is to increase the thermodynamic efficiency of the major system components or the overall CHP system on account of declining cost of the system product, it is necessary to find the extremum of a clearly defined function. This function should be expressed as a quantitative relationship between irreversibility of the thermodynamic processes, the capital investment and the O&M costs of the system components. The obtained results demonstrate exactly the relationship between the monetary costs, exergy destruction and losses within the analyzed CHP system.

Further research should be undertaken to investigate the sensitivity of objective function from changing of the design parameters and after that to perform the optimization of the system, i.e. to search the minimum of the defined above total cost function, $\dot{C}_{tot}$.

ACKNOWLEDGMENTS

The author is grateful to the company along with the technical staff, located in the Varna Wastewater Treatment Plant, Bulgaria for the support and the assistance. The author also would like to thank Assoc. Professor Dr. Dimityr Rusev for his valuable advices, brilliant ideas and constructive comments, which have been very useful in conducting of every one author's research.

NOMENCLATURE

$\dot{C}$ - exergy cost rate, €/h
c - unit exergy cost, €/kWh
CRF - capital recovery factor
$\dot{Ex}$ - exergy flow rate, kW
$\dot{Ex}_D$ - exergy destruction rate, kW
f - exergoeconomic factor, %
n - life time of the system, years
r - relative cost difference, %
$\dot{W}$ - work rate or power, kW
$\dot{Z}$ - levelized capital, operating and maintenance cost per unit time, €/h
φ - relative humidity of the air, %
τ - annual number hours of system operation at 75% load

Subscripts

D – destruction
e - exit
F – fuel
i - inlet
k – *kth* component
L – levelized
L – losses
P - product
q - heat
tot – total
w - work
ACR – air-cooled radiator
BEa,b – biogas engine
CC – carrying charges
CCa,b – cooling charger
CHP – combined heat and power
CI – capital investment
ECA – economic cost analysis
FC – fuel cost

G - generator
GT – gas turbine
HEX – heat exchanger
HS – hydraulic separator
ICE – internal combustion engine
M – mixer
O&M - operating and maintenance
ORC – organic Rankine cycle
OT – oil tank
PEC - purchased equipment cost
PHEX CC – plate heat exchanger from cooling circuit
PHEX PC/SC – plate heat exchanger from secondary circuit
ST- stream turbine
TCa,b – turbocharger
TCC – technological circuit cooled
TEA – thermoeconomic analysis
V - 3-way valve
WP – water pump

REFERENCES

[1] Bhatt, M.S. (2000). Energy audit case studies II – air conditioning (cooling) systems. Journal of Applied Thermal Engineering, 20(3), p. 297-307.

[2] Khurana, S., Banerjee, R., Gaitonde, U. (2005). Energy balance and cogeneration for a cement plant. Journal of Applied Thermal Engineering, 22(5), p. 485-494.

[3] Kabir, G., Abubakar, A.I., El-Nafaty, U.A. (2010). Energy audit and conservation opportunities for pyroprocessing unit of a typical dry process cement plant. Journal of Energy, 35(3), p. 1237-1243.

[4] Sun, Z. S. (2008). Energy efficiency and economic feasibility analysis of cogeneration system driven by gas engine, Journal of Energy and Building, 40, (2), 2008, pp. 126-130.

[5] Moran M., Shapiro H. (2006). Fundamentals of Engineering Thermodynamics. 5th ed. England: John Wiley & Sons Ltd.

[6] Doseva, N., Chakyrova, D. (2015). Energy and exergy analysis of a cogeneration system with biogas engines. Journal of Thermal Engineering, 1, (3), p. 391-401.

[7] Abusoglu A., Kanoglu M. (2009). Exergetic and thermoeconomic analyses of diesel engine powered cogeneration: Part 2 – Application. Journal of Applied Thermal Engineering, 29(2-3), p. 242-249.

[8] Mohammadkhani, F., Khalilarya, S, Mirazaee, I. (2013). Effect of ambient temperature on exergetic and exergoeconomic parameters of a CHP System. Journal of Environmentally Friendly Processes; 1(3), p. 28-37.

[9] Seyyedvalilua, M.H., Mohammadkhani, F., Khalilaryac, S. (2015). A parametric study on exergy and exergoeconomic analysis of a diesel engine based combined heat and power system. International Journal of Engineering, 28(4), p.608-617.

[10] Cavalcanti, E., Motta, H. (2015). Exergoeconomic analysis of a solar-powered / fuel assisted Rankine cycle for power generation. Journal of Energy, 88, p.555-562.

[11] Tempesti, D., Fiaschi, D., Gabuzzini (2012). Thermo-economic assessment of a micro CHP system fuelled by geothermal and solar energy. Proceedings of the 25th international conference ECOS, Italy, Volume III, p.321-359.

[12] Palomino, R. G., Nebra, S. Energetic, exergetic and exergetic cost analysis for a cogeneration system integrated by an internal combustion engine, HRSG and an absorption refrigeration system. Proceedings of ESDA04 7th Biennial ASME Conference Engineering Systems Design and Analysis, July 19-22, 2004, Manchester, United Kingdom.

[13] Temir, G., Bilge, D. (2004). Thermoeconomic analysis of a trigeneration system. Journal of Applied Thermal Engineering, 24(17-18), p. 2689-2699.

[14] Balli O., Aras H. (2010). Thermodynamic and thermoeconomic analyses of a trigeneration (TRIGEN) system with a gas-diesel engine: Part II – An Application. Journal of Energy Conversion and Management, 51(11), p. 2260-2271.

[15] Bagdanavicius, A., Sansom, R., Jenkins, N. et al. (2012). Economic and exergoeconomic analysis of micro GT and ORC cogeneration system. Proceedings of *the 25th International conference ECOS, Italy*, Volume III, p.98-108.

[16] Mert, M., Dilmaç, Ö., Özkan, S. et al. (2012). Exergoeconomic analysis of cogeneration plant in an iron and steel factory. Journal of Energy, 46(1), p.78-84.

[17] Coplan, C.O., Yesin, T. (2006). Energetic, exergetic and thermoeconomic analysis of Bilkent combined cycle cogeneration plant. International Journal of Energy Research, 30, p. 875-889.

[18] Athari, H., Soltani, S. Mohammad, S. et al. (2014). Exergoeconomic analysis of a biomass post-firing combined-cycle power plant. Journal of Energy, 77, p.553-561.

[19] Bagdanavicius, A., Jenkins, N. (2011). Exergoeconomic evaluation of small scale CHP systems. Proceedings of the 8th International Conference environmental engineering, May 19–20, Vilnius, Lithuania, p. 727-734.

[20] Sotomonte, C., Venturini, O. (2010). Exergoeconomic analysis of small-scale biomass steam cogeneration. Proceedings of 13th Brazilian Congress of Thermal Sciences and Engineering December 05-10, 2010, Uberlandia, MG, Brazil.

[21] Colmenar-Santos, A., Zarzuelo-Puch, G., Borge-Diez,,D., García-Dieguez, C. (2016). Thermodynamic and exergoeconomic analysis of energy recovery system of biogas from a wastewater treatment plant and use in a Stirling engine. Journal of Renewable Energy, 88, p.171-184.

[22] Sung, T., Kim, S., Kim, K. (2017). Thermoeconomic analysis of a biogas-fueled micro-gas turbine with a bottoming organic Rankine cycle for a sewage sludge and food waste treatment plant in the Republic of Korea, Journal of Applied Thermal Engineering, 127, p. 963-974.

[23] Abusoglu, A., Demir, S., Kanoglu, M. (2013). Thermoeconomic analysis of a biogas engine powered cogeneration system. Journal of Thermal Science and Technology, 33(2), p. 9-21.

[24] Bejan A, Tsatsaronis G, Moran M. (1996). Thermal design and optimization. Wiley: New York.

[25]Doseva, N. (2017). Thermoeconomic analysis and optimization of energy systems. Doctoral thesis, Varna Technical University, Bulgaria.

[26] Tsatsaronis, G. (1984). Combination of exergetic and economic analysis in energy conversion processes. Proceedings, European Congress on Economic and Management of Energy in Industry, Portugal, Apr. 2-5, Vol. 1, pp.151-157.

[27] Bonnet, S., Alaphilippe, Stouffs, P. (2005). Energy, exergy and cost analysis of a micro-cogeneration system based on an Ericsson engine. International Journal of Thermal Sciences, 44, p. 1161-1168.

[28] Aras, H., Balli, O., Hepbasli, A. (2008). Exergoeconomic analysis of a combined heat and power system with the micro gas turbine (MGTCHP). Journal of Energy Explore Exploit, 26(1), p.53-70.

[29] Balli, O., Aras, H., Hepbasli, A. (2007). Exergoeconomic analysis of a combined heat and power (CHP) system. Journal of Energy Research, 32, p.273-289.

[30] Shokati, N., Ranjabar, F., Yari, M. (2004). A comparative analysis of Rankine and absorption power cycle from exergoeconomic viewpoint. Journal of Energy Conversion Management, 88, p.675-668.

[31] Kim, S.M., Oh, S.D., Kwon, Y.H. et al. (1998). Exergoeconomic analysis pf thermal systems. Journal of Energy, 23(5), p.393-406.

[32] Arriola, D., Olivera Junior, S. (2013). Tetra-combined trigeneration system. Thermoeconomic analysis. Proceedings of 17th International Congress of Mechanical Engineering.

[33] Doseva, N. (2014). Advanced exergetic analysis of cogeneration system with a biogas engine. Proceedings of the 14th International Multidisciplinary Scientific GeoConference SGEM 2014, Book 4, Vol. 1, p. 11-18.

4

EFFECT OF DIFFERENT LEVELS OF HYDROGEN + LPG ADDITION ON EMISSIONS AND PERFORMANCE OF A COMPRESSION IGNITION ENGINE

Yasin Karagöz [1], Hasan Köten [2,*]

ABSTRACT

Compelling emission regulations and diesel engines damage on environment have led researchers to work on alternative fuels such as LPG and hydrogen in recent years. Diesel engine costs are also increasing, especially on the increased costs of after-treatment equipment used in diesel engines. It is also known that diesel engines release NO_x and smoke emissions, which are highly harmful to the environment and to the living health. The effect of the emissions of LPG + H_2 (0%, 15%, 30% and 45%) with different energy contents at full load and same torque (70 Nm) at constant engine speed of 1200 rpm is investigated on CO, THC, NO_x and Smoke emissions. From the results obtained, there was a slight increase in brake thermal efficiency, CO and THC emissions. However, significant improvement in NO_x and smoke emissions has been achieved. The results show that using LPG + H_2 in diesel engines, older technology and lower cost after-treatment equipment can be used, and that NO_x and smoke emissions of diesel engines can be reduced significantly. Also, using small amount of hydrogen (20% of total gas mixture) and LPG, an important improvement can be achieved, and the harmful effects of diesel engines can be suppressed thanks to unique properties of hydrogen fuel.

Keywords: *Liquefied Petroleum Gas, LPG, LPG + H_2, Common-Rail*

INTRODUCTION

Many researchers have focused on the use of gas fuels as alternative fuels in engines [1]. When gas fuels are used in diesel engines in as a dual fuel, some improvement in engine performance is observed, while smoke emissions are reduced [2]. They exhibit low heat and high thermal efficiency, especially under full load or high load operating conditions (peak loads) [3]. Gas fuels can be used in diesel engines in a variety of ways. These are: timed spraying in the manifold, direct spraying into the cylinder, and continuous spraying into the manifold [3]. The method of spraying directly into the combustion chamber is a method of injecting gases into the combustion chamber directly which may lead injectors to corrode [3]. In addition, more special injectors and some modifications to the cylinder head are required. In continuous manifold spraying method; flame spatter, knocking, pre-ignition problems may be encountered. With the aid of an ECU, the injection of gas into the air sucked from the intake manifold by injectors and the ignition of the gas fuel by some pilot injection are the most effective methods, and almost all of the studies in the literature have used gas fuels in diesel engines with this method.

LPG (liquefied petroleum gas) is a clean energy source for future diesel engines and can be produced from alternative energy sources [4]. LPG is also an excellent and economical fuel for exhaust emissions and engine performance [1]. Thanks to the high self-ignition temperature of LPG fuels, they can be easily used as a dual fuel by spraying in manifolds in diesel engines [5]. The easiest and most reliable method of using LPG or other gas fuels in diesel engines is to inject the gas fuel into the intake manifold and then carry out the flammability by pilot injection with the engine in small quantities [3]. Because LPG has a self-ignition temperature of 788 K, it requires an ignition source [6]. Diesel fuel is well suited for ignition with a self-ignition temperature of 525 K, making it the most effective method in practice for pilot injection[3]. LPG and diesel fuel have flame velocities of 4.8 cm / s and 5.12 cm / s at 1.6 air excess coefficient, respectively. With high flame speed and diffusion coefficient, the use of LPG fuel in engines with a small amount of pilot injection will increase combustion efficiency as it will consume more O_2 during combustion and cause more oxidation [3]. In view of the literature results, it is generally evident that reduced HC and CO emissions is the result [3].

This paper was recommended for publication in revised form by Regional Editor Tolga Taner
[1]Yildiz Technical University, Mechanical Engineering Department, Automotive Division, İstanbul, Turkey.
[2]Medeniyet University, Mechanical Engineering Department, Istanbul, Turkey.
**E-mail address: hasan.koten@medeniyet.edu.tr*

LPG is a clean fuel consisting of butane and propane mixture gases [1]. However, from the literature, LPG's gas composition changes significantly affect emissions and engine efficiency [7]. The propane / butane ratio of LPG fuel varies from country to country. Propane / butane mixtures at different ratios affect the combustion process and exhaust emissions [8].

Lata et al. [3] used a 4-cylinder, turbocharged and intercooler diesel engine generator setting with 62.5 kw power. LPG and H_2 / LPG fuels of various gas ratios were sent from the intake manifold and ignited by pilot injection. In thermal efficiency, a maximum 6% efficiency increase was observed with the use of LPG instead of diesel as 40% energy content. Compared to diesel only, HC and CO increased a little; NO_x and smoke emissions have decreased somewhat. When LPG / H_2 (70% LPG, 30% H_2 by volume) was used as energy content instead of 40% diesel, a 27% increase in thermal efficiency and a 68% decrease in HC were observed [3]. It has been found that thermal efficiency is increased when using gas LPG or LPG / H_2 fuels only in situations above 10% load [3]. The reason for this is the increase in ignition delay and the increase in flame-out zones, which reduces the combustion efficiency and decreases the thermal efficiency. Also, at high loads (> 60%), use of over 70% of LPG is defined as the knock limit.

Poonia et al. [7] experimentally examined the effect of intake air temperature, the amount of pilot injected diesel and the amount of EGR on the performance and emissions of the diesel operating with the engine / LPG mixture fuel. It was found that the HC emission increased as the pilot injection amount increased. Optimization in the heating of EGR and intake air led to a slight increase in thermal efficiency and a decrease in HC. Vijayabalan et al. [9] studied the single-cylinder, air-cooled engine by spraying the pilot diesel and by spraying LPG from the intake manifold. The decrease in NO_x emissions is observed at all load stages. In dual fuel usage, higher emissions of CO and HC emissions are evident and thermal efficiency is reduced due to poor ignition under low load conditions. By putting glow plugs in the combustion chamber, this problem is solved which became evident at low loads. In this case, under low load conditions, a 3% efficiency increase, a decrease in 69% HC, 50% CO and 9% smoke emissions was found, but no change in NO_x emissions was observed. Zhili has found that low NO_x and high thermal efficiency can be achieved by adding LPG and di-methyl ether to a diesel engine. Selim [10] et al. tested LPG and CNG using the diesel fuel and Jojoba methyl ester (JME) as pilot fuel in the Ricardo E6 CFR setane engine. As a result of the studies; the maximum pressure increase rate was found to be higher in LPG or CNG additional fuel engines. The maximum pressure increase was found to be as follows: diesel / LPG, diesel / CNG, JME / LGP, JME / CNG. Emission values of CO and HC emissions were found to be as follows: JME / LPG, JME / CNG, diesel / LPG, diesel / CNG. Furthermore, a high amount of LPG fuel was observed to be knocking at high compression ratios and under full load. Qi [5] used mixtures of LPG at 0%, 10%, 20%, 30% and 40% mass in a direct injection diesel engine, the effect of LPG as an additional fuel on emissions and engine performance is investigated. In the test results, it was found that with the increase of LPG, the decrease of the pressure inside the peak cylinder and the increase of the ignition delay is obtained. As LPG is used as additional fuel at low speed, there is a slight increase in heat emission rate and a slight decrease in combustion duration; using over 20% LPG at high speeds; the heat dissipation rate is reduced and the combustion time is prolonged. A slight increase in HC and CO emissions has been observed when there is some decrease in the use of LPG as an additional fuel in NO_x and smoke emissions. A slight increase in specific fuel consumption has also been observed, but LPG is projected to be a more economical alternative fuel with a lower price [5].

Hydrogen, carbon-free, with superior combustion properties and being an environmentally friendly fuel. Like electricity, it is an excellent energy carrier in H_2 [11]. Moreover, H_2 can be produced from renewable energy sources. Gasoline engines are well suited for H_2 conversion, but can also be used in H_2 diesel engines with some simple modifications [12]. Due to the high self-ignition temperature of H_2 (576°C), it is difficult to use directly in diesel engines. They need a number of ignition sources [13]. H_2 is injected into the air sucked from the intake manifold in the diesel engines and the ignition is carried out by pilot injection with a small amount of diesel fuel. H_2 is the easiest and most reliable method of application for diesel engines [11].

H_2 has a wide flammability range and a high combustion rate, making diesel / H_2 dual fuel more superior [3]. Moreover, since H_2 has a high adiabatic flame rate, thermodynamics allows the cycle to approach the ideal conditions more closely [14]. Due to the high diffusion rate of H_2 (0.61 cm^2 / s), air and H_2 can form a better mixture.

In diesel engines, H_2 fuel is usually injected during the intake stroke in the intake manifold, and as described above for the use of LPG fuel in diesel engines, the pilot diesel ignition is the most reliable and simple method [15]. Saravan et al. [16] used H_2 fuel by pilot injection with a stationary diesel engine, engine fuel. It has been observed that a lower amount of water, particulate matter and NO_x are released. With the increase in thermal efficiency, less fuel consumption is achieved. Liew et al. [17] have examined the use of H_2 as an additional fuel in a diesel engine. It has been observed that by using H_2 in high amounts and at high loads, the intracellular pressures and heat dissipation are increased. Furthermore, the 2-step combustion process has been found to turn into a 3-phase combustion process by using H_2 as an additional fuel. In another study by Saravan [18] and others, they sprayed H_2 on a single-cylinder diesel engine intake manifold. At a 75% load condition, they found that the H_2 / diesel increased the efficiency of the dual fuel by 15%. In general, significant reductions were observed in emissions, but only an increase of 1-2% was observed in NO_x emissions. Bretti [19] modified the diesel engine to deliver diesel injectors both H_2 and diesel. The average effective pressure increased from 25 bar to 35 bar and the thermal efficiency increased by 40%. Lilik [20] et al. have investigated the use of H_2 as an additional fuel in a 4-cylinder, turbocharged, common-rail engine at 0%, 2.5%, 5%, 7.5%, 10% and 15% of the energy content. Emissions and engine performance at engine speeds of 1800 and 3600 rpm were investigated in a 25% and 75% load range. It has been observed that the engine has a significant amount of injection delay in its own ECU. All emissions, especially NO_x, have declined significantly. It is also found that the NO / NO_2 ratio of the released NO_x emissions changes, the NO emissions decrease, and the NO_2 emissions increase. Gomes Antunos et al. [21] observed an increase in thermal efficiency of 43% and a decrease in NO_x of 20% with the use of H_2 as an additional fuel for diesel engines. Bose [11] and others have sent a single-cylinder, four-stroke, diesel engine with solenoids from the intake manifold as 0.15 kg / h of additional fuel. (0.15 kg / h) of additional H_2 is sent at constant load conditions of 20%, 40%, 60% and 80%. A slight increase in NO_x was observed under all operating conditions. On top of that, 10% and 20% EGR were used to reduce NO_x emissions. In the case of using EGR with H_2, the thermal efficiency is reduced, but it is still higher than that of diesel fuel. While significant reductions in HC, CO, CO_2 and smoke emissions were observed, only NO_x emissions were observed to increase, but NO_x emissions were also controlled by EGR. Masood [22] and others have sent the diesel engine at different ratios of H_2, and have fired the pilot by spraying. Problems such as difficult ignition, low power output has been overcome by the improvement of the flame extinction zone. Heffel [23] and others have sent H_2 as an additional fuel to the to four-cylinder, Ford, ZETEC engine. 0.78 kg / h and 1.63 kg / h H_2 at 1500 rpm engine speed with EGR. Experimental results have been observed to improve on all emissions, primarily NO_x. Saravan [24] and others have reported that they have been working together with EGR in a diesel engine, sending 20 L / min H_2 as additional fuel. Decrease in all emissions was observed, especially in the case of smoke, particulate, NO_x emissions. Varde and Frame [25] on their direct-injection, single-cylinder, diesel engine, sent H_2 as an additional fuel and reduced work emissions. Using 12.5% hydrogen (energy basis), thermal efficiency increased from 30.5% to 33.7%. However, when H_2 with a small energy content of 5% is added, the thermal efficiency decreases. The reason is thought to be flame propagation is not fully realized because of the very poor H_2 / air mixture. By using H_2 as an additional fuel, the thermal efficiency is increased by reducing the combustion phase and by shortening it to a shorter period of time. Shirk et al. [26] and others have investigated the effect of diesel replacement of 5% and 10% energy content on the common-rail engine on thermal efficiency and exhaust emissions. All emissions, especially NO_x, have been observed to decrease.

In most of the studies carried out, only LPG fuel was used as supplementary gas fuel, or only hydrogen was used as supplementary gas fuel. The effect of diesel, LPG (30% propane and 70% butane) and hydrogen combustion were experimentally investigated in the study. On the other hand, the addition a very low amount of hydrogen fuel into LPG fuel (20% hydrogen by volume of the total gas fuel) and its effect on CO, THC, NO_x and smoke emissions are investigated. Thanks to unique properties of hydrogen, the combustion efficiency was improved in the study. Also, the engine brake torque was kept at 70 Nm and the brake engine speed was fixed at 1200 rpm. For this reason, the effect of different levels of gaseous fuel (LPG + H_2) addition was investigated at constant brake engine speed and constant brake engine torque differently from other studies.

TEST SETUP

Figure 1 gives a schematic view of the test setup. Tests were carried out on a CI engine with single cylinder, 4 strokes, naturally aspirated, common rail fuel system. The CI engine, which is normally a mechanical fuel system, has been turned into a common rail system by using the engine control unit developed by us and placing appropriate electromagnetic injectors. In the tests, the rail pressure of the common-rail system was fixed at a pressure of 1000 bar. The diesel fuel used meets EN590 standards. The LPG fuel used is 30% propane and

70% butane gas. A flow rate of LPG + H_2 gas fuel mixture was measured using a thermal gas flow meter. AVL Digas 4000 emission analyzer and AVL 415S smoke ether is used. All tests were carried out at Yildiz Technical University, Internal Combustion Engines Laboratory. All data obtained were collected with the NI 6215 daq card. An interface has been developed using LabView software and the data can be recorded as desired.

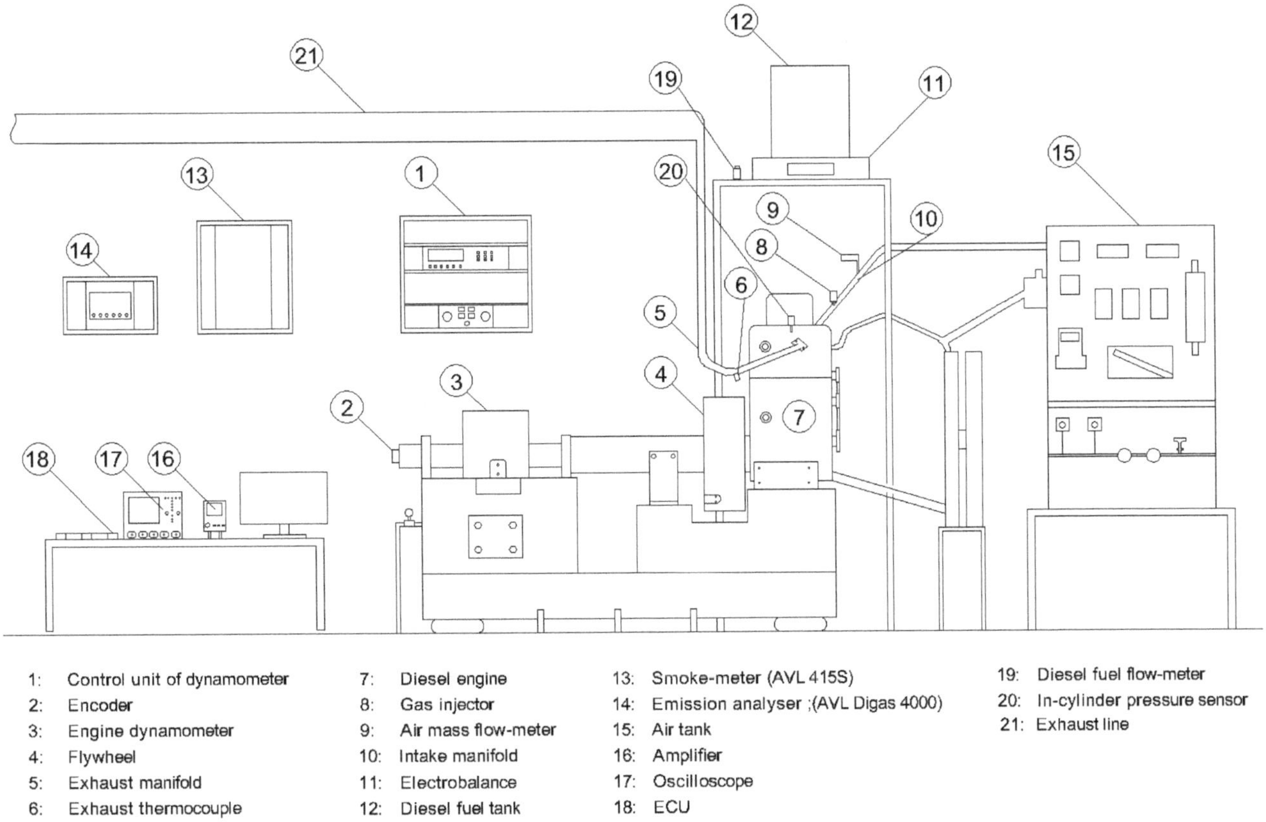

Figure 1. Test setup schematics

CI ENGINE AND EDDY CURRENT DYNAMOMETER

Table 1 lists the test engine specifications and dynamometer properties. Single stroke, 4 stroke, 1160 cc cylinder capacity, Erin Motor brand, direct injection engine is used. In maximum 40 kW load capacity, water cooling type, CI engine was installed using Eddy current dyno. All experiments were carried out at Yildiz Technical University, Internal Combustion Engine Laboratory.

Table 1. The test engine specs and dynamometer properties

Engine brand	Erin motor
Aspiration method	Natural
Max power	18 kW@2700 rpm
Number of cylinders	1
Bore × stroke [mm]	108 × 127
Cylinder volume [cm^3]	1163
Compression ratio	14.7
Speed range min-max [rpm]	800 – 2700
Number of intake & exhaust valves	2 & 2
Cooling type	Water cooled
Dyno & power [kW]	Eddy current & 40

LPG + H_2 FUEL SYSTEM

The schematic of the LPG + H_2 gas fuel system is given in Figure 2. Hydrogen and LPG are supplied separately from high pressure tanks. The hydrogen rate was fixed 20% of total gas mixture fuel (hydrogen + LPG). In all gas mixture addition cases, hydrogen rate was kept constant as 20% of total gas mixture as volume basis. Also, using needle valves after the high pressure tanks, the flow rates for both of gases were adjusted to provide 20% hydrogen rate of total gas mixture. Then, both gases are reduced to 2 bar by pressure regulator, even mixed. The flow rates of both gases are fixed with a needle valve. The flow rate of the hydrogen was fixed to be 20% by volume of the total fuel. A line regulator is installed. Then, a shut-off valve is placed on the line. What's more, even the check valve is placed on the line just before the engine, despite a possible backfiring situation. In addition, it is aimed to prevent a possible line pressure increase by placing a relief valve. In addition, it is aimed that the gas mixture flow will be made more stable by placing one high pressure resistant buffer tank. One thermal flow meter and one rotameter are located on the line. At the end of the line, a Keihin brand LPG-CNG injector was installed to send gas fuel to the engine intake manifold.

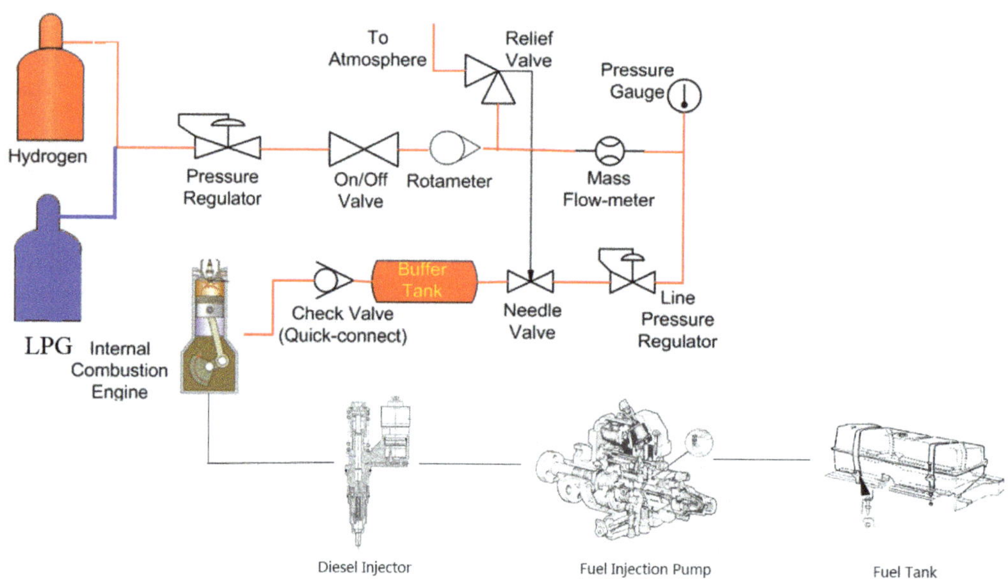

Figure 2. LPG + H_2 and diesel fuel system

THE ENGINE CONTROL SYSTEM

With the developed electronic control system, both diesel and gas injectors can be controlled. By means of an incremental encoder installed on the crankshaft of the engine, the engine speed and the piston position are determined, and the injectors are controlled in the determined advance and the injection quantities, and the engine is supplied with gas fuel (LPG + H_2) or diesel fuel. The gas fuel priming advance is set to TDC in suction stroke while the diesel fuel injection advance is set to 28°BTDC.

EMISSION MEASUREMENT SYSTEMS

Carbon monoxide, oxides of nitrogen and total unburned hydrocarbons emissions were measured with AVL Digas 4000. With the AVL 415S, smoke emission values are measured as FSN. Since the engine power and the engine speed are kept constant during operation, no specific emission is required.

THE TEST CONDITIONS

The CI engine was loaded at a constant engine speed of 1200 rpm and at full engine load. Test conditions are given in Table 2. The effects of LPG + H_2 gas fuel addition on engine performance and emissions were investigated in different gas fuel energy contents (0%, 15%, 30% and 45%). Brake engine torque value is fixed at 70 Nm in all fuel conditions. The diesel spraying rate is fixed at 28°BTDC. During the work, no engine knock or backfiring problem has been encountered. The uncertainty and total accuracy values are presented in Table 3. EN 590 diesel automotive fuel was used in the study. Also, the energy content means that the energy

Table 2. The test properties summary.

Engine brake torque	Engine speed	Start of diesel injection	LPG+H_2 energy fraction	Diesel energy fraction
70 Nm	1200 rpm	28°BTDC	0%	100%
70 Nm	1200 rpm	28°BTDC	15%	85%
70 Nm	1200 rpm	28°BTDC	30%	70%
70Nm	1200 rpm	28°BTDC	45%	55%

Table 3. Measurement Accuracies and Total Uncertainty

Parameter	Device	Accuracy
Engine torque	Load cell	±0.05 Nm
Engine speed	Incremental Encoder	±5 rpm
Cylinder pressure	Kistler 6253C	±0.5 %
Diesel flow rate	Biotech VZS-005	±1 % [of reading]
LPG + H_2 flow rate	New-Flow TMF	±1 % [F.S.]
CO	AVL DiCom 4000	0.01 % Vol.
THC	AVL DiCom 4000	1 ppm
NO_x	AVL DiCom 4000	1 ppm
Smoke	AVL 415S	0.4 % Vol.
Calculated results		Uncertainty value
Engine power	±0.28 %	
Thermal efficiency	±1.10 ÷ 1.51 % [0% ÷ 45% LPG + H_2]	

ratio of gas mixture to the total fuel mixture (diesel + gas mixture (LPG + Hydrogen)). Using lower heating values of fuels, the energy flow rates of fuels was calculated.

RESULTS AND DISCUSSION

The effect of the LPG + H_2 gas mixture addition (0%, 15%, 30% and 45%) on brake thermal efficiency, CO, THC, NO_x and smoke emissions in different energy contents at 1200 rpm constant engine speed and 100% brake engine load investigated experimentally on the work [30-32]. The gas fuel mixture was sent through the suction manifold through the LPG-CNG injector at the suction stroke. On the other hand, the CI engine converted to a common-rail fuel system, the injection rate of the electromagnetic diesel injector is controlled by a self-developed ECU like a gas injector.

The brake thermal efficiency is obtained by dividing the obtained engine power value by the fuel energy consumed by the engine. The obtained brake thermal efficiency values are given in Figure 3 depending on the different LPG + H_2 gas fuel contents. The brake thermal efficiency values decreased by 0.8%, 2.1% and 2.9%, respectively, with the addition of 15%, 30% and 45% of gas fuel, respectively. The main reason for this decline in thermal efficiency is the fact that the CI engine is not designed for LPG and that the diesel injection advance is not changed. However, due to the high diffusion coefficient and high flame velocity of hydrogen in the gas fuel, this decline has remained slight. The result of the study in terms of brake thermal efficiency is similar with the results of Lata and Misra [33] in low and medium operation conditions.

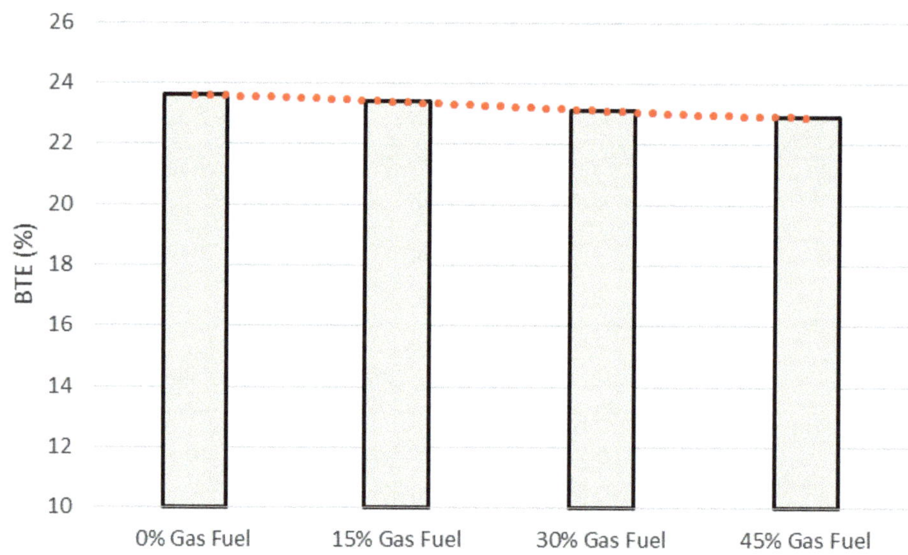

Figure 3. Effect of various level of LPG+H_2 addition on brake thermal efficiency

CO is a toxic pollutant and is quite harmful for environment and living health [27]. The main reason for the rise in CO emissions is the local lack of oxygen [27]. In Figure 4, the addition of LPG + H_2 gas fuel energy content at different ratios has been influenced on CO emissions. The CO and NO_x emission levels are high for a diesel engine since a naturally aspirated CI engine was used and any after-treatment system wasn't used in the experimental study. With the addition of 15%, 30% and 45% gas fuel, only 4%, 12% and 16% increase in diesel fuel has been achieved. Normally, with the increase of gas fuels, it is expected that the regional mixture is better. However, this causes replacing gas mixtures and low volumetric efficiency especially in high speed. The greatest reason for this increase is the lack of oxygen entering the combustion chamber at the same power value due to the volume occupancy of the LPG due to the fact that hydrogen is also a gas fuel. Low oxygen content in cylinder causes to decrease combustion efficiency and CO emissions were increased with gaseous fuel addition. Another reason to increase in CO emissions is the crevices in the combustion chamber. Since the engine was designed according to diesel fuel, when using engine with gaseous fuel, the crevices causes to increase in CO emissions. The results of the study in terms of CO emission is similar with the results of the study of Lata et al. **[34]**. Also, Lata et al. **[34]** found that LPG+H_2 addition up to 60%, the CO emission levels were higher than LPG and H_2 addition at 80% engine load operating condition.

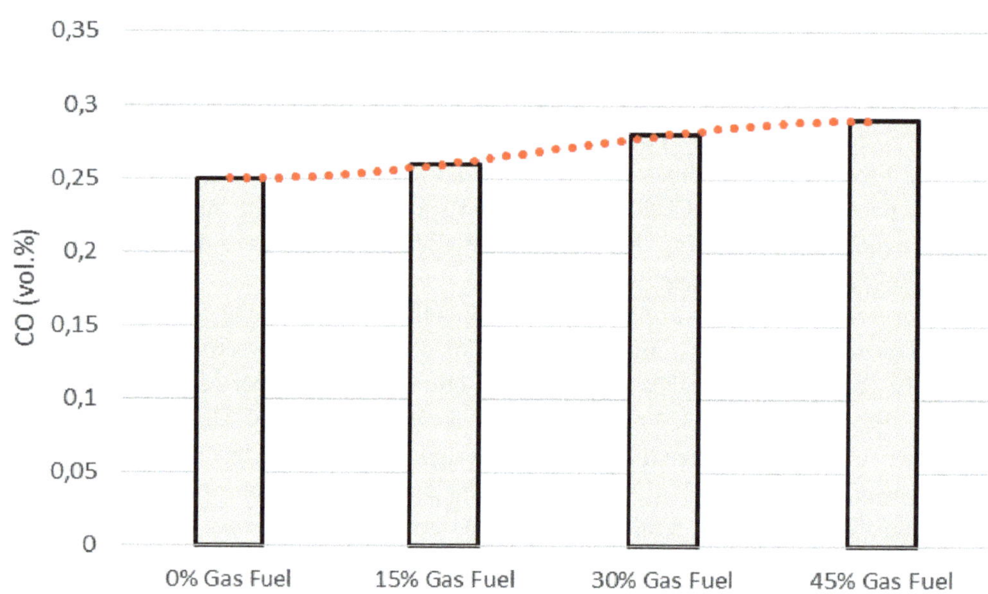

Figure 4. Effect of various level of LPG+H_2 addition on CO emissions

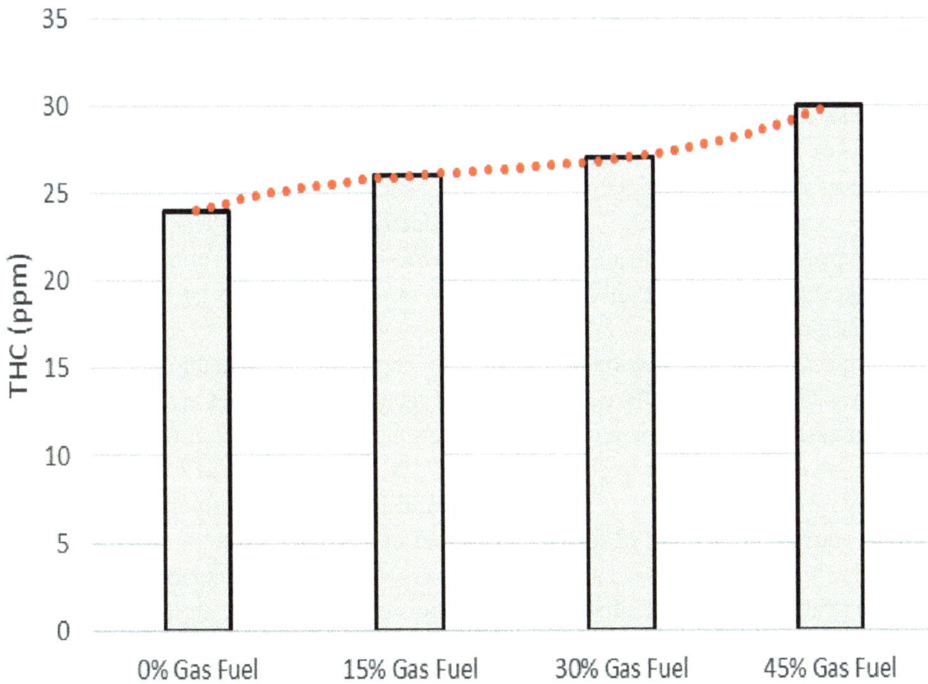

Figure 5. Effect of various level of LPG+H_2 addition on THC emissions

Hydrocarbon-based emissions are derived from the combustion of hydrocarbons in the fuel, either by burning or not burning [8]. Exhausted hydrocarbon emissions as a result of incomplete combustion or combustion in engines are called total hydrocarbons [8]. Figure 5 shows the effect of addition of gas fuel (LPG + H_2) with different energy contents (0%, 15%, 30% and 45%) on THC emissions. With the addition of 15%, 30% and 45% LPG + H_2, the results obtained show an increase of 8.3%, 12.5% and 25% of THC emissions. The greatest reason for this increase in THC emissions is that the engine is not designed for LPG and H_2 gas fuels and THC emissions are accumulated there during combustion and are thrown out of the engine due to the excess of the engine crevice. Lata et al. **[34]** found that the addition of LPG+H2 causes to increase THC emissions. For this reason, the result of the study is similar with the results of Lata et al. **[34]** for low engine load. Also, in high engine load an instability was observed on the study of Late et al. **[34]** in terms of THC emissions.

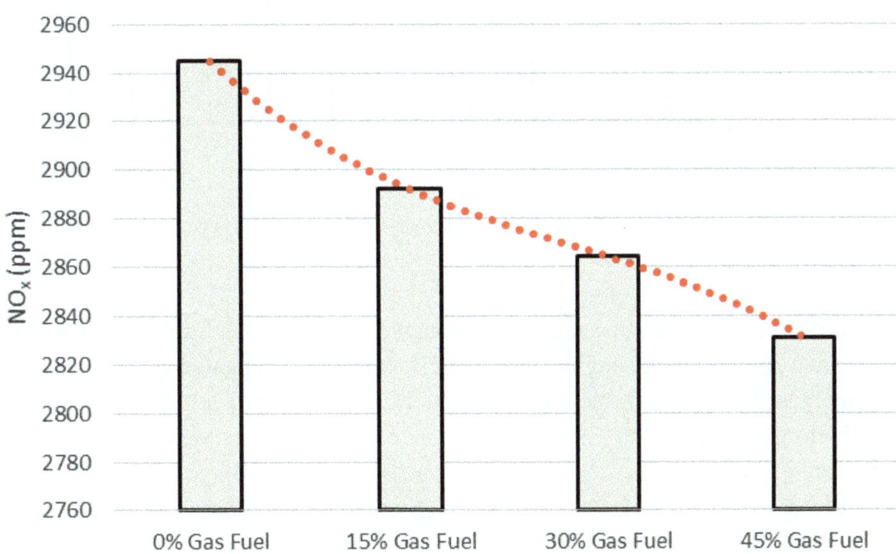

Figure 6. Effect of various level of LPG+H_2 addition on NO_x emission

Nitrogen oxides are formed by nitric monoxide and nitrogen dioxide gases. It is quite poisonous and harmful. They are explained by two different mechanisms [28, 29]. These are Zeldovich mechanism and Fenimore mechanism. In diesel engines, NO emissions are high. The main reason for this is that the diesel engines operate in lean conditions and the pressure inside the cylinder and therefore the temperature is high. The effect of the addition of gas fuel (LPG + H_2) at different ratios (0%, 15%, 30% and 45%) on NO_x emissions is given in Figure 6. From the results obtained, NO_x emissions decreased by 1.8%, 2.8% and 3.9% with addition of LPG + H_2 as energy content of 15%, 30% and 45%, respectively. This result is a decrease in the amount of oxygen due to the air entering the combustion chamber due to the gas fuel covering the volume in the cylinder. Gaseous fuel acts like EGR due to lack of oxygen. Lata et al. **[34]** found similar results at high loads with the study in terms of NO_x emissions.

One of the most important emission problems in CI engines is the formation of carbonaceous-originated smoke emission. They are also generally derived from oxygen deficiency and are one of the most demanding emission types for engine manufacturers in emission regulation [8]. The variation in smoke emissions due to the addition of different LPG + H_2 gas fuels is given in Figure 7. With the addition of 15%, 30% and 45% LPG + H_2, the results were only 13.6%, 20.3% and 28.8% lower than diesel fuel, respectively. The improvement in these smoke emissions is caused by the high diffusion coefficient of gas fuels. In addition, the absence of hydrogen carbon atoms and high flame speed caused smoke emissions to be released. The addition of LPG + H_2 mixtures to diesel fuel can be studied at different rates to see the effects of gas mixtures on the performance and emission as an additional study. Lata et al. **[35]** found that the combustion efficiency was increased with LPG + H2 addition. Improvement in combustion efficiency causes to decrease in smoke emissions. The results of Lata et al. **[34]** are similar with the study in terms of smoke emissions. Lata et al. [34] found that increasing amount of gaseous fuel (LPG, LPG + H_2, H_2) causes to decrease of smoke emissions.

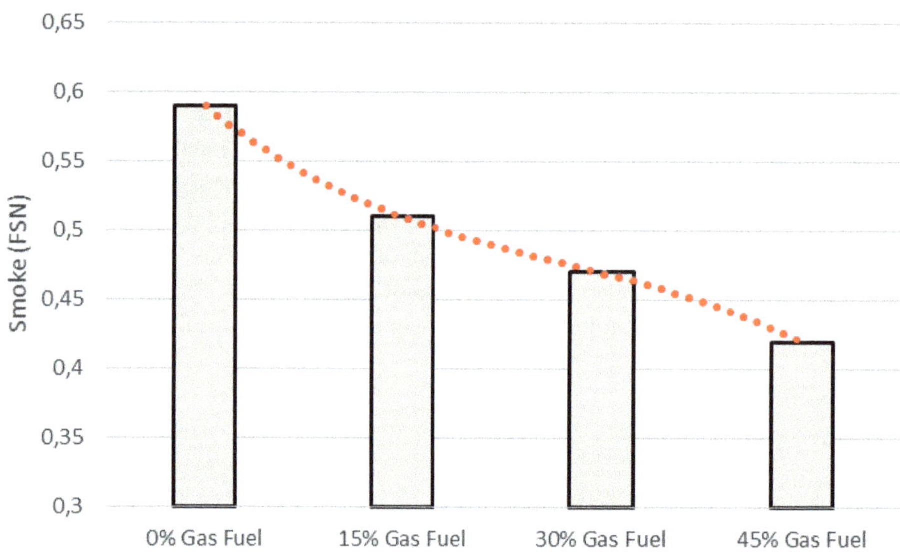

Figure 7. Effect of various level of LPG+H_2 hydrogen addition on smoke emission

CONCLUSION

The effect of LPG + H_2 addition on engine performance and emissions in different energy contents (0%, 15%, 30% and 45%) at a constant brake engine torque of 70 Nm was investigated at 1200 rpm engine speed. The hydrogen ratio in the gas fuel mixture was kept constant 20% of the total gas fuel by volume. The effect of different levels of hydrogen addition on engine performance and emissions were experimentally investigated. The results obtained are given below.

- The BTE decreased by 0.8%, 2.1% and 2.9% respectively with addition of 15%, 30% and 45% LPG + hydrogen compared to neat diesel fuel.
- CO emissions increased by 4%, 12% and 16%, respectively with 15%, 30% and 45% LPG + hydrogen addition compared to neat diesel fuel.

- THC emissions increased by 8.3%, 12.5% and 25% with 15%, 30% and 45% LPG + hydrogen addition compared to neat diesel.
- NO_x emissions improved by 1.8%, 2.8% and 3.9% with 15%, 40% and 75% LPG + hydrogen addition.
- Smoke emissions improved by 13.6%, 20.3% and 28.8% with 15%, 40% and 75% LPG + hydrogen addition compared with neat diesel fuel.
- Using LPG + Hydrogen as additional fuel, the inversely proportional NO_x emissions and smoke emission were decreased simultaneously.
- Lastly, using small amount of hydrogen in gas mixture (LPG + Hydrogen) as additional fuel in diesel engines can be a solution to extinguish detrimental effects of diesel engines.

In the future works, the effect of micro pilot diesel injection and high rates of gas fuel mixture (LPG + Hydrogen) addition on emissions, performance and combustion characteristics of diesel engine will be studied.

ACKNOWLEDGEMENT

This research was supported by TÜBITAK (Scientific and Technological Research Council of Turkey) with 1512 project. Project Number: 2150175. The author is also indebted to Şahin Metal A.Ş. and Erin Motor for test apparatus and equipment donation.

NOMENCLATURE

CI	Compression ignition
CNG	Compressed natural gas
CO	Carbon monoxide
CO_2	Carbon dioxide
ECU	Electronic control unit
H_2	Hydrogen
HC	Hydrocarbons
LPG	Liquefied petroleum gas
NO_x	Oxides of nitrogen
THC	Total unburned hydrocarbons

REFERENCES

[1] Saleh, H. E. (2008). Effect of variation in LPG composition on emissions and performance in a dual fuel diesel engine. Fuel, 87(13-14), 3031-3039.

[2] Kumar, M. S., Ramesh, A., & Nagalingam, B. (2003). Use of hydrogen to enhance the performance of a vegetable oil fuelled compression ignition engine. International Journal of Hydrogen Energy, 28(10), 1143-1154.

[3] Lata, D. B., Misra, A., & Medhekar, S. (2012). Effect of hydrogen and LPG addition on the efficiency and emissions of a dual fuel diesel engine. International Journal of Hydrogen Energy, 37(7), 6084-6096.

[4] Beroun, S., & Martins, J. (2001). The development of gas (CNG, LPG and H2) engines for buses and trucks and their emission and cycle variability characteristics (No. 2001-01-0144). SAE Technical Paper.

[5] Qi, D. H., Bian, Y. Z., Ma, Z. Y., Zhang, C. H., & Liu, S. Q. (2007). Combustion and exhaust emission characteristics of a compression ignition engine using liquefied petroleum gas–diesel blended fuel. Energy Conversion and Management, 48(2), 500-509.

[6] Sahoo, B. B., Sahoo, N., & Saha, U. K. (2009). Effect of engine parameters and type of gaseous fuel on the performance of dual-fuel gas diesel engines—A critical review. Renewable and Sustainable Energy Reviews, 13(6-7), 1151-1184.

[7] Poonia, M. P., Ramesh, A., & Gaur, R. R. (1999). Experimental investigation of the factors affecting the performance of a LPG-diesel dual fuel engine. SAE transactions, 499-508.

[8] Heywood JB. Internal combustion egine fundementals. (1988).New York:McGraw-Hill.

[9] Vijayabalan, P., & Nagarajan, G. (2009). Performance, emission and combustion of LPG diesel dual fuel engine using glow plug. JJMIE, 2, 105-110.

[10] Selim, M. Y., Radwan, M. S., & Saleh, H. E. (2008). Improving the performance of dual fuel engines running on natural gas/LPG by using pilot fuel derived from jojoba seeds. Renewable Energy, 33(6), 1173-1185.

[11] Bose, P. K., & Maji, D. (2009). An experimental investigation on engine performance and emissions of a single cylinder diesel engine using hydrogen as inducted fuel and diesel as injected fuel with exhaust gas recirculation. International Journal of Hydrogen Energy, 34(11), 4847-4854.

[12] Das, L. M. (2002). Hydrogen engine: research and development (R&D) programmes in Indian Institute of Technology (IIT), Delhi. International journal of hydrogen energy, 27(9), 953-965.

[13] Das, L. M. (2002). Near-term introduction of hydrogen engines for automotive and agricultural application. International Journal of Hydrogen Energy, 27(5), 479-487.

[14] Boretti, A. (2011). Advantages of the direct injection of both diesel and hydrogen in dual fuel H2ICE. International Journal of Hydrogen Energy, 36(15), 9312-9317.

[15] Liew, C., Li, H., Nuszkowski, J., Liu, S., Gatts, T., Atkinson, R., & Clark, N. (2010). An experimental investigation of the combustion process of a heavy-duty diesel engine enriched with H2. International Journal of Hydrogen Energy, 35(20), 11357-11365.

[16] Saravanan, N., & Nagarajan, G. (2008). An experimental investigation of hydrogen-enriched air induction in a diesel engine system. International Journal of Hydrogen Energy, 33(6), 1769-1775.

[17] Liew, C., Li, H., Nuszkowski, J., Liu, S., Gatts, T., Atkinson, R., & Clark, N. (2010). An experimental investigation of the combustion process of a heavy-duty diesel engine enriched with H2. International Journal of Hydrogen Energy, 35(20), 11357-11365.

[18] Saravanan, N., & Nagarajan, G. (2010). Performance and emission studies on port injection of hydrogen with varied flow rates with Diesel as an ignition source. Applied Energy, 87(7), 2218-2229.

[19] Szwaja, S., & Grab-Rogalinski, K. (2009). Hydrogen combustion in a compression ignition diesel engine. International Journal of Hydrogen Energy, 34(10), 4413-4421.

[20] Lilik, G. K., Zhang, H., Herreros, J. M., Haworth, D. C., & Boehman, A. L. (2010). Hydrogen assisted diesel combustion. International Journal of Hydrogen Energy, 35(9), 4382-4398.

[21] Antunes, J. G., Mikalsen, R., & Roskilly, A. P. (2009). An experimental study of a direct injection compression ignition hydrogen engine. International journal of hydrogen energy, 34(15), 6516-6522.

[22] Masood, M., Ishrat, M. M., & Reddy, A. S. (2007). Computational combustion and emission analysis of hydrogen–diesel blends with experimental verification. International Journal of Hydrogen Energy, 32(13), 2539-2547.

[23] Heffel, J. W. (2003). NOx emission and performance data for a hydrogen fueled internal combustion engine at 1500rpm using exhaust gas recirculation. International Journal of Hydrogen Energy, 28(8), 901-908.

[24] Saravanan, N., Nagarajan, G., Kalaiselvan, K. M., & Dhanasekaran, C. (2008). An experimental investigation on hydrogen as a dual fuel for diesel engine system with exhaust gas recirculation technique. Renewable Energy, 33(3), 422-427.

[25] Varde, K. S., & Frame, G. A. (1983). Hydrogen aspiration in a direct injection type diesel engine-its effects on smoke and other engine performance parameters. International Journal of Hydrogen Energy, 8(7), 549-555.

[26] Shirk, M. G., McGuire, T. P., Neal, G. L., & Haworth, D. C. (2008). Investigation of a hydrogen-assisted combustion system for a light-duty diesel vehicle. International Journal of Hydrogen Energy, 33(23), 7237-7244.

[27] Wei, L., & Geng, P. (2016). A review on natural gas/diesel dual fuel combustion, emissions and performance. Fuel Processing Technology, 142, 264-278.

[28] Fenimore, C. P. (1971, January). Formation of nitric oxide in premixed hydrocarbon flames. In Symposium (International) on Combustion (Vol. 13, No. 1, pp. 373-380). Elsevier.

[29] Stiesch, G. (2003). Modeling engine spray and combustion processes. Springer Science & Business Media.

[30] Karagöz, Y., Sandalcı, T., Koylu, U. O., Dalkılıç, A. S., & Wongwises, S. (2016). Effect of the use of natural gas–diesel fuel mixture on performance, emissions, and combustion characteristics of a compression ignition engine. Advances in Mechanical Engineering, 8(4), 1687814016643228.

[31] Karagöz, Y., Güler, İ., Sandalcı, T., Yüksek, L., Dalkılıç, A. S., & Wongwises, S. (2016). Effects of hydrogen and methane addition on combustion characteristics, emissions, and performance of a CI engine. International Journal of Hydrogen Energy, 41(2), 1313-1325.

[32] Karagöz, Y. (2018). Effect of hydrogen addition at different levels on emissions and performance of a diesel engine. Journal of Thermal Engineering, 4(2), 1780-1790.

[33] Lata, D. B., & Misra, A. (2010). Theoretical and experimental investigations on the performance of dual fuel diesel engine with hydrogen and LPG as secondary fuels. International Journal of Hydrogen Energy, 35(21), 11918-11931.

[34] Lata, D. B., Misra, A., & Medhekar, S. (2012). Effect of hydrogen and LPG addition on the efficiency and emissions of a dual fuel diesel engine. International Journal of Hydrogen Energy, 37(7), 6084-6096.
[35] Lata, D. B., Misra, A., & Medhekar, S. (2011). Investigations on the combustion parameters of a dual fuel diesel engine with hydrogen and LPG as secondary fuels. International Journal of Hydrogen Energy, 36(21), 13808-13819.

5

CFD ANALYSIS OF LAMINAR FORCED CONVECTIVE HEAT TRANSFER FOR TiO2/WATER NANOFLUID IN A SEMI-CIRCULAR CROSS-SECTIONED MICRO-CHANNEL

H. Kaya[1], R. Ekiciler[2,*], K. Arslan[3]

ABSTRACT

In this study, forced convection flow and heat transfer characteristics of TiO2/water nanofluid flow with different nanoparticle volume fractions (1.0%, 2.0%, 3.0% and 4.0%) in semi – circular cross – sectioned micro – channel was numerically investigated. The three - dimensional study was conducted under steady state laminar flow condition where Reynolds number changing from 100 to 1000. CFD model has been generated by using ANSYS FLUENT 15.0 software based on finite volume method. The flow was under hydrodynamically and thermally developing flow condition. Uniform surface heat flux boundary condition was applied at the bottom surface of the micro – channel. The average and local Nusselt number and Darcy friction factor values were obtained using numerical results. Also, the effects of using nanofluid on local values of Nusselt number and Darcy friction factor were investigated. Numerical results indicate that the increasing of nanoparticle volume fraction of nanofluid, the average Nusselt number increases; however, there is no significant variation in average Darcy friction factor.

Keywords: *CFD, Semi – Circular Cross – Sectioned Micro – Channel, Forced Convection, Nanofluid*

INTRODUCTION

Micro and mini – channel applications play very important role for developing cooling capacity in microelectronic equipment such as micro – reactors, refrigeration systems and micro fuel cells. Micro – channels are identified to be one of the fundamental components to transport fluid within a tiny area. Increasing heat transfer rate in microelectronic components becomes very important for scientists. Hence, the researches on this area are mostly carried out for obtaining more efficient working fluids. Nowadays, the most efficient developed working fluids are nanofluids. The dispersed nanoparticle into a base fluid process was explored and called nanofluid. Nanofluids consist of solid metallic or nonmetallic nanoparticles and base fluid such as water, ethylene glycol etc. There is significant enhancement of convective heat transfer using nanofluids as working fluids.

In the past numerical and experimental studies, it was obtained that the flow characteristics in micro – channels are mostly different from macro – channels. The friction factor in micro – channel under laminar flow condition is higher than in macro – channel [1]. An extensive review of forced convection flow in circular and non – circular duct cross – sections were presented by Shah and London [2], Kakaç et al. [3] and Kakaç and Liu [4]. An experimental investigation was made by Berbish et al. [5] for turbulent forced convection heat transfer and pressure drop characteristics of air flow inside a macro – level horizontal semi – circular cross – sectioned duct. Throughout the duct, variations of surface and mean temperatures, local heat transfer coefficient, local Nusselt number and local friction factor were presented. Empirical correlations were obtained for the average Nusselt number and Darcy friction factor as a function of Reynolds number. Manglik and Bergles [6] analyzed numerically constant property, laminar flow heat transfer in a semi – circular tube with a uniform wall temperature. A steady state turbulent flow in a smooth semi – circular cross – sectioned duct was numerically investigated by Arslan [7]. Flow was hydrodynamically and thermally developing under uniform surface heat flux with uniform peripheral wall heat flux in that study. Local heat transfer coefficient and local Darcy friction factor were obtained through the duct length. Languri and Hooman [8] investigated numerically a slip flow in micro – channel for various cross – sections under forced convection condition. Velocity slip and temperature

This paper was recommended for publication in revised form by Regional Editor Tolga Taner
[1] Department of Mechanical Engineering, Bartın University, Bartın, Turkey, hkaya@bartin.edu.tr
[2] Department of Mechanical Engineering, Gazi University, Ankara, Turkey, recepekiciler@gazi.edu.tr
[3] Department of Mechanical Engineering, Karabük University, Karabük, Turkey, kamilarslan@karabuk.edu.tr
**E-mail address: recepekiciler@gazi.edu.tr*

jump boundary conditions were applied at the uniformly heated walls. A numerical analysis has been conducted by Geyer et al. [9] for fully developed laminar flow and heat transfer in periodic semi – circular cross – sectioned trapezoidal channels. It is obtained that the semi – circular geometry provides high rate of heat transfer with relatively small pressure drop with respect to straight pipe.

Investigations conducted on nanofluids are mostly about their thermophysical properties; however, experimental and numerical investigations have been performed for flow and heat transfer characteristics of nanofluids. The investigations in literature about nanofluids using TiO_2 nanoparticle are presented here. Hussein et al. [10] numerically analyzed the effect of cross - sectional area of tube for friction factor and heat transfer using water with TiO_2 nanoparticles. Volume fractions of nanoparticles were 1.0%, 1.5%, 2.0% and 2.5%, respectively. Circular, elliptical and flat tubes were used and the results were compared with experimental data in literature. Parallel results were obtained with deviation of 2.0%. An experimental investigation was conducted by Duangthongsuk and Wongwises [11] about forced convection heat transfer and flow characteristics of TiO_2/water (0.2%) nanofluid in a double – tube counter flow heat exchanger. It was obtained that the convective heat transfer coefficient was higher about 6 – 11% than the base fluid and also it was seen that variation in pressure drop of nanofluid was very little. Kayhani et al. [12] experimentally investigated the convective heat transfer and pressure drop of TiO_2/water (nanoparticle volume fractions of 0.1%, 0.5%, 1.0%, 1.5% and 2.0%) nanofluid flow through a circular tube. The tube was flat, uniformly heated and the flow is under turbulent flow condition. An increase was observed in Nusselt number about 8% for 2.0% nanoparticle volume fraction of nanofluid. CFD analysis was conducted by Moraveji et al. [13] about the effect of nanofluid on cooling performance and pressure drop in mini – channel heat sink. TiO_2 and SiC nanoparticles were used with 0.8, 1.6, 2.4, 3.2 and 4.0% nanoparticle volume fractions for five inlet velocities and acceptable correlations were obtained for Nusselt number and friction factor. Celen et al. [14] studied on a two – dimensional numerical model of TiO_2/water nanofluid flow in plain and enhanced pipes in order to investigate pressure drop. They compared the numerical results with their previous experimental data. The results were presented with graphically for local and average values of temperature, pressure and velocity distribution. Also, pressure drop results were given according to nanoparticle volume fractions and tube types. An experimental analysis of heat transfer characteristics and friction factor of TiO_2 and SiO_2 water based nanofluids was carried out by Azmi et al. [15]. They determined the effect of nanoparticle volume fractions of nanoparticles of TiO_2 and SiO_2 on viscosity and thermal conductivity. The results indicate the enhancements in viscosity and thermal conductivity of TiO_2/water nanofluid were greater than SiO_2/water nanofluid. In addition, they observed a maximum enhancement of 26% in heat transfer coefficients with TiO_2/water nanofluid at 1.0% nanoparticle volume fraction, while there was 33% enhancement of SiO_2/water nanofluid at 3.0% nanoparticle volume fraction. Arani and Amani [16] investigated experimentally convective heat transfer and pressure drop of TiO_2/water nanofluid with nanoparticle volume fraction of 0.002 – 0.02 and the range of Reynolds number was 8000 – 51000. They observed that increasing the Reynolds number or nanoparticle volume fraction, the Nusselt number increases. It was obtained that at the higher Reynolds number greater than 30,000, there was no significant variation in Nusselt number with changing nanoparticle volume fractions. Also, effect of nanoparticle diameter on heat transfer performance was experimentally studied [17]. Experiments were carried out for fully developed turbulent flow condition using TiO_2/water nanofluid. The diameters of TiO_2 nanoparticles were 10, 20, 30 and 50 nm with the nanoparticle volume fractions were from 0.01 to 0.02. It was observed that there was no increment in Nusselt number, with decreasing the diameter of nanoparticles. The highest value of amount of heat transfer was observed for 20 nm nanoparticle diameter. A numerical study was carried out by Peng et al. [18] for heat transfer characteristics under turbulent flow condition in a horizontal circular tube using TiO_2/water nanofluids. Single – phase and multi – phase models were used for numerical computations. The new empirical correlation of nanoparticle viscosity for multi – phase model was suggested and indicated that it was more accurate than traditional ones. An experimental analysis about forced convective heat transfer characteristics and pressure drop of TiO_2/water nanofluid has been reported by Kahani et al. [19]. The geometry of the model was helical coiled tube and nanoparticle volume fractions of nanofluid was between 0.25 – 2.0% with a range of Reynolds number of 500 – 4500. It was observed that the heat transfer coefficient increases with increasing volume fraction and Reynolds number. The maximum thermal performance was obtained at 2.0% nanoparticle volume fraction and Reynolds number was 1750. Uysal et al. was numerically investigated flow characteristics of ZnO/Ethylene glycol nanofluid with φ=1.0 – 4.0% volumetric concentrations in rectangular micro-chanells. Laminar flow conditions were used and the dimensionless properties temperatures, convective heat transfer coefficient, Nusselt number, pressure drop, and

Darcy friction factor values were detrmined [20]. A numerical study was conducted by Makinde [21] about nanofluid flow over a plate with Cu/water, TiO_2/water and Al_2O_3/water nanofluids with low volume fractions. The effects of viscous dissipation and Newtonian heating on boundary flow were numerically determined by using Runge-Kutta-Fehlberg method. He has beern reported that the maximum heat transfer rate on the plate surface had obtained with Cu/water nanofluid. Khamis et al. [22] has reported a numerical analysis about water-based Cu and Al_2O_3 nanofluids flowing in a permeable cylindrical pipe with Navier slip. Unsteady flow conditions were taken into account with variable thermophysical properties along the saturated porous medium in the cylinder. They obtain the increase in volume fraction and Grashof number cause the temperature and velocity enhancement.

Energy and momentum analysis in semi – circular cross – sectioned micro - channels are a little complicated. In order to designing of mini or micro thermal equipment, fundamental knowledge of flow and heat transfer in semi – circular cross - sectioned micro - channel is needed. This geometry is used in heat exchanger in general [7]. In other respects, semi – circular geometry for laminar flow has not been sufficiently investigated in detail. The geometrical properties are very significant for heat removing systems. The channels on the CPU's or any electronic chips need to be heat rejection under higher heat flux. This paper proposes using nanofluids these types of micro-chanells. The novelty of this study is the numerical approach for nanofluid flow in a semi-circular cross sectioned micro-channel for influence of geometry. In this paper, TiO_2/water nanofluid flow inside a horizontal straight semi – circular cross – sectioned micro – channel under hydrodynamically and thermally developing laminar flow condition was numerically analyzed. TiO_2/water nanofluids with nanoparticle volume fractions of 1.0%, 2.0%, 3.0% and 4.0% were used as the heat transfer medium. The significant enhancements in heat transfer for nanofluids take place at under 5.0% nanoparticle volume fraction. It was determined that the enhancements in heat transfer above this value have remained lower values because of the viscosity and shear stress [27-29]. The continuity, momentum and energy equations for three - dimensional flow in semi – circular cross - sectioned micro – channel were solved using finite volume method. The variation of average and local Nusselt number and Darcy friction factor values with Reynolds number and channel length were investigated. Applicable empirical engineering correlations for average Nusselt number and average Darcy friction factor were presented. This study was carried out in order to emphasize the effect of channel geometry and nanoparticle volume fraction on flow and thermal performance of semi – circular micro – channel geometry.

MATHEMATICAL MODEL AND VALIDATION

The micro-channel is thought to be located over a hot surface like a CPU or microchip to disappear heat generation as mentioned before. In order to allow us to figure out more clearly the structure of the case the schematic diagram illustrating the geometry and coordinate system with boundary conditions is given in Figure. 1. Mathematical modeling of semi – circular cross – sectioned micro – channel was performed to solve the case. Three – dimensional Navier – Stokes and energy equations were used to identify the flow and heat transfer in micro – channel. The length and the diameter of the micro - channel are $L = 3x10^4$ µm and $D_i = 1.5x10^2$ µm, respectively. The thermophysical properties of nanofluids were obtained at the bulk temperature.

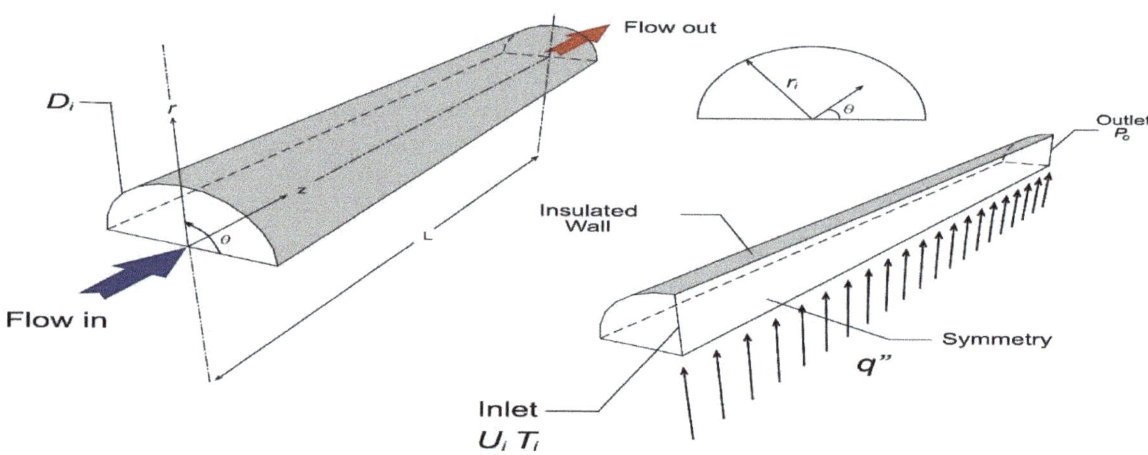

Figure 1. Boundary conditions and coordinate system of the semi – circular cross – sectioned micro – channel

The continuity, momentum and energy equations in cylindrical coordinate system [23]:

Continuity equation:

$$\nabla \vec{V}^* = 0 \tag{1}$$

Momentum equation

$$(\vec{V}^* \nabla^*)\vec{V}^* = -Eu\nabla^* P^* + \mathrm{Re}\nabla^{*2}\vec{V}^* \tag{2}$$

Energy equation:

$$\frac{1}{\mathrm{Re}\,\mathrm{Pr}}\nabla^{*2}T^* + \frac{E_c}{\mathrm{Re}}\Phi^* = 0 \tag{3}$$

where the dimensionless parameters in the Eqns. (1) – (3);

$$\vec{V}^* = \frac{\vec{V}}{U_m}, \quad \vec{\nabla}^* = \vec{\nabla} D_h, \quad \mathrm{Re} = \frac{\rho V D_h}{\mu}, \quad P^* = \frac{P - P_\infty}{P_0 - P_\infty}, \quad g^* = \frac{\vec{g}}{g}, \quad T^* = \frac{T - T_\infty}{T_w - T_\infty}, \quad \mathrm{Pr} = \frac{\mu c_p}{k},$$

$$Ec = \frac{U_m^2}{c_p (T_0 - T_m)}, \quad Eu = \frac{P_0 - P_\infty}{\rho V^2}$$

The ANSYS FLUENT 15.0 solves the problems by computing the Navier-Stokes equations according to designated flow type and boundary conditions. Thus continuity, momentum and energy equations were solved using this software. The momentum equation is used for calculating the pressure drop, friction coefficient and shear stress for the flow. Also, the energy equation is used for obtaining the outlet temperature values according to given heat flux and inlet temperature. In order to solve the case, uniform velocity and temperature profiles of TiO_2/water nanofluid was performed at the inlet of the micro-channel. The no slip boundary condition was applied on the micro - channel walls. The bottom surface of the micro-channel is affected by the heat generation of an electronic device. Therefore, a uniform heat flux was applied on the bottom surface of the micro-channel, while the rest of the micro – channel was thermally isolated from the environmental effects since the working conditions of microchips and micro heat exchangers. These types of micro-channels work under constant heat flux effect in general [20]. Also, the thermophysical properties of nanofluids are completely dependent with temperature. Thus, given heat flux were used to avoid from higher temperature differences. Velocity inlet boundary condition was used for indicating the flow regime that was mentioned as laminar before. Moreover, nanofluid flow enters the micro - channel at T_i=300 K and ANSYS FLUENT 15.0 pressure outlet boundary condition was performed at the outlet of the micro - channel. The symmetry boundary condition was applied on the symmetry plane of the micro-channel.

Single phase approach is considered in this study since the nanoparticle diameters of TiO_2/water nanofluid used in this study are smaller than 100 nm (d_p = 25 - 50 nm) [16]. The fluid flow has been modelled as continuous phase. The flow is assumed to be three-dimensional and steady state. Also, the following assumptions were made that, incompressible and laminar flow conditions, temperature independent thermophysical properties and negligible buoyancy and radiation effects.

The thermophysical properties of TiO_2/water nanofluid were obtained using the expressions as follows [10]:

$$\rho_{nf} = \left(\frac{\phi}{100}\right)\rho_p + \left(1 - \frac{\phi}{100}\right)\rho_f \tag{4}$$

$$C_{nf} = \frac{\frac{\phi}{100}(\rho C)_p + \left(1 - \frac{\phi}{100}\right)(\rho C)_f}{\rho_{nf}} \tag{5}$$

$$k_{nf} = k_f(0.8938)\left(1+\frac{\phi}{100}\right)^{1.37}\left(1+\frac{T_{nf}}{70}\right)^{0.2777}\left(1+\frac{d_p}{150}\right)^{-0.0336}\left(\frac{\alpha_p}{\alpha_f}\right)^{0.01737} \quad (6)$$

$$\mu_{nf} = \mu_f\left(1+\frac{\phi}{100}\right)^{11.3}\left(1+\frac{T_{nf}}{70}\right)^{-0.038}\left(1+\frac{d_p}{170}\right)^{-0.061} \quad (7)$$

The thermophysical properties of püre water and TiO_2 nanoparticle are presented in Table 1 [24].

Table 1. Thermophysical properties of water and TiO_2 nanoparticle

	Pure Water	**TiO_2**
k [W/m.K]	0.613	8.4
c_p [J/kg.K]	4179	710
ρ [kg/m^3]	997	4157
μ*10^6 [kg/m.s]	855	-

The convective heat transfer coefficient, average Nusselt number, average Darcy friction factor and Reynolds number were determined as follows:

$$h = \frac{\dot{m}C_{nf}(T_o - T_i)}{A_s(T_w - T_b)} \quad (8)$$

$$\mathrm{Re} = \frac{4\dot{m}}{\pi D_h \mu} \quad (9)$$

$$Nu = \frac{hD_h}{k} \quad (10)$$

$$f = \frac{2\Delta p\left(\frac{D_h}{L}\right)}{\rho U^2} \quad (11)$$

In order to compute the non-dimensional properties, the hydraulic diameter, D_h, for the semi – circular micro-channel was chosen as the characteristic length and obtained from the Eq. (12) [7]. Average heat transfer coefficient was also obtained from Eq. (13) [23].

$$D_h = \frac{\pi D_i}{\pi + 2} \quad (12)$$

$$h = \frac{q''}{T_w - T_b} \quad (13)$$

Furthermore, all nanofluid properties were taken at the bulk temperature of nanofluid [25]:

$$T_b = (T_i + T_o)/2 \quad (14)$$

In this study, in order to perform the numerical analysis, a general finite – volume based commercial CFD software ANSYS FLUENT 15.0 was used. The code provides the mesh flexibility by structured and unstructured meshes. Computations were conducted under laminar flow condition and the energy equation was solved by neglecting radiative effects. The Navier – Stokes and energy equations were solved numerically until the residuals lower than 10-6 through the iterative process.

In order to prove the reliability of the numerical model the grid independence study was conducted since the mesh structure has vital importance in CFD investigations. For this reason, mesh structure was constituted with the following specifications. Tetrahedron cells were generated with a fine mesh near the plate walls. To obtain fine mesh distribution, boundary layer mesh type was used adjacent to the surfaces of the micro – channel. As depicted in Figure 2, the non – uniform grid distribution was performed to the plane perpendicular to the main flow direction. In order to enhance the resolution and accuracy, the number of grid points close to each wall was increased.

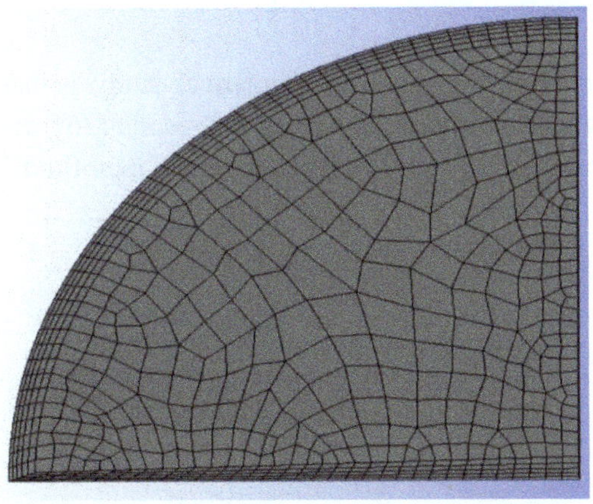

Figure 2. Mesh distribution of the computational domain

The mesh independence study was carried out until the variation in both average Nusselt number and average Darcy friction factor were less than 1%. To provide the certainty of results, a grid independence study was performed for pure water flow using ten different grid sizes changing from $1.36x10^5$ to $1.74x10^6$ for Re = 1000. This process gave the effects of grid size on flow and heat transfer. It is illustrated in Figure 3 that there was no remarkable variation on average Nusselt number and average Darcy friction factor between $9.07x10^5$ and $1.74x10^6$ grid sizes. According the mesh independence process, the grid size $9.07x10^5$ was chosen and used for Reynolds number 100 – 1000. Also, same procedure was applied for other nanoparticle volume fractions.

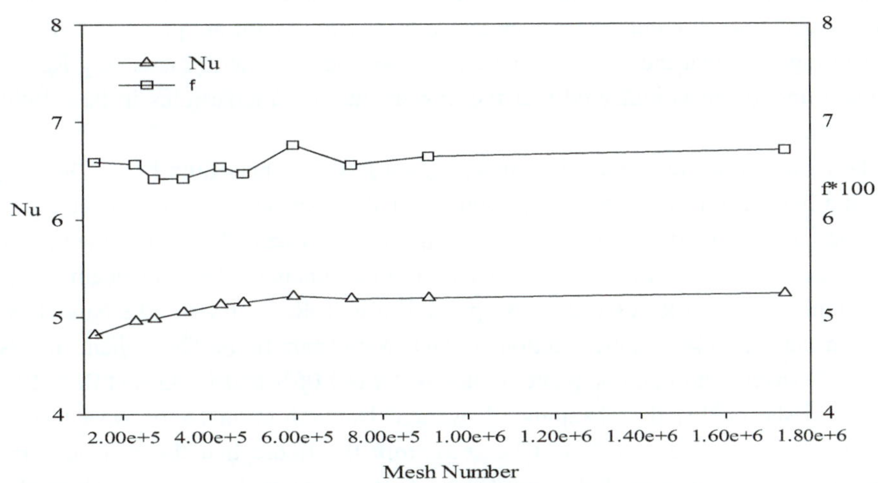

Figure 3. Variation of average Nusselt number and average Darcy friction factor with mesh number

RESULTS AND DISCUSSION

In this study, the fluid flow and heat transfer characteristics of TiO_2/water nanofluid flow in a semi circular cross – sectioned micro-channel was numerically analyzed as mentioned before. Different nanoparticle volumetric concentrations (ϕ= 1.0%, 2.0%, 3.0% and 4.0%) were used in the computations. The Reynolds number was changed from 100 to 1000. The results were presented with average and local Nusselt number and Darcy friction factor. Moreover, the velocity and temperature profiles in the micro-channel were given with contur graphical representation.

In order to test the numerical code, results for numerical analysis were compared with the Hausen equation in the literature. The results of pure water flow in micro - channel were compared with Hausen equation which was obtained by means of experimental studies [26], used for calculating average Nusselt number under laminar flow condition in pipes, given with Eq. (15).

$$Nu = 3.66 + \frac{0.0668(D/L)\mathrm{Re}\,\mathrm{Pr}}{1 + 0.04\left[(D/L)\mathrm{Re}\,\mathrm{Pr}\right]^{2/3}} \tag{15}$$

It is seen in Figure 4 that the current numerical study is harmonious with Hausen equation. The maximum deviation between the numerical and literature results [26] was obtained nearly ±7%. The validation was assumed to be in acceptable range and then calculations of nanofluid flow was carried out.

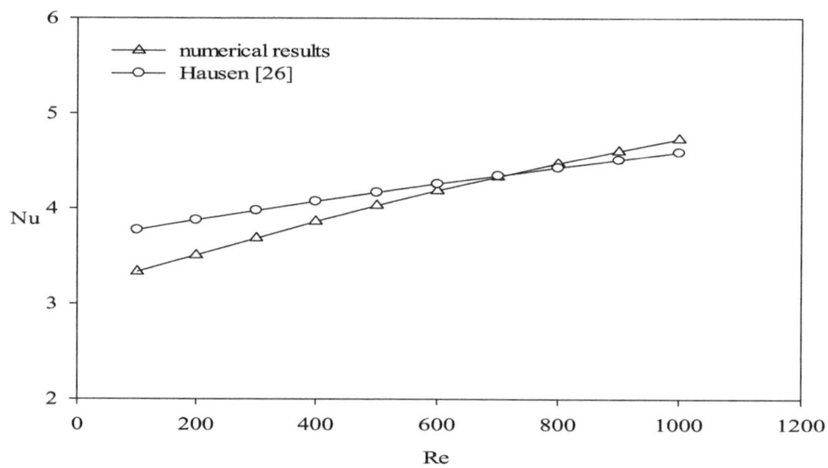

Figure 4. Comparison of current numerical analysis and Hausen equation in literature

Nusselt number is a measure of convective heat transfer coefficient and it is reliable to comparison of heat transfer characteristics since it is dimensionless. In this case the effect of volume fractions (1.0 – 4.0%) of the nanofluid on average Nusselt number and average Darcy friction factor are presented in Figure 5 and Figure 6. It can be seen in Figure 5 that the average Nusselt number increases with increasing Reynolds number and nanoparticle volume fraction. It is indicated that the movements of nanoparticles in base fluid provide higher energy exchange rates.

The highest value of average Nusselt number was observed in nanofluid has 4.0% nanoparticle volume fraction at the same boundary conditions. It represents that the convection heat transfer rate of nanofluid for nanoparticle volume fraction of 4.0% is more effective. Figure 5 represents that increasing the volume fraction of nanofluid increases the convective heat transfer capacity in micro-channel. The enhancement in Nusselt number at low Reynolds number remains lower values as opposed to high Re numbers. The Nusselt number increases with the increase in nanoparticle volume fraction of TiO_2/water nanofluid. The enhancements in the Nusselt number for 4.0% TiO_2/water nanofluid compared to pure water is 3.06% and 17.85% at Re = 100 and Re = 1000, respectively. Figure 6 shows that the changing of average Darcy friction factor with Reynolds number for different nanoparticle volume fractions. It can be seen from the figure that there is no significant effect of changing volume fractions of nanoparticles on Darcy friction factor in micro – channel as reported for Al_2O_3/Water and Al_2O_3/Water-EG nanofluids by Ref.[30-32]. Because at lower velocity values of fluid, the shear stress near the walls of micro-channel is higher values. The value of obtained Darcy friction factor at Re=100 is almost 10 times higher than at Re=1000. Adding nanoparticle to base fluid absolutely increases the pressure drop but the friction factor indicates dimensionless pressure drop.

The velocity and temperature profiles in channels are important for identifying the flow in the channel. The velocity and temperature distributions in micro - channel are shown in Figure 7 – 10. The contours of flow velocity magnitude for 4.0% vol. TiO_2/water nanofluid with different Reynolds numbers are given in Figure 7. It is seen that the velocity profile does not change with Reynolds number. Maximum velocity is obtained at the center of the micro-channel and diminishing through the radial direction since the shear stress is maximum close the walls. Additionally, the temperature distributions of the TiO_2/water nanofluid flow at the outlet of the channel for different Reynolds number are given in Figure 8. Temperature profiles are almost the same distribution for each Reynolds number. Gradually reduction of the temperature values is observed from bottom surface to top of the micro-channel since the bottom surface is exposed to heat flux.

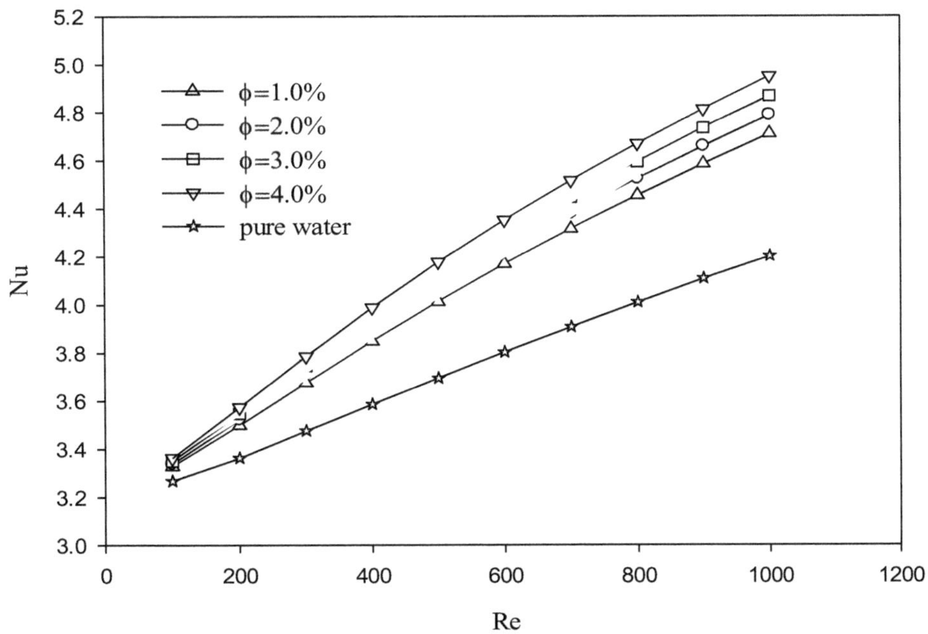

Figure 5. Variation of average Nusselt number with Reynolds number for different nanoparticle volume fractions

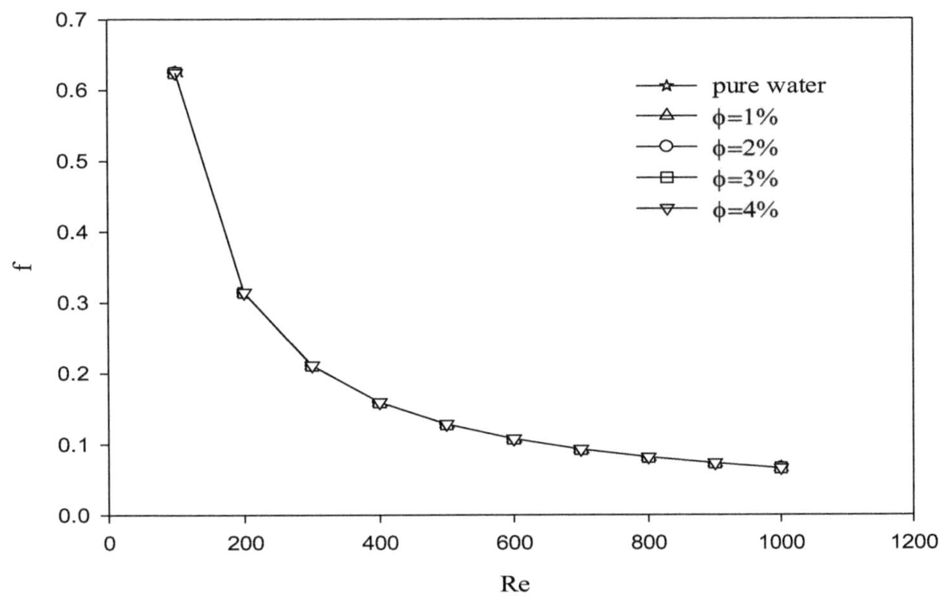

Figure 6. Variation of Darcy friction factor with Reynolds number for different nanoparticle volume fractions

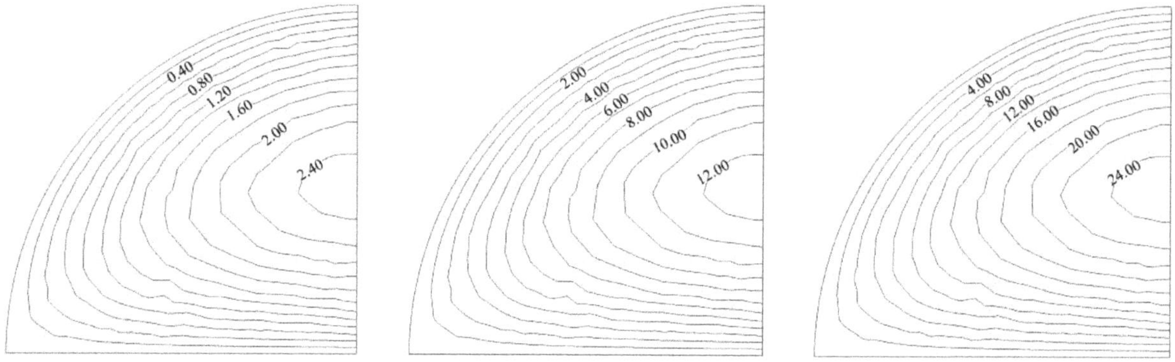

(a) (b) (c)

Figure 7. Velocity distributions at the outlet of the micro-channel for 4.0% nanoparticle volume fraction of TiO_2/water nanofluid (a) Re=100, (b) Re=500, (c) Re=1000

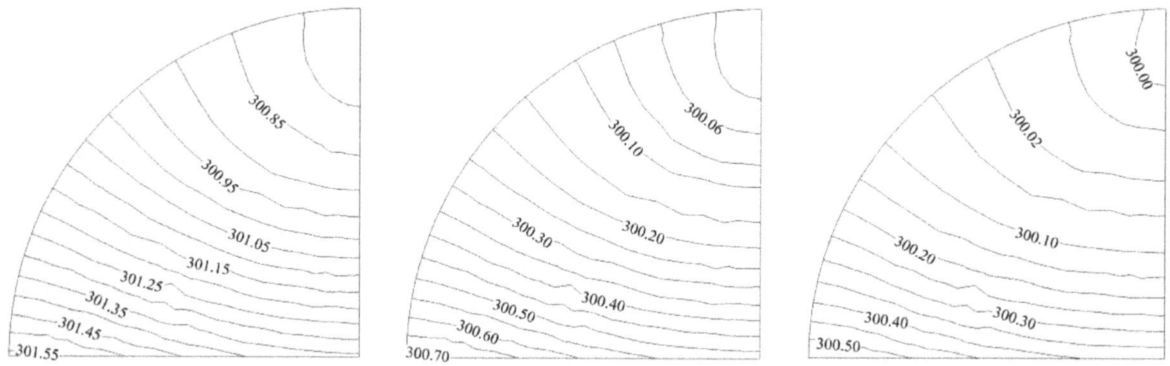

(a) (b) (c)

Figure 8. Temperature distributions at the outlet of the micro-channel for 4.0% nanoparticle volume fraction of TiO_2/water nanofluid (a) Re=100, (b) Re=500, (c) Re=1000

In order to determine the effect of nanoparticle volume concentration on temperature and velocity profiles in the micro-channel contours was constituted at Re=500. Figure 9 shows the flow velocity magnitude with different nanoparticle volume fractions at the outlet of the micro-channel and it is seen that the same profile is obtained. The center velocity of 4.0% volume fraction TiO_2/water nanofluid is calculated as 12 m/s which is 26.3% higher than 1.0% volume fraction TiO2/water nanofluid. The temperature contours at the outlet of the micro-channel are presented in Figure 10 for different nanoparticle volume fractions of nanofluids at Re = 500. It is obtained from the figure that the temperature distribution is not affected too much by changing nanoparticle volume fraction.

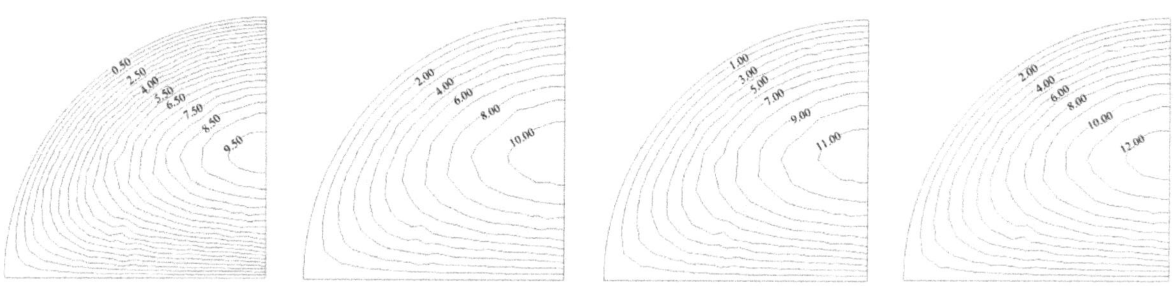

(a) (b) (c) (d)

Figure 9. Velocity distributions at the outlet of the micro - channel for different nanoparticle volume fractions of TiO_2/water nanofluid at Re=500 (a) 1.0% (b) 2.0% (c) 3.0% (d) 4.0%

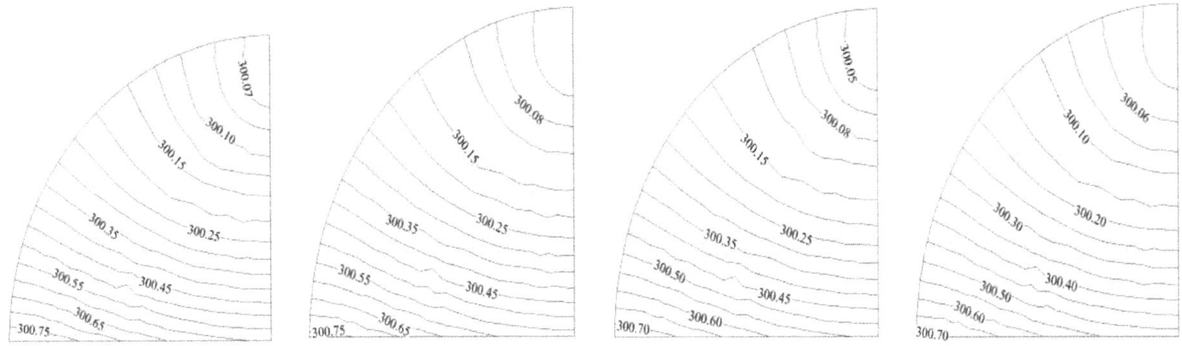

(a) (b) (c) (d)

Figure 10. Temperature distributions at the outlet of the micro-channel for different nanoparticle volume fractions of TiO_2/water nanofluid at Re=500 (a) 1.0% (b) 2.0% (c) 3.0% (d) 4.0%

After the all computations for this case, the applicable correlations of average Nusselt number and average Darcy friction factor changing with Reynolds number and nanoparticle volume fraction were obtained from the numerical results of TiO_2/water nanofluid flow in a semi – circular cross – sectioned micro-channel. The empirical correlations are given in Eq. (16) and (17) for average Nusselt number and average Darcy friction factor, respectively. Average Nusselt number is directly dependent with Reynolds number and volume fraction of nanofluid as indicated in Eq. (16). Prandtl number is not used as a variable parameter in the Nusselt number equation. The reason of absence of the Prandtl number in the equation (16) is using only one type of nanofluid in the calculations.

$$Nu = 1.58\, Re^{0.17}\, \phi^{0.03} \tag{16}$$

$$f = 62.88 / Re \tag{17}$$

Local values of dimensionless parameters are very significant for observing the variation of flow characteristics along the channels. Therefore, variation of local Nusselt number according to dimensionless length of micro - channel for different nanoparticle volume fractions and Reynolds numbers is presented on Figure 11. It is determined that the local Nusselt number increases with increasing nanoparticle volume fraction; however, it decreases along with micro-channel length. Also, it can be seen that flow reaches thermally fully developed condition in the micro – channel for lower Reynolds numbers. At higher Reynolds numbers, flow is under thermally developing condition.

Figure 12 represents that the variation of local Darcy friction factor with dimensionless micro - channel length for different nanoparticle volume fractions and Reynolds numbers. Local Darcy friction factor can be evaluated as follows:

$$f_x = \frac{8\tau_x}{\rho U_{in}^2} \tag{18}$$

where τ_x is the local shear stress.

The maximum values of local Darcy friction factor were obtained at the inlet of the microchannel at each Reynolds number. Local Darcy friction factor values decreases until about x/D_h=30 and then become constant along the micro - channel length. Also, flow reaches hydrodynamically developing condition at x/D_h=30. The values of flow velocity and local shear stress increase with increase in Reynolds number, but the enhancement in velocity is higher than local shear stress. Hence, it decreases with increasing Reynolds number as seen in Figure 12. As a result, it can be said that the nanofluid flow reaches hydrodynamically fully developed flow condition inside the micro - channel.

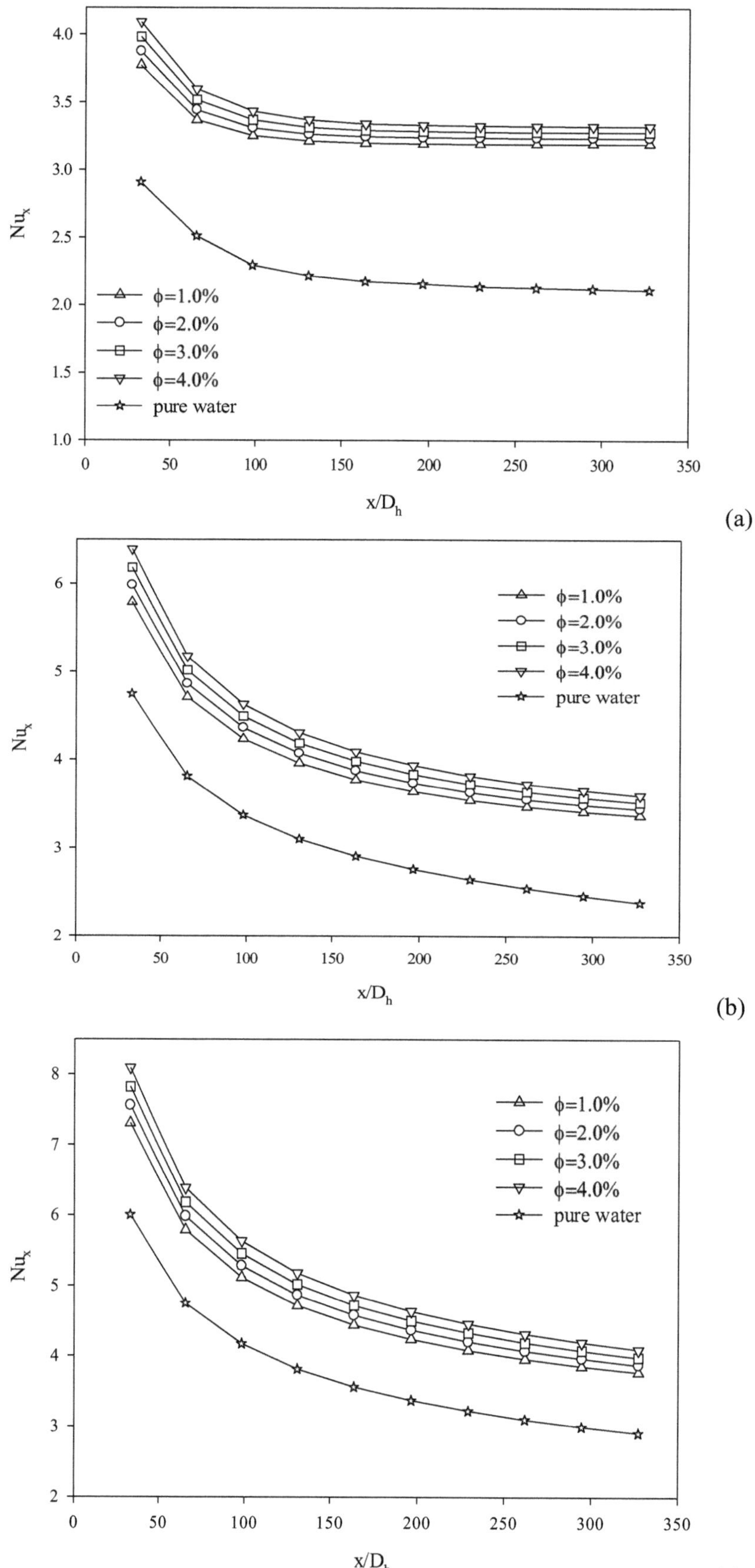

Figure 11. The variation of local Nusselt Number with dimensionless length of micro - channel for (a) Re=100, (b) Re=500, (c) Re=1000

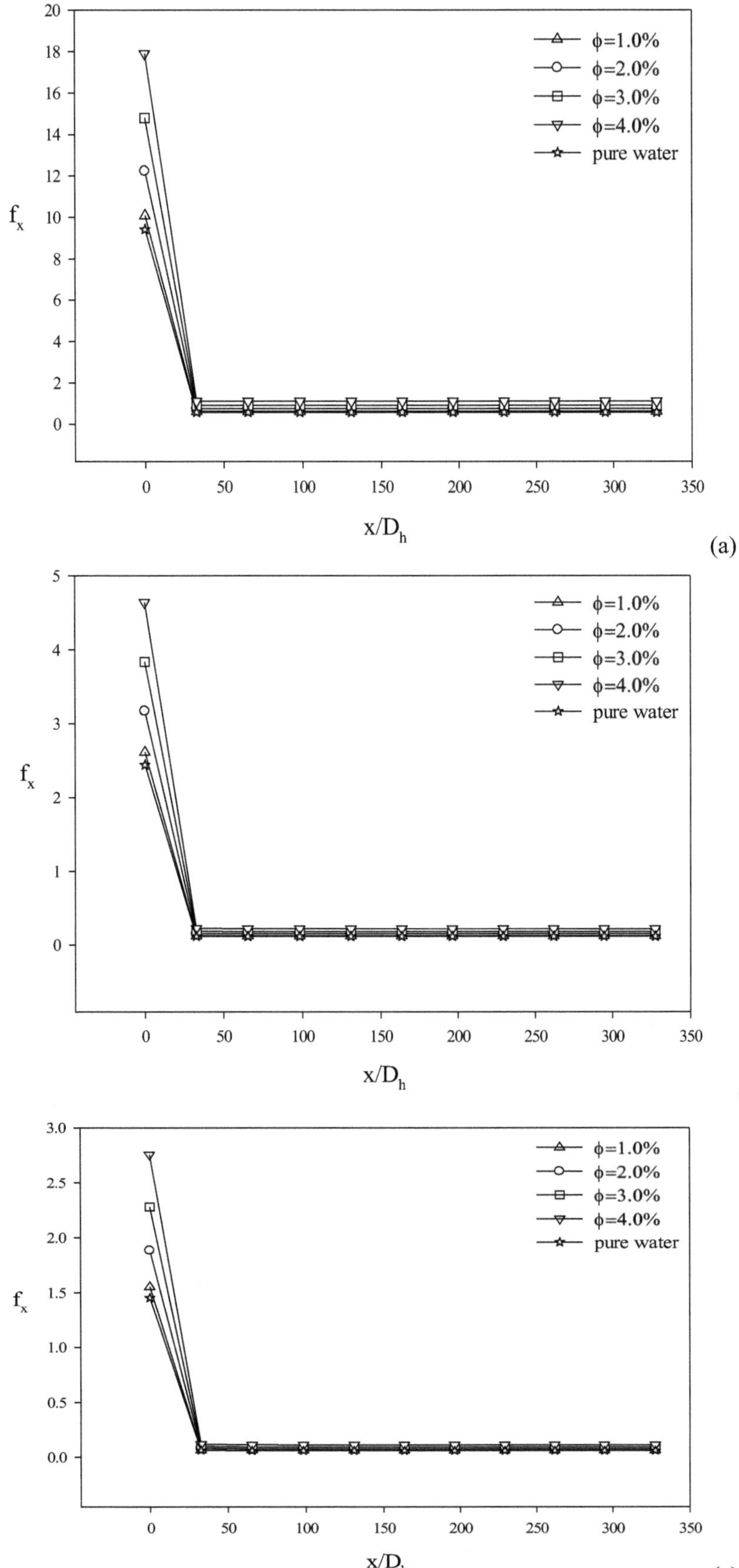

Figure 12. The variation of local friction factor with dimensionless length of micro - channel for (a) Re=100, (b) Re=500, (c) Re=1000

CONCLUDING REMARKS

In the present three – dimensional numerical investigation, TiO_2/water nanofluid flow in a semi – circular cross – sectioned micro – channel for different nanoparticle volume fractions was investigated. Flow characteristics were under hydrodynamically and thermally developing, steady laminar flow condition. Computations were conducted for different Reynolds numbers ($100 \leq Re \leq 1000$) and different nanoparticle volume fractions (1.0%, 2.0%, 3.0% and 4.0%). The results of numerical calculations showed that the increase of nanoparticle volume fraction increases the average Nusselt number up to 10% according to base fluid. However, there is no significant variation for average Darcy friction factor with changing nanoparticle volume fraction. Furthermore, the increment is observed for average Nusselt number with increasing Reynolds number, notwithstanding, average Darcy friction factor decreases. Velocity and temperature distribution at the outlet of the micro - channel were also presented graphically for different Reynolds number and different nanoparticle volume fractions of nanofluid. According to the numerical computations of laminar flow in semi – circular micro - channel, new engineering correlations were determined for average Nusselt number and average Darcy friction factor depends on Reynolds number and nanoparticle volume fraction for TiO_2/water nanofluid flow in semi – circular micro – channel.

NOMENCLATURE

C	specific heat [J/kgK]
ϕ	volumetric concentration
ρ	density [kg/m^3]
k	thermal conductivity [W/mK]
T	temperature [K]
d	particle diameter [m]
α	thermal diffusivity [m^2/s]
μ	viscosity [kg/ms]
$\dot{m}$	mass flow rate [kg/s]
U	velocity [m/s]
h	convection heat transfer coefficient [W/m^{2}K]
Nu	Nusselt number
f	friction factor
p	pressure [Pa]
L	length [m]
q	heat flux [W/m^2]
Re	Reynolds number
Pr	Prandtl number
D	channel diameter [m]
τ	shear stress [Pa]
Ec	Eckert number
Eu	Euler number
Φ	viscous dissipation
x	local values
nf	nanofluid
p	solid particle
f	liquid phase
h	hydraulic diameter
i	inlet
o	outlet
w	wall
b	bulk

REFERENCES

[1] Guobing, Z., Shi – Chune, Y. (2011). Effect of surface roughness on laminar liquid flow in micro – channels, Applied Thermal Engineering, 31(2-3), 228 – 234.

[2] Shah, R. K.., London, A. L. (1978). Laminar flow forced convection in ducts, Academic Press, New York.

[3] Kakaç, S., Shah, R. K., Aung W. (1987). Handbook of single – phase convective heat transfer, John Wiley and Sons.

[4] Kakaç, S., Liu H. (1998). Heat exchanger selection, rating and thermal design, CRC Press, USA.

[5] Berbish, N. S., Moawed, M., Ammar M. (2011). Heat transfer and friction factor of turbulent flow through a horizontal semi – circular duct, Heat and Mass Transfer, 47(4) 377 – 384.
[6] Manglik, R. M., Bergles, A. E. (1998). Laminar flow heat transfer in a semi – circular duct with uniform wall temperature, International Journal of Heat and Mass Transfer, 31 (3) 625 – 636.
[7] Arslan, K. (2014). Three-dimensional numerical investigation of turbulent flow and heat transfer inside a horizontal semi – circular cross – sectioned duct, 18(4), 1145 – 1158.
[8] Languri, E. M., Hooman, K. (2011). Slip flow convection in a microchannel with semi-circular cross-section, International Communications in Heat and Mass Transfer, 38(2), 138-143.
[9] Geyer, P. E., Fletcher, D. F., Haynes, B. S. (2007). Laminar flow and heat transfer in a periodic trapezoidal channel with semi – circular cross – section, International Journal of Heat and Mass Transfer, 50(17-18), 3471 – 3480.
[10] Hussein, A. M., Sharma, K. V., Bakar, R. A., Kadirgama K. (2013). The effect of cross sectional area of tube on friction factor and heat transfer nanofluid turbulent flow, International Communications in Heat and Mass Transfer, 47, 49 – 55.
[11] Duangthongsuk, W., Wongwises, S. (2009). Heat transfer enhancement and pressure drop characteristics of TiO_2 – water nanofluid in a double – tube counter flow heat exchanger, International Journal of Heat and Mass Transfer, 52(7-8), 2059 – 2067.
[12] Kayhani, M. H., Soltanzadeh, H., Heyhat, M. M., Kowsary, F. (2012). Experimental study of convective heat transfer and pressure drop of TiO_2/water nanofluid, International Communications in Heat and Mass Transfer, 39(3), 456 – 462.
[13] Moraveji, M. K., Ardehali, R. M., Jam, A. (2013). CFD investigation of nanofluid effects (cooling performance and pressure drop) in mini – channel heat sink, International Communications in Heat and Mass Transfer, 40, 58 – 66.
[14] Çelen, A., Kayacı, N., Çebi, A., Demir, H., Dalkılıç, A. S., Wongwises, S. (2014). Numerical investigation for the calculation of TiO_2 – water nanofluids' pressure drop in plain and enhanced pipes, International Communications in Heat and Mass Transfer, 53, 98 – 108.
[15] Azmi, W. H., Sharma, K. V., Sarma, P. K., Mamat, R., Najafi, G. (2014). Heat transfer and friction factor of water based TiO_2 and SiO_2 nanofluids under turbulent flow in a tube, International Communications in Heat and Mass Transfer, 59, 30 – 38.
[16] Arani, A. A., Amani, J. (2012). Experimental study on the effect of TiO_2 – water nanofluid on heat transfer and pressure drop, Experimental Thermal and Fluid Science, 42, 107 – 115.
[17] Arani, A. A., Amani, J. (2013). Experimental investigation of diameter effect on heat transfer performance and pressure drop of TiO_2 – water nanofluid, Experimental Thermal and Fluid Science, 44, 520 – 533.
[18] Peng, W., Minli, B., Jizu, L., Chengzi, H., Yuyan, W. (2014). Numerical investigation on the turbulent flow characteristic of nanofluids in a horizontal tube, Numerical Heat Transfer Part A: Applications, 66, 646 – 668.
[19] Kahani, M., Heris, S. Z., Mousavi, S. M. (2014). Experimental investigation of TiO_2/water nanofluid laminar forced convective heat transfer through a helical coiled tube, Heat and Mass Transfer, 50, 1563 – 1573.
[20] Uysal, C., Arslan, K., Kurt, H. (2016). A Numerical Analysis of Fluid Flow and Heat Transfer Characteristics of ZnO-Ethylene Glycol Nanofluid in Rectangular Microchannels, Strojniški vestnik - Journal of Mechanical Engineering 62(10), 603-613.
[21] Makinde, O. D. (2013). Effects of viscous dissipation and Newtonian heating on boundary layer flow of nanofluids over a flat plate. International Journal of Numerical Methods for Heat and Fluid flow, 23(8), 1291-1303.
[22] Khamis, S., Makinde, O. D., Nkansah-Gyekye Y. (2015). Buoyancy – driven heat transfer of water based nanofluid in a permeable cylindrical pipe with Navier slip through a saturated porous medium. Journal of Porous Media, 18(12), 1169-1180.
[23] Bejan, A. (2004). Convective Heat Transfer, Wiley, New York.
[24] Incropera, F. P., DeWitt, D. P. (2002). Fundamentals of Heat and Mass Transfer, John Wiley & Sons, New York.
[25] Çengel, Y. A. (1988). Heat Transfer a Practical Approach, McGraw – Hill New York.
[26] Hausen, H., (1959), "Neue Gleichungen fur die Wameiibertragung bei Freieroder Erzwungerner Stromung," Allg. Warmetchn., 9, 75–79.
[27] Li, C. H., Peterson, G. P. (2006). Experimental investigation of temperature and volume fraction variations on the effective thermal conductivity of nanoparticle suspensions (nanofluids), Journal of Applied Physics, 99, 1–8.
[28] Verma, S. K., Tiwari, A. K. (2015). Progress of nanofluid application in solar collectors: A review, Energy Conversion and Management, 100, 3324-346
[29] Zhang, X., Gu, H., Fujii, M. (2006). Experimental Study on the Effective Thermal Conductivity and Thermal Diffusivity of Nanofluids, International Journal of Thermophysics, 27(2), 569-580
[30] Jung, J.Y., Oh, H. S. Kwak, H. Y. (2009). Forced convective heat transfer of nanofluids in microchannels. International Journal of Heat and Mass Transfer, 52, 466-472.

[31] Ekiciler, R., Arslan, K. (2018) CuO/water Nanofluid flow over microscale backward-facing step and heat transfer performance analysis. Heat Transfer Research 49:1489-1505.
[32] Kaya, H., Ekiciler, R., Arslan, K. (2018) Entropy generation analysis of forced convection flow in a semi-circular shaped microchannel with TiO_2/water nanofluid. Heat Transfer Research doi: 10.1615/HeatTransRes.2018025888.

6

GEOMETRY EFFECTS ON THERMOHYDRAULIC BEHAVIOR OF FLUID FLOW IN A SQUARE ENCLOSURE WITH AN INNER CIRCULAR TUBE

A. Ahmadi Nadooshan[1*], A. Jahanbakhshi[1], M. Bayareh[1]

ABSTRACT

Convective heat transfer in non-circular channels are of interest in many industrial applications. In the present work, fluid flow in the space between a square channel and a circular tube with different positions of the holder rigid plate is investigated numerically. The use of holder plates is applicable in industrial applications. Holder plates allow different flows with different thermal and hydrodynamic behaviors in a channel at constant Reynolds numbers. Six geometries selected to explore the effect of the position of holder plate on heat transfer rate. Results demonstrated that the plate position has significant effects on fluid flow behavior. It is found that hydrodynamic and thermal behavior affected by the plate position for different Reynolds numbers. For example, for the case that the circular tube is positioned in the center of the square channel with two inclined plates, average convective heat transfer outside and between the two plates is 344 $[\frac{w}{m^2k}]$ and 465 $[\frac{w}{m^2k}]$, respectively.

Keywords: *Square Channel, Forced Convection, Numerical Simulation, Holder Rigid Plate, Reynolds Number*

INTRODUCTION

Regarding great application of non-circular sections in many engineering areas, some researchers investigated heat transfer in laminar fluid flows in such cross-sections. Due to limited contact surface area between the fluid and the wall, heat transfer is low in these channels. As a result, the rate of shear stress from contact between the fluid and channel wall is lower in the flowing fluid. Therefore, pressure drop in these channels is lower than that in circular ones. Low pressure drop in these sections emphasizes their importance in industrial applications. Moreover, reduced contact surface area between the fluid and walls reduces convective heat transfer. Today, increasing heat transfer is investigated to design smaller heat exchangers. Since non-circular sections, such as square and triangular channels, are associated with pressure drop reduction in passing fluids, they should be considered for investigation. Choi [1] was the first one who used fluids containing a suspension of nanoparticles and investigated their significant thermal properties through measuring convective heat transfer. Murshed and Castro [2] investigated forced convection for various nanofluids and addressed different base fluids, such as water, ethylene glycol, and engine oil, and found out that nanofluids have higher heat transfer coefficient in comparison with the base fluids.

Mirmasoumi and Behzadmehr [3] investigated laminar mixed convection of water-aluminum oxide in a horizontal circular tube using a two-phase mixture model. Results indicated that although nanoparticles have no effect on hydrodynamic parameters of flow in the fully developed region, they significantly affect thermal parameters. Pishkar and Ghasemi [4] investigated mixed convection of nano and pure fluids in a horizontal channel containing a heat source using a single-phase numerical method. They showed that the application of nanofluid improves the distribution of heat and increases mean temperature of the heat source. Vasefi and Alizadeh [5] investigated the effect of CuO-water nanofluid on forced convection of water flow in horizontal channels with different cross-sections using the two-phase Euler-Lagrange method. Their results indicated that heat transfer increases due to the use of nanoparticles.

Zeinali et al. [6] investigated heat transfer from nanofluids in channels with triangular and equilateral sections and found that lower pressure drop occurs in triangular cross-section in comparison with other cross-sections. In addition, their results showed that heat transfer intensity depends on the volume fraction of nanoparticles. Effects of estimation error in effective dynamic viscosity and conductivity of aluminum-water nanofluid on free convective heat transfer in a heated square enclosure was studied by Ho et al. [7]. They showed that the heat transfer intensity in the middle of the enclosure (cavity) could be determined using dynamic

This paper was recommended for publication in revised form by Regional Editor Omid Mahian
[1]Department of Mechanical Engineering, Shahrekord University, Shahrekord, IRAN
**E-mail address: ahmadi@eng.sku.ac.ir, Akram.jahanbakhshi@gmail.com, m.bayareh@eng.sku.ac.ir*

viscosity equation by referring to the base fluid. Mansour et al. [8] simultaneously investigated the laminar-mixed convection and heat transfer of aluminum oxide-water nanofluid in a tube with its wall exposed to uniform heat flux. Nguyen et al. [9] performed a study on convective heat transfer of aluminum oxide-water nanofluid in the process of cooling electronic components. They showed that a nanofluid with 36 nm particle size provides higher convective heat transfer than the one with 47 nm particle size. Heris et al. [10] numerically investigated laminar convective heat transfer of a nanofluid in a tube under a constant temperature boundary condition. Shah and London [11] studied heat transfer properties of a laminar flow in channels with different shapes, including a channel with an equilateral triangle cross-section, a channel with an equilateral triangle cross-section and rounded edges, and channels with equilateral triangle and right triangle cross-sections in a various range of heat boundary conditions. Zhang [12] studied convective heat transfer from laminar fully-developed flows at different Reynolds numbers from hydrodynamic and thermal perspectives in ducts with triangle and equilateral triangle cross-sections, uniform wall temperature, and apex angles from 30° to 120°. Results demonstrated that the Nusselt number of nanofluid increases by increasing the concentration of nanoparticles and reducing their diameter. Heyhat and Kowsary [13] investigated the effect of non-uniform distribution of nanoparticles on flow and thermal parameters of water-alumina nanofluid in a circular tube. They indicated that non-uniform distribution of nanoparticles improves the heat transfer. Kalteh and Mehrzad [14] numerically investigated laminar flow heat transfer of a nanofluid in a space between a square tube and a rigid tube located in its center, and concluded that average Nusselt number is affected by the volume fraction of nanoparticles in the base fluid. The major subject in some other studies, as well as the current one, is study of thermal and hydrodynamic behavior of water-AL_2O_3 nanofluid flow in a channel with non-circular cross-section, containing a middle rigid body and holder plate. In addition, the effects of Reynolds number, Prandtl number, concentration of nanoparticles, and inlet temperature are addressed.

In the present work, water-AL_2O_3 nanofluid flow is considered in a rectangular cross-section channel (H×H, H=0.05 m) with an inner circular tube (D=0.025m). Table 1 indicates the thermophysical properties of AL_2O_3 nanoparticles. Figure 1 shows the schematic of the problem. The flow is assumed to be laminar, incompressible and steady with constant properties. The channel walls are exposed to a constant thermal flux of 20,000 w/m^2. Inlet fluid temperature is 293 K. Moreover, holder plates and middle rigid tubes are considered to be isolated. The Reynolds number is Re=100, 250, and 500. In addition, the flow is uniform in the inlet and is hydro-dynamically fully-developed along the tube direction. Figure 2 presents the geometries are used in the current study. This numerical study aims to compare the effects of holder plates on fluid convective heat transfer and flow pattern

Table 1. Thermophysical properties of AL_2O_3 nanoparticles

Average particle diameter[nm]	density[kg/m^3]	C_p [Kj/kg.K]	K [W/m.K]
20	3970	765	25

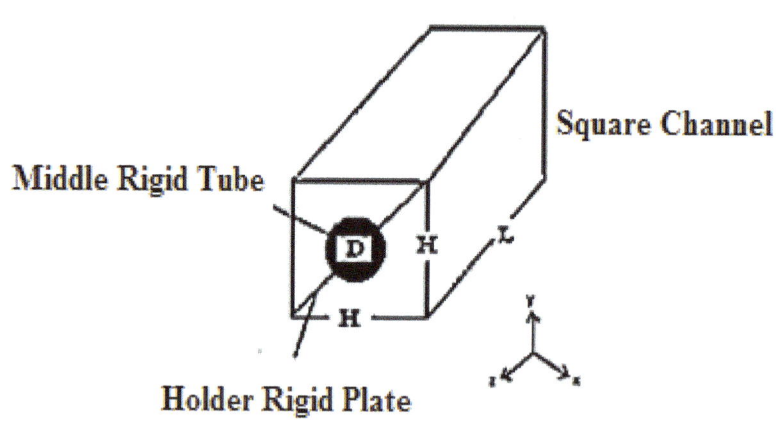

Figure 1. Schematic of the problem

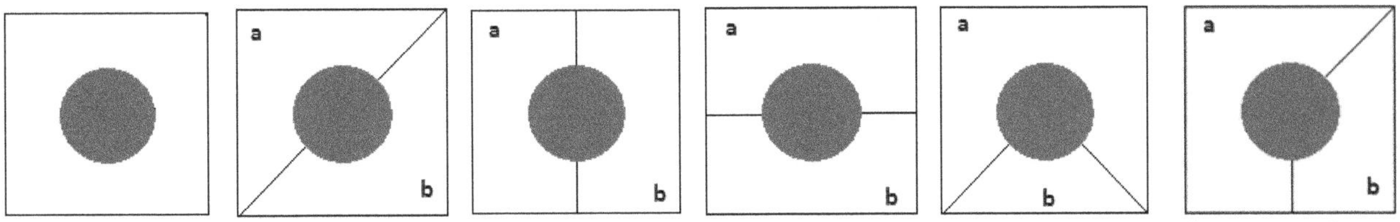

Geometry 1 **Geometry 2** **Geometry 3** **Geometry 4** **Geometry 5** **Geometry 6**

Figure 2. Different positions of holder plates.

MATHEMATICAL MODEL AND VALIDATION

By neglecting thermal loss, viscous loss, radiation, and gravity and assuming Newtonian and incompressible fluid in this study, the governing equations for three dimensional laminar flow inside a channel are as follows [14-16]:

Continuity equation:

$$\frac{\partial u}{\partial x}+\frac{\partial v}{\partial y}+\frac{\partial w}{\partial z}=0 \tag{1}$$

Momentum equation in the X- direction:

$$u\frac{\partial u}{\partial x}+v\frac{\partial u}{\partial y}+w\frac{\partial u}{\partial z} = -\frac{1}{\rho}\frac{\partial p}{\partial x}+\frac{\mu}{\rho}\left(\frac{\partial^2 u}{\partial x^2}+\frac{\partial^2 u}{\partial y^2}+\frac{\partial^2 u}{\partial z^2}\right) \tag{2}$$

Momentum equation in the Y- direction:

$$u\frac{\partial v}{\partial x}+v\frac{\partial v}{\partial y}+w\frac{\partial v}{\partial z} = -\frac{1}{\rho}\frac{\partial p}{\partial y}+\frac{\mu}{\rho}\left(\frac{\partial^2 v}{\partial x^2}+\frac{\partial^2 v}{\partial y^2}+\frac{\partial^2 v}{\partial z^2}\right) \tag{3}$$

Momentum in the Z- direction:

$$u\frac{\partial w}{\partial x}+v\frac{\partial w}{\partial y}+w\frac{\partial w}{\partial z} = -\frac{1}{\rho}\frac{\partial p}{\partial y}+\frac{\mu}{\rho}\left(\frac{\partial^2 w}{\partial x^2}+\frac{\partial^2 w}{\partial y^2}+\frac{\partial^2 w}{\partial z^2}\right) \tag{4}$$

Energy equation:

$$u\frac{\partial T}{\partial x}+v\frac{\partial T}{\partial y}+w\frac{\partial T}{\partial z}=\alpha\left(\frac{\partial^2 T}{\partial x^2}+\frac{\partial^2 T}{\partial y^2}+\frac{\partial^2 T}{\partial z^2}\right) \tag{5}$$

Fluid thermal diffusivity is defined as follows [17]:

$$\alpha=\frac{K}{\rho C_p} \tag{6}$$

Reynolds and Prandtl numbers are defined as follows [18]:

$$Re=\frac{\rho V D_H}{\mu} \tag{7}$$

$$pr=\frac{\mu}{\rho\alpha} \tag{8}$$

Hydraulic diameter is the characteristic length employed in open channels or channels with non-circular cross-sections. The hydraulic diameter is calculated by the following equation:

$$D_H=\frac{4A}{P} \tag{9}$$

$$A=\frac{1}{2}[(H\times H)-(\pi r^2)] \tag{10}$$

$$P = [(2 \times H) + (\pi \times r) + \sqrt{H^2 + H^2} - 2 \times r] \tag{11}$$

Where A is wetted cross-section, P is wetted perimeter of the cross-section, H is edge size of the cross-section, and r is the radius of the inner tube. Equation 12 is used to specify the mean temperature of the fluid.

$$T_{bulk} = \frac{(q''z\,p)}{(\dot{m}c_p)} + T_{in} \tag{12}$$

Also, mass flow rate, convective heat transfer coefficient, Nusselt number, friction factor and inlet velocity are defined as follows:

$$\dot{m} = \rho A V \tag{13}$$

$$h = \frac{q''}{T_{wall} - T_{bulk}} \tag{14}$$

$$Nu = \frac{h D_H}{k} \tag{15}$$

$$f = \frac{\tau_{wall}}{\frac{1}{2}\rho v^2} \tag{16}$$

$$V = \frac{Re\mu}{\rho D_H} \tag{17}$$

Structure grid is employed to examine the grid resolution test. The grid is uniform along the channel and proportionally finer in the inlet, outlet, corners (near walls), and curved surface. To investigate the dependency of numerical solution results from grid point's number, four types of meshes, namely coarse, medium, fine and very fine are studied (Fig. 3). The figure shows that two grids with 135000 and 156000 cells lead to similar results. To decrease the calculation costs, a rectangular mesh with 1,350,000 cells and 1,395,436 nodes is selected.

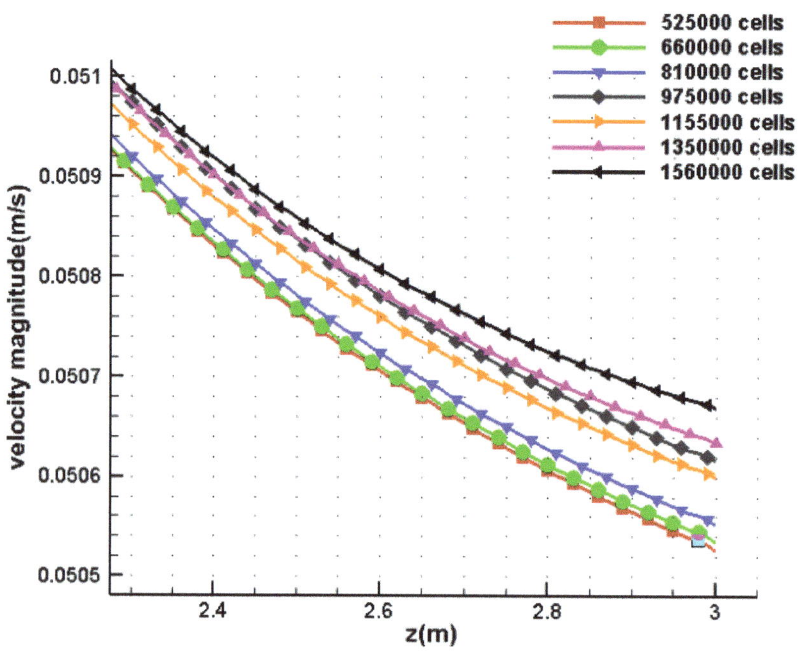

Figure 3. Velocity along the channel for different number of grid cells

The results are compared to the reference [11] and [14] for validation. Obtained average Nusselt number and Re×f are compared with their results (Table 2). It is found that our simulations have sufficient accuracy to predict thermal and hydrodynamic behavior of the fluid flow.

Table 2. Average Nusselt number and Re×f for pure water flow through a channel

Percentage error	Present work	Mehrzad &Kalteh [14]	Shah & London [11]
3.328	Nu=5.14	Nu=5.6615	Nu=5.317
1.316	f.Re=22.31	f.Re=24.054	f.Re=22.02

RESULTS AND DISCUSSION

According to figures 4, 5, and 6, the maximum velocity is observed in the middle of channel cross section, away from holder plates. This is due to low shear stress in this region. In addition, different geometries are introduced for discussion.

Since gravity effect is neglected in the current study, similar results are obtained for geometries 3 and 4. The holder plate in the channel divides the volume of flowing fluid into two parts; however, the Reynolds number and flow regime are similar for all cases.

It can be seen that when the holder plate is in line with the symmetry line, higher velocity is obtained for smaller volume size of the duct, such as geometries 5 (a, b) and 6 (a, b). In all Reynolds number in this study, the highest and lowest velocities were obtained in the middle of Geometries 5 (b) and 1, respectively. Regarding the provided contours, they have different values at different Reynolds number.

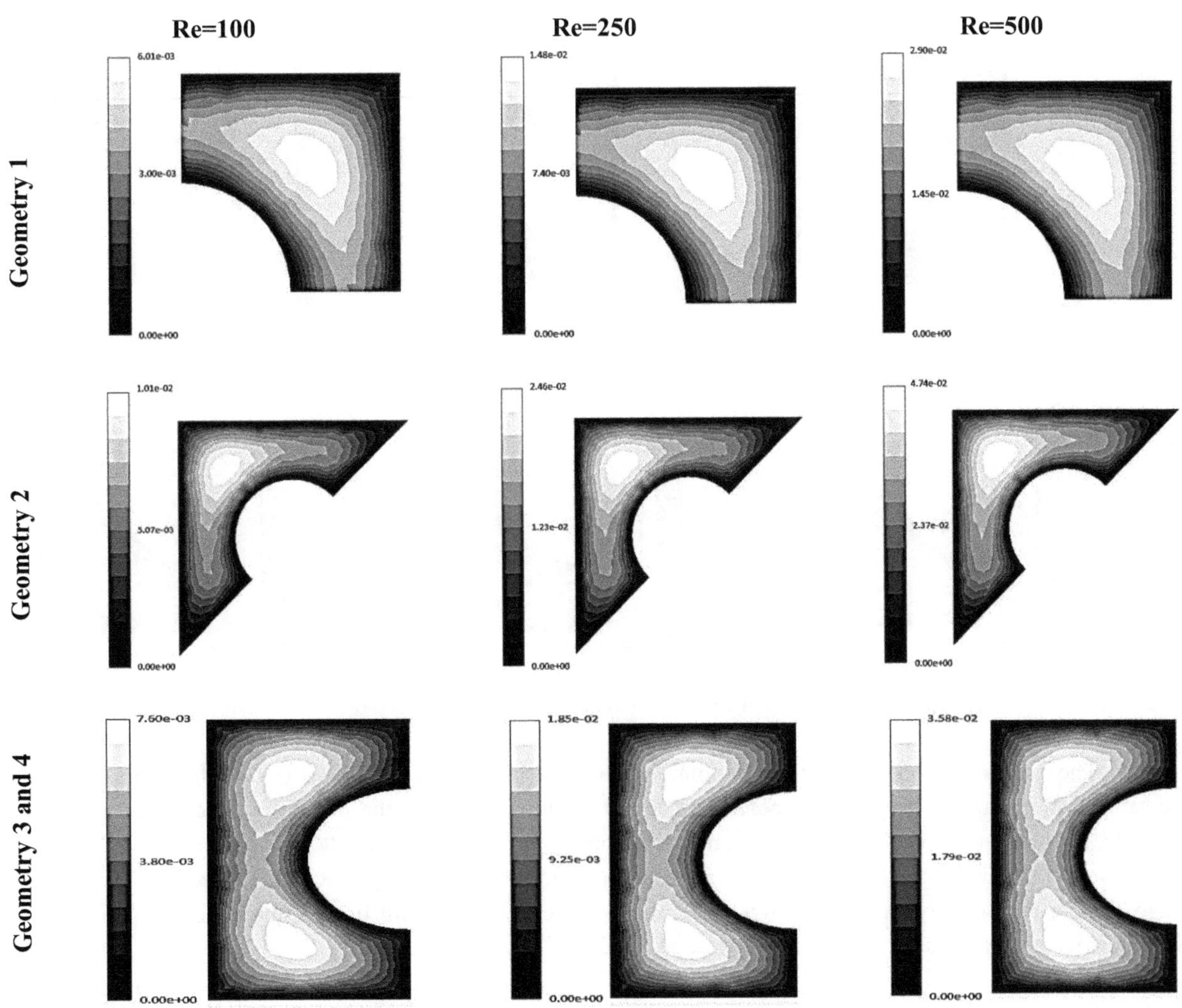

Figure 4. Streamlines for geometries 1, 2, 3, and 4

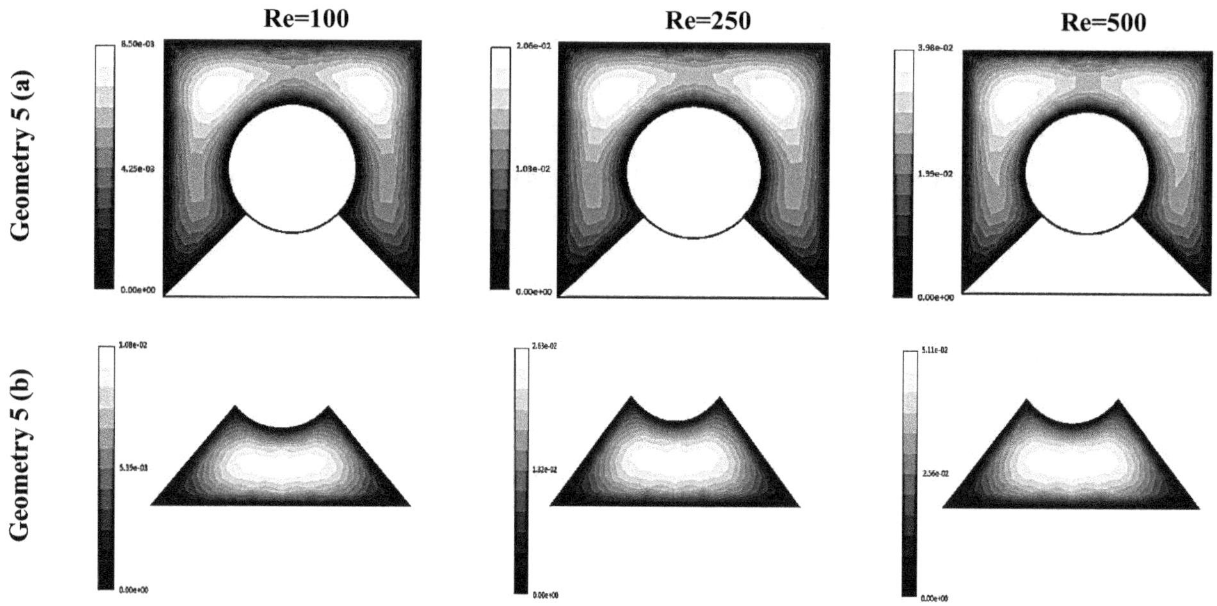

Figure 5. Streamlines for geometry 5

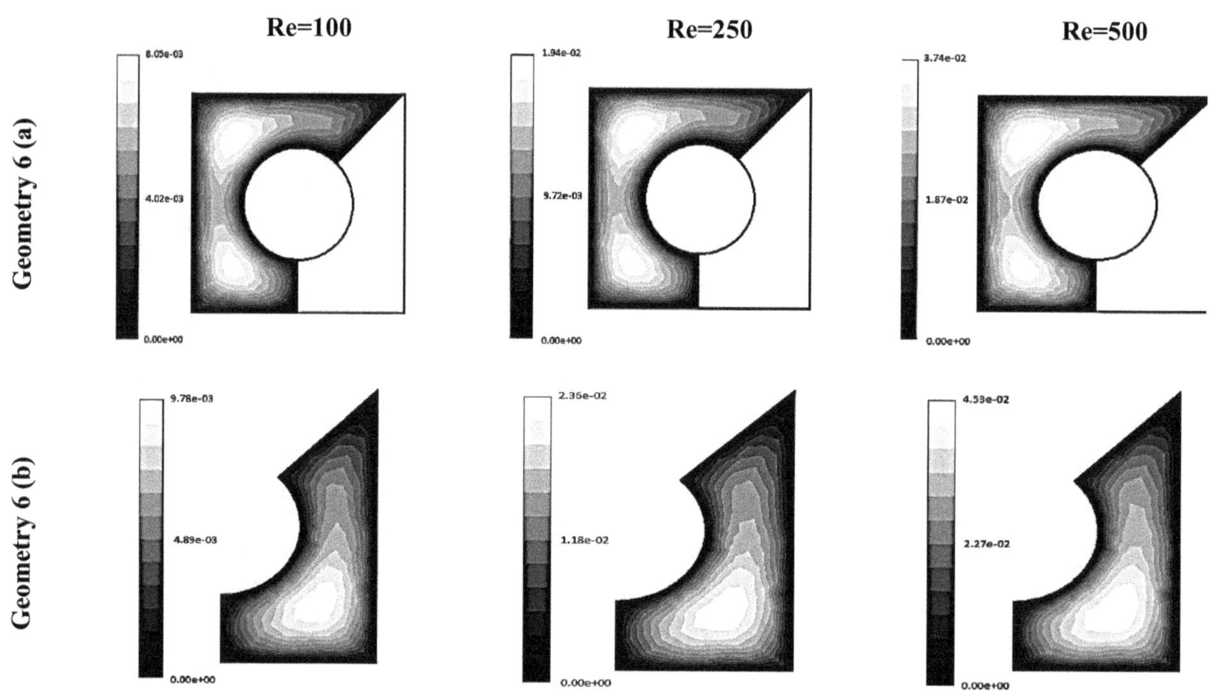

Figure 6. Streamlines for geometry 6

The interesting point is that different outlet velocities can be obtained in this case at the same time under the same flow regime at constant Reynolds numbers. Its rate depends on the position of the holder plate. For example, the highest velocity for sections *a* and *b* of geometry 6 at Re=100 are 0.00805 (m/s) and 0.00978 (m/s), respectively. In addition, the highest velocity for sections *a* and *b* of geometry 6 at Re=500 are 0.0374 (m/s) and 0.0453 (m/s), respectively. As shown, when the holder plate is in line with the symmetry line of the channel, fluid flow has similar hydrodynamic behavior at both sides of the plate.

Figures 7, 8, and 9 present the isothermal line for the outlet section of the flow for different geometries. According to these figures, holder plate allows for different temperature in the outlet cross-section of the channel at the same time under the same flow regime and in the same duct. Moreover, the highest temperature occurs around the holder plate.

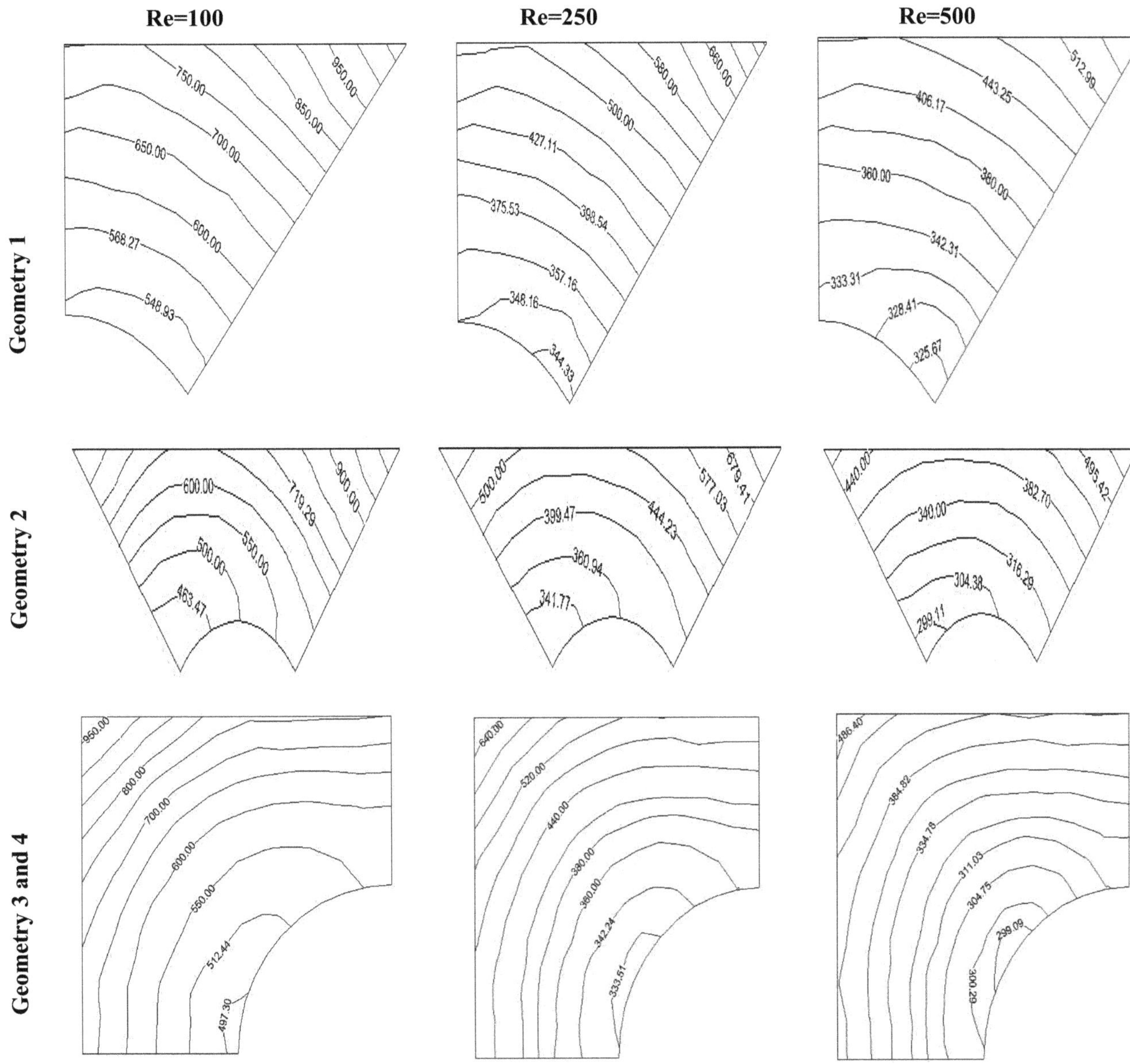

Figure 7. Temperature lines in outlet cross-section of flow for geometries 1, 2, 3, and 4

Figure 10 presents the average convective heat transfer for different Reynolds numbers. At Re=100, the highest and lowest h are observed for Geometries 5 (b) and 1, which were equal to 73.11 and 33.87, respectively. For Geometries 2 (a, b), 3, 4, and 5 (a), the mean values of h are very close to each other and equal to 59.11, 58, 93, and 58.56, respectively.

At Re=250, the highest and lowest mean values of h are observed for Geometries 5 (b) and 2 (a, b), which are equal to 181.2 and 130.01, respectively. The highest and lowest mean values of h occur at Re=500 for the same geometries at Re=100 (Figure 10).

Researchers tend to evaluate the pressure drop required for maintaining internal flows. Figure 11 presents the pressure drop as a function of channel length. The highest and lowest values of the pressure drop occur for the Geometries 5 (b) and 1, respectively.

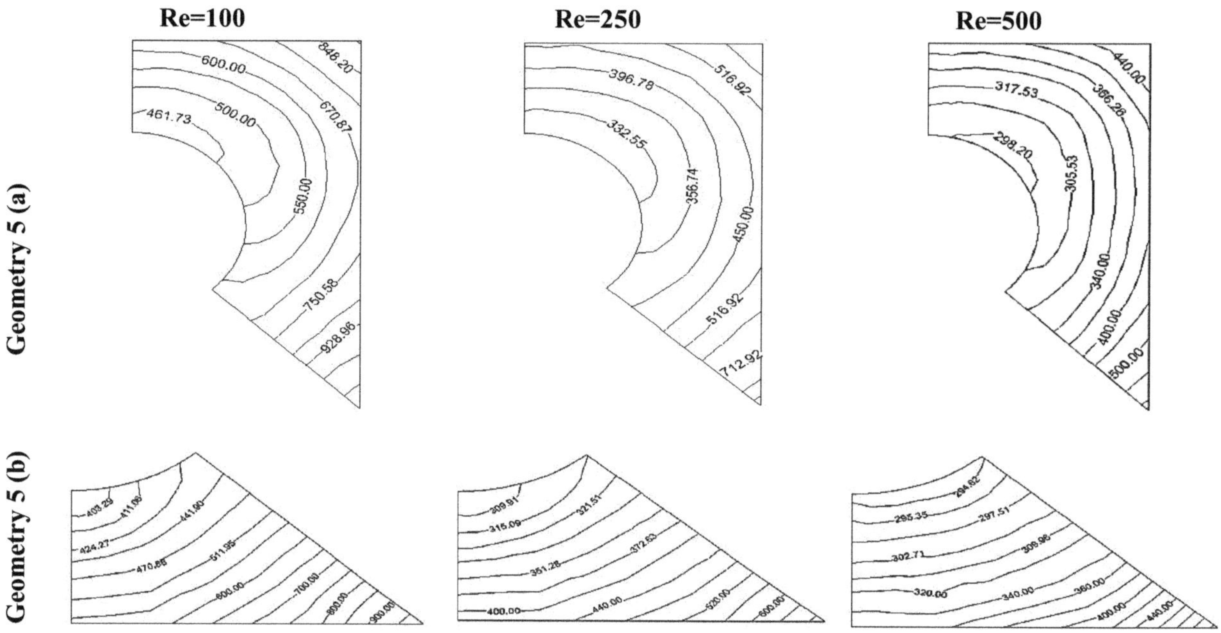

Figure 8. Temperature lines in outlet cross-section of flow for geometry 5

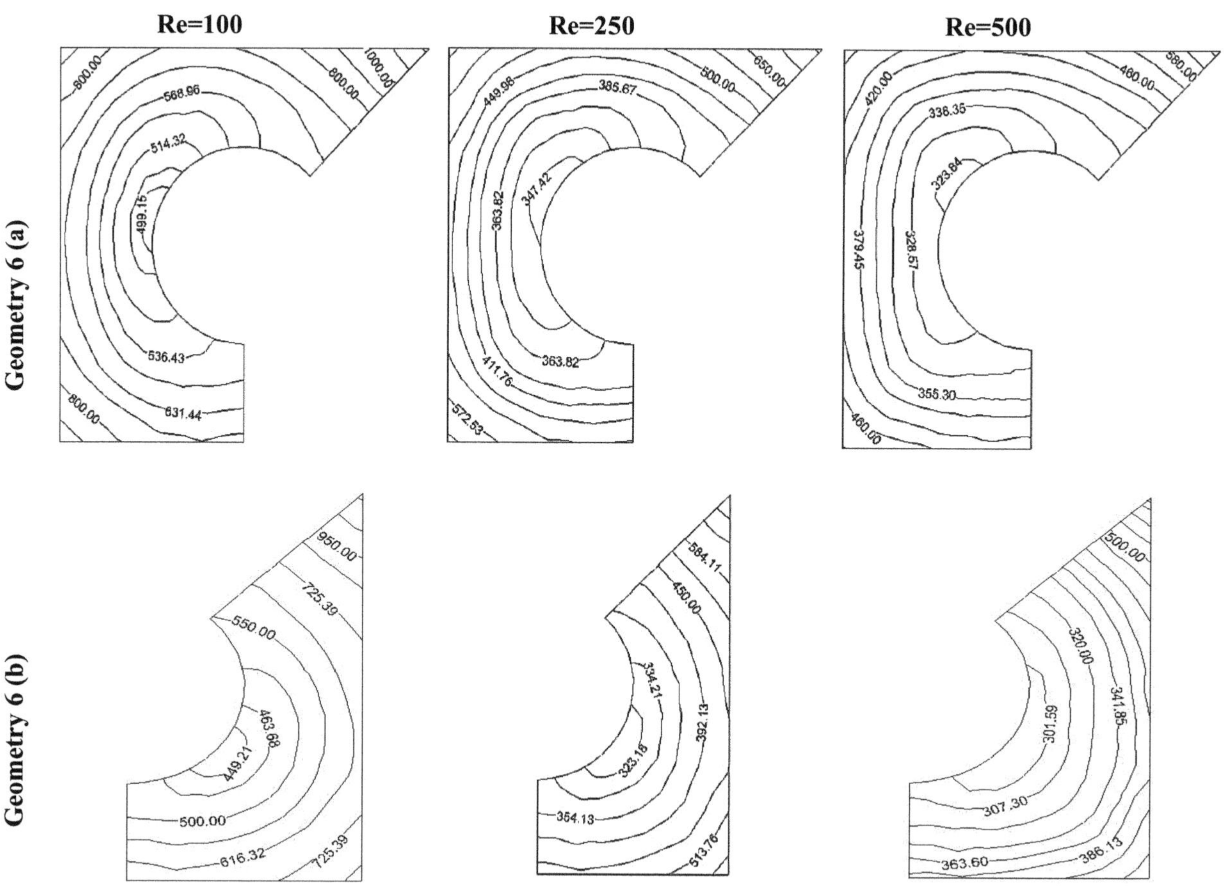

Figure 9. Temperature lines in outlet cross-section of flow for geometry 6

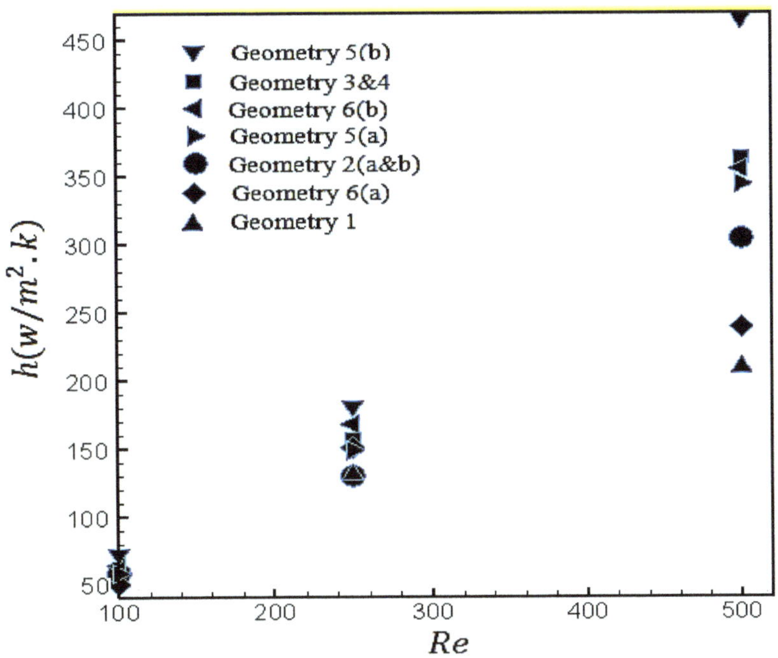

Figure 10. Comparison of mean convective heat transfer for different Reynolds numbers

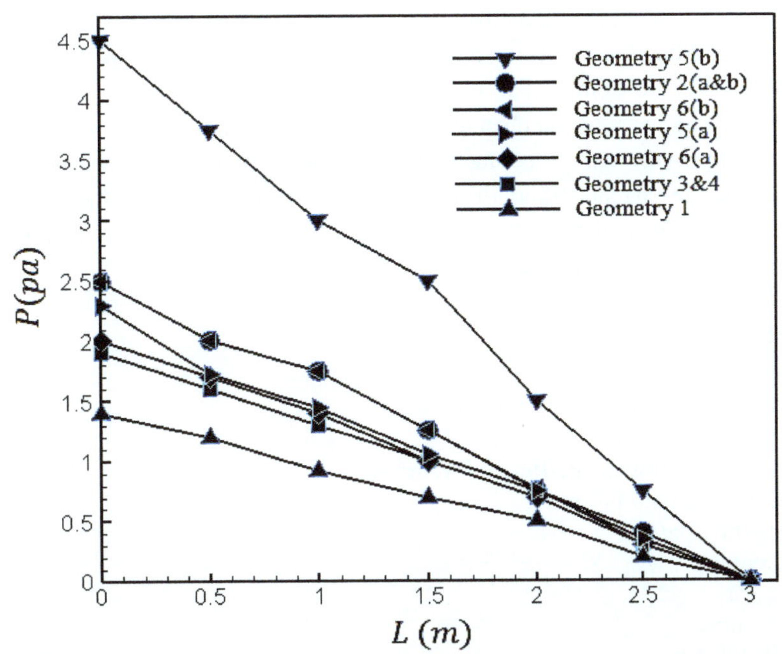

Figure 11. Comparison of the pressure drop versus the channel length

Figure 12 presents fluid mass temperature (bulk temperature, T_{bulk}) at different Reynolds numbers. The mean temperature is used to explain thermal behavior of internal flows in the absence of constant free flow temperature. Bulk temperature is defined in a certain transversal cross-section based on thermal energy transferred by the fluid passing through that section. It is observed that the wall temperature around the perimeter of a non-circular section cross-section changes under a uniform thermal flux conditioning. It is expected that the ratio of fluid temperature to its inlet temperature increases by moving the fluid along the channel. In this problem, the highest T_{bulk} at Re = 100, 250, and 500 corresponds to the Geometry 1, which are equal to 517, 387.42, and 345.313 K, respectively. The lowest temperature at Re=100, 250, and 500 was obtained for the Geometry 5 (b) that are equal to 444.53, 387.42, and 500, respectively.

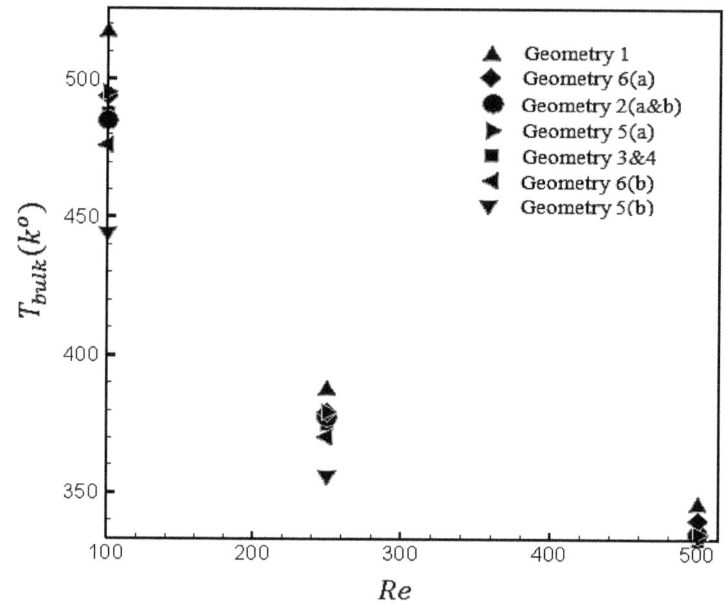

Figure 12. Comparison of fluid mass temperature at different Reynolds numbers

CONCLUDING REMARKS

In the present work, the convective heat transfer of water-AL_2O_3 nanofluid in the space between a square channel and middle rigid plate for different positions of the holder plate was investigated. The results indicated that the position of holder plate has significant effects on the flow behavior. When this plate is not located in the line of symmetry of the channel cross-section, it can produce two flows with different velocities and outlet temperatures at the same time under the same inlet conditions. When the holder plate is in line with the symmetry line of the channel, fluid flow has similar hydrodynamic behavior at both sides of the plate. Finally, the studies of Erdinc and Yilmaz [21] is proposed to compare the present results with the results for the case of converging and diverging channels.

NOMECLATURE

A	Area [m^2]
C_p	Specific thermal coefficient [kJ/kg.K]
D_H	Hydraulic diameter [m]
h	Convective heat transfer coefficient [W/m^2.K]
k	Thermal conductivity [W/m.K]
$\dot{m}$	Mass flow rate [kg/s]
Nu	Nusselt number
p	Pressure [Pa]
Pr	Prandtl number
Re	Reynolds number
T	Temperature [K]
u	x-component of the velocity field [m/s]
v	y-component of the velocity field [m/s]
w	z-component of the velocity field [m/s]
α	Thermal expansion coefficient [1/K]
ρ	Density [W/m^2]
μ	Viscosity [Pa. s]

REFERENCES

[1] Choi. S.U.S., Eastman. J.A. (1995). Enhancing thermal conductivity of fluids S with nanoparticles. In Developments and Applications of Non-Newtonian Flows. Volume FED-231/MD-66. Siginer DA, Wang HP, editor. ASME, New York, 99–105.

[2] Murshed SS, De Castro CN. (2011). Forced convective heat transfer of nanofluids in minichannels. InTwo Phase Flow, Phase Change and Numerical Modeling. InTech.

[3] Mirmasoumi. S , Behzadmehr. A. (2008). Numerical study of laminar mixed convection of a nanofluid in a horizontal tube using two-phase mixture model. Applied Thermal Engineering.;28(7). 717-27.
[4] Pishkar. I, Ghasemi. B. (2012). Effect of nanoparticles on mixed convection heat transfer in a horizontal channel with heat source. Modares Mechanical Engineering. 12(2):95-108.
[5] Vasefi. I, Alizadeh. M. (2013). A numerical investigation of CuO–water nanofluid in different geometries by two-phase Euler–Lagrange method. World Appl Sci J. 26(10). 1323-9.
[6] Heris. SZ, Talaii. E, Noie. SH. (2012). CuO/water nanofluid heat transfer through triangular ducts. Iran. J. Chem. Eng.;9(1). 23-32.
[7] Ho. CJ, Chen. MW, Li. ZW. (2008). Numerical simulation of natural convection of nanofluid in a square enclosure: effects due to uncertainties of viscosity and thermal conductivity. International Journal of Heat and Mass Transfer. 51(17). 4506-16.
[8] Ben Mansour. R, Galanis. N, Nguyen.(2009). CT. Developing laminar mixed convection of nanofluids in an inclined tube with uniform wall heat flux. International Journal of Numerical Methods for Heat & Fluid Flow. 19(2). 146-64.
[9] Nguyen. CT, Roy. G, Gauthier. C, Galanis. N.(2007). Heat transfer enhancement using Al 2 O 3–water nanofluid for an electronic liquid cooling system. Applied Thermal Engineering. 27(8):1501-6.
[10] Heris. SZ, Esfahany. MN, Etemad. G. (2007). Numerical investigation of nanofluid laminar convective heat transfer through a circular tube. Numerical Heat Transfer, Part A: Applications. 52(11). 1043-58.
[11] Shah. RK, London. AL. (2014). Laminar flow forced convection in ducts: a source book for compact heat exchanger analytical data. Academic press.
[12] Zhang. LZ. (2007). Laminar flow and heat transfer in plate-fin triangular ducts in thermally developing entry region. International Journal of Heat and Mass Transfer. 50(7). 1637-40.
[13] Heyhat. MM, Kowsary. F. (2010). Effect of particle migration on flow and convective heat transfer of nanofluids flowing through a circular pipe. Journal of Heat Transfer. 132(6). 062401.
[14] A. Bejan. (2013). Convection Heat Transfer, 4th Edition, Wiley.
[15] Pak. BC, Cho. YI. (1998). Hydrodynamic and heat transfer study of dispersed fluids with submicron metallic oxide particles. Experimental Heat Transfer an International Journal. 11(2). 151-70.
[16] Hamilton. RL, Crosser. OK.(1962). Thermal conductivity of heterogeneous two-component systems. Industrial & Engineering chemistry fundamentals. 1(3). 187-91.
[17] Kalteh. M, Abbassi. A, Saffar-Avval. M, Frijns. A, Darhuber. A, Harting. J. (2012). Experimental and numerical investigation of nanofluid forced convection inside a wide microchannel heat sink. Applied Thermal Engineering. 36. 260-8.
[18] Chon. CH, Kihm. KD, Lee. SP, Choi. SU. (2005). Empirical correlation finding the role of temperature and particle size for nanofluid (Al 2 O 3) thermal conductivity enhancement. Applied Physics Letters. 87(15). 153107.
[19] Nguyen. CT, Desgranges. F, Roy. G, Galanis. N, Maré. T, Boucher. S, Mintsa. HA. (2007). Temperature and particle-size dependent viscosity data for water-based nanofluids–hysteresis phenomenon. International Journal of Heat and Fluid Flow. 28(6):1492-506.
[20] Sieder. EN, Tate. GE. (1936). Heat transfer and pressure drop of liquids in tubes. Industrial & Engineering Chemistry. 28(12). 1429-35.
[21] Erdinc. MT, Yilmaz. T. (2018). Numerical Investigation of Flow and Heat Transfer in Communicating Converging and Diverging Channels, Journal of Thermal Engineering, 4(5), 2318-2332.

7

THERMOECONOMIC ANALYSIS OF A WATER TO WATER HEAT PUMP UNDER DIFFERENT CONDENSER AND EVAPORATOR CONDITIONS

B. Aksu[1,*], C. Uysal[2], H. Kurt[3]

ABSTRACT

A thermoeconomic analysis of a water to water heat pump are performed under different condenser and evaporator conditions. Experiments are realized for different volumetric inlet temperatures of 14.4, 17 and 19 $°C$ and different volumetric flow rates of 50, 100, 150 lt/h for condenser cooling water. Same inlet temperatures with condenser cooling water are used for evaporator water inlet, while constant volumetric flow rate of 100 lt/h is used for each case. Modified Productive Structure Analysis (MOPSA) is used for thermoeconomic analysis. It is found that increases in inlet temperature and in volumetric flow rate cause to decrease in both the unit cost of heat delivered (C_H) and the unit cost of entropy generation (C_S). As a result, in the case of T_{in}=14.4 $°C$ and $\dot{\forall}$ = 50 lt/h, C_H and C_S values are obtained to be 0.0489 \$/kWh and 0.0221 \$/kWh, respectively, while C_H and C_S values are obtained to be 0.0385 \$/kWh and 0.0175 \$/kWh for T_{in}=19 $°C$ and $\dot{\forall}$ = 150 lt/h, respectively.

Keywords: *Water to Water Heat Pump, Condenser Cooling Water, Thermoeconomic Analysis, MOPSA*

INTRODUCTION

Growing energy demand, limited life time and reservoirs of fossil fuels and environmental effects caused by fossil fuel usage such as greenhouse effect and global warming cause to pay a more attention for alternative energy sources. In this sense, heat pump systems can be considered one of these energy sources because of that they provide cooling or heating by using non-fossil based energy sources. Heat pump systems transfer the energy from lower temperature source to higher temperature source. Different energy sources such as air, water and ground can be used by heat pump systems. To be able to provide the energy transfer from lower temperature source to higher one, heat pump systems consume electricity. Although the performance improvement studies in the manner of thermodynamics are important, the studies for economically assessment of this kind system are also important to be able to evaluate the unit cost of products.

Thermoeconomic analysis that combines thermodynamical and economic analysis is an effective tool to evaluate thermal systems. Thermoeconomic analysis provides an information for designers or engineers about the effect of some modifications or malfunctions and dysfunctions on the unit cost of system products. In this sense, several studies about thermoeconomic analysis of heat pump systems have been performed on.

Waheed et al. [1] have performed a study to reduce heat pump size and heat loss developed advanced vapor recompression heat pump models. They compared the thermoeconomic and environmental performance of the developed models with the traditional distillation process. As a result, the developed models provided significant energy savings. The total annual costs for the models were 92259 US\$, 92835US\$, 95454 US\$, and cost savings were 33.6%, 29.7% and 24.4%, respectively. Qureshi and Zubair [2] have investigated the thermoeconomic factors for the inventory of the heat exchanger at finite thermal capacities of irreversible refrigerants and heat pumps. As a result, the use of mathematical expressions would be better for determining the effect of the internal dissipation. Kodal et al. [3-4] have conducted performance analyzes using finite time thermodynamic based on a thermoeconomic objective function for irreversible refrigerators and heat pumps. They analytically derived optimum design parameters for the maximum conditions of the objective functions for refrigerators and heat pumps and have discussed the effects of internal irreversibility, economic parameters and external temperatures on global and optimal performance. Teyber et al. [5] have offered an optimization to minimize the permanent magnet and cooling cost in magnetic heat pumps. As a result, a constant temperature

This paper was recommended for publication in revised form by Regional Editor Ozgen Acikgoz

[1]*Gas and Gas Installation Program, TOBB Vocational School of Technical Sciences, Karabuk Uni., 78050, Karabuk, TURKEY*
[2]*Automotive Technologies Program, TOBB Vocational School of Technical Sciences, Karabuk Uni., 78050, Karabuk, TURKEY*
[3]*Mechanical Engineering Department, Faculty of Engineering & Architecture, Necmettin Erbakan Uni., 42140, Konya, TURKEY*

**E-mail address: baksu@karabuk.edu.tr*

range of 15 K and an optimal magnet configuration for 4.2 kg of gadolinium, using 16.3 kg of permanent magnet material, they obtained cooling power of 3.3 kW with a second law efficiency of 0.33. Esfahani et al. [6] have developed a thermoeconomic model of an existing system which has low and high pressure compressors. The thermoeconomic model has been used to assess the unit cost of the fresh water and cooling to evaluate the flexibility of the system for fuel allocation from different electricity and heat energy sources. The results showed that the system with two compressors had high flexibility to allocate the different energy sources when the availability of the sources was limited for a given value of fresh water and cooling production.Verda et al. [7] have conducted thermoeconomic analysis for district heating networks operating at low temperatures. The analyzes were made on simple samples in order to provide a quantitative evaluation of various cost terms according to user / producer conditions, topology and properties.

Arat and Arslan [8] have investigated geothermal heat pump supported district heating system exergetically and economically by using NPL and LLC method. The results have showed that 13776 houses would be heated with this system and it could be an attractive investment for the region. Erbay and Hepbasli [9] have investigated the ground source heat pump drying system in terms of exergetic and advanced exergoeconomics. Results showed that total costs could be decrease from 4.008 $/h to 2.569 $/h. For this, the improvements had to be made at drying tube and condenser. Sayyadi and Nejatolahi [10] have compared and discussed the results of optimizing by the ways of thermodynamic, thermoeconomic, and both thermodynamic both thermoeconomic according to the total revenue requirement model (TTR) of the cooling tower supported with ground source heat pump. Further, the sensitivity of optimized systems to the interest rate, the annual number of operating hours in cooling mode, the electricity price, and the water price are studied in detail. Erbay and Hepbasli [11] have found that the total cost was 1.347 $/h and the exergoeconomic factor was 0.029, by performing an exergoeconomic analysis of the ground source heat pump food dryer. Mastrullo and Renno [12] have investigated a model of heat pump whose evaporator operates as a photovoltaic collector in the terms of thermoeconomic and have compared the conventional heat pump with the photovoltaic heat pump thermoeconomically in the same operating conditions. Akbulut et al. [13] have experimentally and theoretically investigated exergoeconomic and exergoenvironmental analyses of wall cooling systems fed by a vertical type of ground source heat pump integrated wall cooling system for cooling mode. The results showed that the energy and exergy efficient of the whole system were 74.85% and 29.90%, respectively. The external environmental effects of the whole system were found as 42.6% and the exergoeconomic factor as 77.68%. Qin and Hao [14] have showed that the coefficient of performance of the sewage source heat-pump system approximates to 4.0. They have determined that the cost of the system is around 16.77 ¥/m^2 and the annual cost of saving is around 444000 ¥. When they examined the thermoeconomic cost of the system, they obtained an average value of 7.8 W/W. Erbay and Hepbasli [15] have compared the performance of the pilot scale air source heat pump food dryer with the traditional and exergoeconomic aspects. By the results, they have determined that the most necessary cost to reduce the cost is in the heat recovery unit. Kwak et al. [16] have performed thermoeconomic analysis of a ground-source heat pump system with a vertical or horizontal ground heat exchanger. Modified Productive Structure Analysis (MOPSA) method is used for thermoeconomic analysis. They have reported that the unit cost of heat delivered is 0.063 $/kWh, while the unit cost of electricity supplied to the system is 0.14 $/kWh and the coefficient of performance of the system is 3.27.

In this study, thermoeconomic analysis of a water-to-water heat pump system is performed on under different condenser and evaporator conditions. Same inlet temperatures of water supplied to condenser and evaporator are used in the experiments, which are T_{in}= 14.4 $°C$, 17 $°C$ and 19 $°C$. Moreover, different volumetric flow rates are used for the water stream supplied to the condenser ($\dot{\forall}$ =50 lt/h, 100 lt/h and 150 lt/h), while volumetric flow rates of water stream supplied to evaporator are to be fixed as $\dot{\forall}$ = 100 lt/h in all experiments. Modified Productive Structure Analysis (MOPSA) is used for thermoeconomic analysis of the heat pump system considered in this study. The unit costs of mechanical exergy, thermal exergy, heat delivered and negentropy are presented and the effects of condenser and evaporator conditions on the unit cost of these terms are discussed.

SYSTEM DESCRIPTION AND EXPERIMENTAL SETUP

A schematic diagram of experimental setup for the water-to-water heat pump considered in this study is shown in Figure 1.

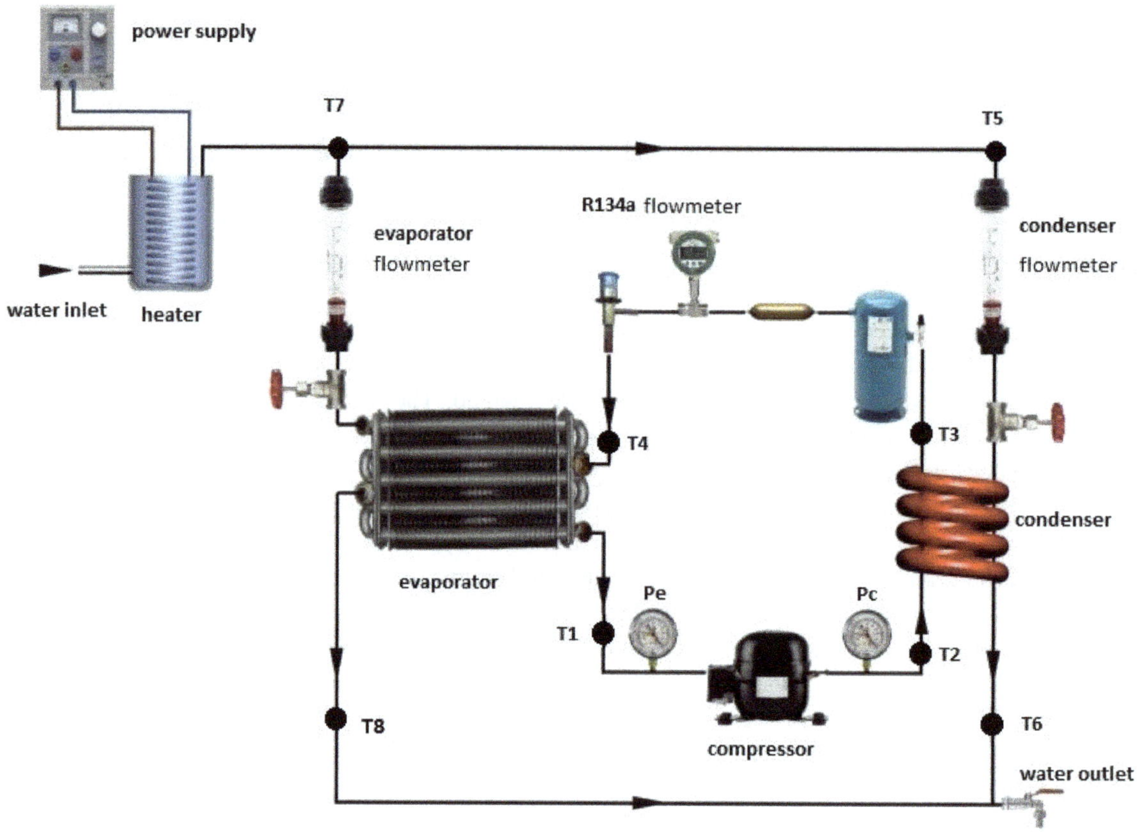

Figure 1. Experimental setup for heat pump system considered in this study

The system consists of four main components: compressor, condenser, expansion valve and evaporator. Refrigerant R-134a is used as working fluid for this study. R-134a enters to compressor as saturated vapor and is pressurized in compressor. R-134a leaving from compressor as superheated and pressurized enters to condenser. The refrigeration delivers heat to water supplied to condenser and leaves from condenser as saturated liquid. The pressure of refrigerant leaving from condenser is reduced in expansion valve to obtain a phase change for low temperatures in evaporator. Heat removed from medium is transferred to refrigerant in evaporator and refrigerant leaves from evaporator as saturated vapor.

An experimental setup is installed and experiments are realized for different inlet temperatures of water supplied to condenser and evaporator and different volumetric flow rates of water supplied to condenser. The volumetric flow rate of 100 lt/h is used for water supplied to evaporator in all experiments for each cases. The inlet temperatures for water supplied to condenser and evaporator are $14.4\,°C$, $17\,°C$ and $19\,°C$. An electrical heater controlled with a thermostat is used to provide the demanded temperatures for water supplied to condenser and evaporator. In temperature measurements for water supplied to condenser and evaporator and for refrigerant circulated in the system, multichannel data logger with PT100 type sensors is used. In pressure measurements for the system states, REFCO MR-205-DS-R22 is used for low pressure line, while REFCO MR-305-DS-R22 is used for high pressure line. The experiments are also realized for different volumetric flow rate of water supplied to the condenser, while constant volumetric flow rate of 100 lt/h is used for the water supplied to the evaporator. The volumetric flow rates used for the water supplied to the condenser are 50 lt/h, 100 lt/h and 150 lt/h. A turbine type flow meter is used for measurements of volumetric flow rate of refrigerant circulated in the system. In the measurement of volumetric flow rates of volumetric flow rate of water supplied to the condenser and evaporator, a rotameter type having measurement range of 40-400 lt/h is used.

The experiments are realized for 9 different cases by changing the inlet temperature of the water supplied to the condenser and evaporator and by changing the volumetric flow rate of water supplied to the condenser. The properties considered for each cases are presented in Table 1.

Table 1. The cases considered for experiments of heat pump system

Cases	Inlet temperature for condenser, T_{in} (°C)	Inlet temperature for evaporator, T_{in} (°C)	Volumetric flow rate for evaporator, $\dot{V}$ (lt/h)	Volumetric flow rate for condenser, $\dot{V}$ (lt/h)
Case I	14.4	14.4	100	50
Case II	14.4	14.4	100	100
Case III	14.4	14.4	100	150
Case IV	17	17	100	50
Case V	17	17	100	100
Case VI	17	17	100	150
Case VII	19	19	100	50
Case VIII	19	19	100	100
Case IX	19	19	100	150

The results obtained from experiments are presented for only Case I and Case IX. Table 2 and 3 show the properties obtained by experiments for each state located in the heat pump system shown in Figure 1 for Case I and Case IX, respectively.

Table 2. The properties for each state located in the heat pump system for Case I

State	$\dot{m}$(kg/s)	T (°C)	P (bar)	h(kJ/kg)	s(kJ/kgK)	ex (kJ/kg)
1	0.006406	1.6	3.1	399.41	1.7258	27.278200
2	0.006406	68.96	15	441.96	1.7548	61.186200
3	0.006406	55.23	15	279.68	1.2616	45.879800
4	0.006406	1.6	3.1	279.68	1.29	37.416600
5	0.013889	14.4	4	60.85	0.21567	1.085940
6	0.013889	28.32	4	119.07	0.41341	0.379420
7	0.027778	14.4	4	60.85	0.21567	1.085940
8	0.027778	9.78	4	41.487	0.14779	1.951180

Table 3. The properties for each state located in the heat pump system for Case IX

State	$\dot{m}$(kg/s)	T (°C)	P (bar)	h(kJ/kg)	s(kJ/kgK)	ex (kJ/kg)
1	0.007968	7.6	3.82	402.84	1.7228	31.6022
2	0.007968	73.36	19.33	443.02	1.7495	63.8256
3	0.007968	60.8	19.33	288.63	1.2884	46.8434
4	0.007968	7.6	3.82	288.63	1.3158	38.6782
5	0.041667	19	4	80.105	0.28211	0.54182
6	0.041667	25.33	4	106.58	0.37175	0.3041
7	0.027778	19	4	80.105	0.28211	0.54182
8	0.027778	12.26	4	51.885	0.18438	1.44536

THERMOECONOMIC ANALYSIS

Thermoeconomic analysis provides the detailed information about unit cost of any thermodynamical parameter, especially for product of thermal systems, to designer or engineer and enables cost allocation when system analyzed has two or more products. There are several thermoeconomic methods in the literature [17-28]. Among these methods, Modified Productive Structure Analysis (MOPSA) method was selected for this study because of that MOPSA method assigns a unit cost for entropy production and provides correct results for unit cost of products for thermal systems [29].

Modified Productive Structure Analysis (MOPSA)

MOPSA method, which was firstly introduced by Kim et al. [23], provides exergy costing without any flow stream cost calculation. In MOPSA method, the component of thermomechanical exergy can be costed in separate form as thermal and mechanical exergy costs. Therefore, general exergy balance equation can be written for MOPSA application as follows:

$$\dot{E}_x^{CHE} + \left(\sum_{inlet} \dot{E}_x^{T} - \sum_{outlet} \dot{E}_x^{T} \right) + \left(\sum_{inlet} \dot{E}_x^{P} - \sum_{outlet} \dot{E}_x^{P} \right) + T_o\left(\sum_{inlet} \dot{S}_i - \sum_{outlet} \dot{S}_i\right) = \dot{E}_x^{W} \quad (1)$$

In Equation 1, first term on the left side express the chemical exergy of fuel supplied to the system. Second and third terms on the left side are thermal and mechanical components of thermomechanical exergy of any stream, respectively. Fourth term on the left side denote exergy destruction of the component considered. The term on the right side of Equation 1 the exergy value of work, which is equal to the value of work.

Thermoeconomics is an engineering branch that combines thermodynamics and economics analysis. Therefore, economic analysis should be applied to realize the thermoeconomical analysis of a thermal system. In this study, the annualized (levelized) cost method proposed by Moran [30] is used to estimate the capital cost of components. The present worth factor (PWF) and the capital recovery factor (CRF) are given as follows, respectively:

$$PWF = \frac{1}{(1+i)^n} \quad (2)$$

$$CRF = \frac{i(1+i)^n}{(1+i)^n - 1} \quad (3)$$

where i and n terms denote the interest rate and the life time of system, respectively. The salvage value of the system can be calculated as follows:

$$SV = \mu TCI \quad (4)$$

where μ and TCI represent the salvage value ratio and the total capital investment of the system. The present worth (PW) and the annual capital cost of the system ($A\dot{C}$) are given as follows, respectively.

$$PW = TCI - SV \times PWF \quad (5)$$

$$A\dot{C} = PW \times CRF \quad (6)$$

The capital investment cost of overall system is calculated as follows:

$$\dot{Z}_{system} = \phi \frac{A\dot{C}}{\tau} \quad (7)$$

where ϕ and τ denote the maintenance factor and the total annual number of operating hours of the system at full load capacity. The capital investment cost rate of kth component of the system can be calculated as follows:

$$\dot{Z}_k = \dot{Z}_{system} \frac{PEC_k}{TCI} \quad (8)$$

where *PEC* express the purchased equipment cost of kth component in the system.

The general cost balance equation for MOPSA based adding unit cost of related terms in the exergy balance equation given in Equation 9 can be written as follows:

$$\dot{Ex}^{CHE}C_0 + \left(\sum_{inlet} \dot{Ex}_{in}^{T} - \sum_{outlet} \dot{Ex}_{out}^{T}\right)C_T + \left(\sum_{inlet} \dot{Ex}_{in}^{P} - \sum_{outlet} \dot{Ex}_{out}^{P}\right)C_P$$

$$+T_0\left(\sum_{inlet} \dot{S}_{in} - \sum_{outlet} \dot{S}_{out}\right)C_S + \dot{Z}_k = \dot{Ex}^{W}C_W \qquad (9)$$

where T and P superscripts represent thermal and mechanical components of thermomechanical exergy, respectively. C_0, C_S and C_W are the unit exergy cost of fuel, entropy and work, respectively. C_T and C_P are also the unit exergy cost of thermal and mechanical components of thermomechanical exergy, respectively.

MOPSA Application for Water-to-Water Heat Pump

The exergy balance equations are obtained by applying the general exergy balance equation given in Equation 1 to each components of the heat pump system shown in Figure 1. The obtained equations are given as follows:

Compressor:

$$(\dot{Ex}_1^T - \dot{Ex}_2^T) + (\dot{Ex}_1^P - \dot{Ex}_2^P) + T_0(\dot{S}_1 - \dot{S}_2) = \dot{Ex}^W \qquad (10)$$

Condenser:

$$(\dot{Ex}_2^T - \dot{Ex}_3^T + \dot{Ex}_5^T - \dot{Ex}_6^T) + (\dot{Ex}_2^P - \dot{Ex}_3^P + \dot{Ex}_5^P - \dot{Ex}_6^P)$$

$$+T_0(\dot{S}_2 - \dot{S}_3 + \dot{S}_5 - \dot{S}_6) = 0 \qquad (11)$$

Expansion valve:

$$(\dot{Ex}_3^T - \dot{Ex}_4^T) + (\dot{Ex}_3^P - \dot{Ex}_4^P) + T_0(\dot{S}_3 - \dot{S}_4) = 0 \qquad (12)$$

Evaporator:

$$(\dot{Ex}_4^T - \dot{Ex}_1^T + \dot{Ex}_7^T - \dot{Ex}_8^T)C_T + (\dot{Ex}_4^P - \dot{Ex}_1^P + \dot{Ex}_7^P - \dot{Ex}_8^P)$$
$$+T_0(\dot{S}_4 - \dot{S}_1 + \dot{S}_7 - \dot{S}_8) = 0 \qquad (13)$$

In this study, for economic analysis, the interest ratio (i) and the life time of the system (n) are assumed to be 10% and 5 years, respectively. In addition; the salvage value ratio (μ), the maintenance factor (ϕ) and the total annual number of operating hours of the system at full load capacity () are selected to be 12%, 1.06 and 2880 hours, respectively.

The cost balance equations are obtained by applying the general cost balance equation given in Equation 9 to each components of the heat pump system illustrated in Figure 1 and are given as follows:

Compressor:

$$(\dot{Ex}_1^T - \dot{Ex}_2^T)C_T + (\dot{Ex}_1^P - \dot{Ex}_2^P)C_P + [T_0(\dot{S}_1 - \dot{S}_2)]C_S + \dot{Z}_{com} = \dot{Ex}^W C_W \qquad (14)$$

Condenser:

$$(\dot{Ex}_2^T - \dot{Ex}_3^T)C_T + (\dot{Ex}_2^P - \dot{Ex}_3^P)C_P - \dot{Q}_{room}C_H + [T_0(\dot{S}_2 - \dot{S}_3)]C_S + \dot{Z}_{cnd} = 0 \qquad (15)$$

Expansion valve:

$$(\dot{Ex}_3^T - \dot{Ex}_4^T)C_T + (\dot{Ex}_3^P - \dot{Ex}_4^P)C_P + [T_0(\dot{S}_3 - \dot{S}_4)]C_S + \dot{Z}_{exp} = 0 \qquad (16)$$

Evaporator:

$$(E\dot{x}_4^T - E\dot{x}_1^T)C_T + (E\dot{x}_4^P - E\dot{x}_1^P)C_P + (E\dot{x}_7 - E\dot{x}_8).0 + [T_0(\dot{S}_4 - \dot{S}_1)]C_S + \dot{Z}_{eva} = 0 \quad (17)$$

where $\dot{Q}_{room}$ term represents the heat transfer rate delivered to medium heated and it is formulated as follows:

$$\dot{Q}_{room} = (E\dot{x}^T{}_6 - E\dot{x}^T{}_5) + (E\dot{x}^P{}_6 - E\dot{x}^P{}_5) + T_0(\dot{S}_6 - \dot{S}_5) \quad (18)$$

With writing from Equation 14 to Equation 17, all cost balance equations required to calculate the unknown unit cost parameters assigned as C_T, C_P, C_S and C_H are obtained. The value of the unit cost of electricity supplied to the system (C_W) is 0.0586 \$/kWh [31]. The unit cost of heat gained by evaporator is assumed to be 0, similarly with Reference [16].

The overall cost balance equation is obtained by summing Equation 14 to Equation 17 as follows:

$$-\dot{Q}_{room}C_H + \sum \dot{Z}_k = E\dot{x}^W C_W \quad (19)$$

The unit cost of heat obtained by solving Equation 14 to Equation 17 should be same with the result obtained by Equation 19.

RESULTS AND DISCUSSION

The effect of volumetric flow rate and inlet temperature of condenser cooling water on the unit cost of heat delivered to medium by a water-to-water heat pump system is investigated. Three different volumetric volume rates ($\dot{\forall}$ = 50 lt/h, 100 lt/h and 150 lt/h) and three different inlet temperatures (T_{in} = 14 °C , 17 °C and 19 °C) of condenser cooling water are considered. For evaporator, same inlet temperatures are used with that of condenser for each case. However, constant volumetric flow rate of 100 lt/h are used for each case. By this way, the exergy and thermoeconomic analyses are applied for 9 different cases of the heat pump system. However, the results obtained for only Case I and Case IX are presented in this paper.

Coefficient of performance (COP) and second law efficiency (η_{II}) values of the water-to-water heat pump system for each cases are given in Table 4.

Table 4. COP and η_{II} values of water-to-water heat pump system for each case

	Case I	Case II	Case III	Case IV	Case V	Case VI	Case VII	Case VIII	Case IX
COP	2.98	3.21	3.45	2.94	3.19	3.46	2.90	3.13	3.44
η_{II}	0.137	0.080	0.061	0.151	0.089	0.069	0.164	0.096	0.073

It is found that COP value of the system increases with increase in volumetric flow rate of inlet water supplied to condenser, while it is not almost affected by inlet temperature of water supplied to condenser and evaporator. However, the second law efficiency of the system decreases with increasing volumetric flow rate of inlet water supplied to condenser. In addition, increase in inlet temperature of water supplied to condenser and evaporator causes to increase in the second law efficiency of the system. The highest second law efficiency value of the system is obtained to be 16.4% for $\dot{\forall}$ = 50 lt/h and T = 19 °C , while the lowest one is obtained to be 6.1% for $\dot{\forall}$ = 150 lt/h and T = 14.4 °C .

Exergy balance values obtained for each component with applying Equations 10 to Equation 13 to related components by using the data presented in Table 2 and 3 for Case I and Case IX are illustrated in Table 5 and Table 6, respectively.

Table 5. Exergy balance for the system components of the heat pump system for Case I

Component	Refrigerant (kW)	Water stream (kW)	Heat flow (kW)	Lost work rate (kW)	Work input rate (kW)
Compressor	0.217215			0.055361	-0.27205
Condenser	-0.098053		0.808618	-0.941513	
Exp. valve	-0.054215			0.054215	
Evaporator	-0.064947	-0.026689		0.831937	
Total	0.0	-0.026689	0.808618	0.0	-0.27205

Table 6. Exergy balance for the system components of the heat pump system for Case IX

Component	Refrigerant (kW)	Water stream (kW)	Heat flow (kW)	Lost work rate (kW)	Work input rate (kW)
Compressor	0.256756			-0.0634	-0.32136
Condenser	-0.135314		1.10313	1.09487	
Exp. valve	-0.065060			-0.06506	
Evaporator	-0.056382	-0.025099		-0.96641	
Total	0.0	-0.025099	1.10313	0.0	-0.32136

The negative sign for exergies given in Table 5 and 6 expresses the fuel concept of the component, while the positive sign expresses the product concept of the component. As can be seen from Table 5 and 6, the maximum exergy destruction is observed in condenser. The sign of lost work rate of condenser is positive, differently with other components. It is due to that condenser is boundary of the system, and negative sign is added for boundary terms in MOPSA method.

The capital investment costs for the system components obtained with economic analysis are presented in Table 7.

Table 7. The purchased equipment cost and capital investment cost values for the system components

Component	PEC ($)	$\dot{Z}$ ($/h)
Compressor	94.41	0.00847
Condenser	54.28	0.00487
Expansion valve	11.8	0.00105
Evaporator	102.55	0.00921
Total	263.04	0.02360

The purchased equipment cost of overall system is 263.04 $. The capital investment cost of overall system is calculated to be 0.02360 $/h.

The cost balance equations given by Equation 14 to Equation 17 should be solved with using the data presented in Table 2 and 5 and Table 3 and 6 for Case I and Case IX, respectively. By this way, the unit cost of unknown parameters can be found.

The results obtained for the unit cost of thermal exergy are presented in Table 8 for all cases given in Table 1.

Table 8. The change of unit cost of thermal exergy C_T ($/kWh) with temperature and volumetric flow rate of water supplied to the condenser

	$\dot{\forall}$ = 50 lt/h	$\dot{\forall}$ = 100 lt/h	$\dot{\forall}$ = 150 lt/h
T=14 °C	0.1413	0.1416	0.1424
T=17 °C	0.1374	0.1372	0.1379
T=19 °C	0.1347	0.1349	0.1362

As can be seen from Table 8, the unit cost of thermal exergy decreases with increase in inlet temperature of the water supplied to the condenser and evaporator. In addition, it is found that increase in mass flow rate of water supplied to the condenser causes to an increase in the unit cost of thermal exergy.

Table 9 shows the obtained results for the unit cost of mechanical exergy for different volumetric flow rates of water supplied to the condenser and for different inlet temperatures of water supplied to the condenser and evaporator.

Table 9. The change of unit cost of mechanical exergy C_P (\$/kWh) with temperature and volumetric flow rate of water supplied to the condenser

	$\dot{\forall}$ = 50 lt/h	$\dot{\forall}$ = 100 lt/h	$\dot{\forall}$ = 150 lt/h
T=14 °C	0.0734	0.0732	0.0733
T=17 °C	-0.0276	0.0627	0.0576
T=19 °C	0.0518	0.0498	0.0509

As can be seen from Table 9, it may be said that the unit cost of mechanical exergy has a decreasing trend with increasing inlet temperature of water supplied to the condenser and evaporator. Moreover, it may be said that the obtained values for unit cost of mechanical exergy for $\dot{\forall}$ = 100 lt/h are lower compared to that of other volumetric flow rates.

The variation of unit cost of negentropy with the inlet temperature of water supplied to the condenser and evaporator and the volumetric flow rate of water supplied to the condenser is given in Table 10.

Table 10. The change of unit cost of negentropy C_S (\$/kWh) with temperature and volumetric flow rate of water supplied to the condenser

	$\dot{\forall}$ = 50 lt/h	$\dot{\forall}$ = 100 lt/h	$\dot{\forall}$ = 150 lt/h
T=14 °C	0.0221	0.0207	0.0195
T=17 °C	0.0211	0.0196	0.0178
T=19 °C	0.0203	0.0186	0.0175

It is found that the unit cost of negentropy decreases with increase in the inlet temperature of water supplied to the system and with increase in the volumetric flow rate of water supplied to the condenser. The minimum unit cost of negentropy is obtained to be 0.0175 \$/kWh for $\dot{\forall}$ = 150 lt/h and T= 19 °C.
The results obtained for the unit cost of heat delivered are presented in Table 11 for all cases given in Table 1.

Table 11. The change of unit cost of heat delivered C_H (\$/kWh) with temperature and volumetric flow rate of water supplied to the condenser

	$\dot{\forall}$ = 50 lt/h	$\dot{\forall}$ = 100 lt/h	$\dot{\forall}$ = 150 lt/h
T=14 °C	0.0489	0.0460	0.0433
T=17 °C	0.0459	0.0429	0.0401
T=19 °C	0.0434	0.0406	0.0385

It is found that the unit cost of heat delivered decreases with increase in the inlet temperature of water supplied to the condenser and evaporator and with increase in the volumetric flow rate of water supplied to the condenser. The lowest unit cost of heat delivered is obtained to be 0.0385 \$/kWh for Case IX, while the highest one is obtained to be 0.0489 \$/kWh for Case I.

The cost flow rates of the system components for Case I and Case IX are presented in Table 12 and 13, respectively.

For Case I, the cost flow rate of heat delivered for the condenser is found to be -0.039195 \$/kWh, while the cost flow rate of electricity supplied to the system is 0.015581 \$/kWh. Whereas, for Case IX, the cost flow rate of heat delivered for the condenser is found to be -0.42471 \$/kWh, while the cost flow rate of electricity supplied to the system is 0.018831 \$/kWh. The highest values of cost flow rate for negentropy are obtained for the condenser.

Table 12. The cost flow rates of the system components for Case I

Component	C_T **(\$/h)**	C_P **(\$/h)**	C_H **(\$/h)**	C_S **(\$/h)**	C_W **(\$/h)**	$\dot{Z}$ **(\$/h)**
Compressor	-0.015240	-0.007690	0.000000	-0.001112	0.015581	0.008470
Condenser	0.013955	0.000000	-0.039195	0.020326	0.000000	0.004870
Expansion valve	-0.007689	0.007690	0.000000	-0.001051	0.000000	0.001050
Evaporator	0.008975	0.000000	0.000000	-0.018163	0.000000	0.009210
TOTAL	0.000000	0.000000	-0.039195	0.000000	0.015581	0.023600

Table 13. The cost flow rates of the system components for Case IX

Component	C_T **(\$/h)**	C_P **(\$/h)**	C_H **(\$/h)**	C_S **(\$/h)**	C_W **(\$/h)**	$\dot{Z}$ **(\$/h)**
Compressor	-0.020953	-0.005238	0.000000	-0.001110	0.018831	0.008470
Condenser	0.018429	0.000000	-0.042471	0.019160	0.000000	0.004870
Expansion valve	-0.005155	0.005238	0.000000	-0.001139	0.000000	0.001050
Evaporator	0.007679	0.000000	0.000000	-0.016912	0.000000	0.009210
TOTAL	0.000000	0.000000	-0.042471	0.000000	0.018831	0.023600

CONCLUDING REMARKS

Thermoeconomic analysis of a water-to-water heat pump is realized for different evaporator and condenser conditions. Different inlet temperature of $T_{in} = 14.4\,°C$, $17\,°C$ and $19\,°C$ are used for water supplied to the condenser and evaporator and different volumetric flow rates of $\dot{\forall} = 50$ lt/h, 100 lt/h and 150 lt/h are used for water supplied to the condenser. Constant volumetric flow rate of $\dot{\forall} = 100$ lt/h for water supplied to the evaporator is used. Modified Productive Structure Analysis (MOPSA) method is used for thermoeconomic analysis. Results showed that unit costs of heat delivered and of negentropy increases with increase in the volumetric flow rate of water supplied to the condenser and with increase in the inlet temperature of water supplied to the condenser. By this way, the unit costs of heat delivered and of negentropy for Case IX are lower than that of Case I. However, the result obtained for cost flow rates of heat delivered for Case IX is higher than that of Case I. It is due to that the electricity consumption increases with increasing volumetric flow rate of water supplied to the condenser and the amount of heat delivered. Moreover, it is found that the unit cost of negentropy decreases with increase in the inlet temperature of water supplied to the condenser and with increase in the volumetric flow rate of water supplied to the condenser and evaporator.

AKNOWLEGMENT

This work was supported by Scientific Research Projects Coordination Unit of Karabuk University. Project Number: KBU-BAP-14/1-DR-001

NOMENCLATURE

$A\dot{C}$	annual capital cost [\$/year]
C	unit cost [\$/kWh]
ex	specific exergy [kJ/kg]
$\dot{E}_x$	exergy rate [kW]
h	specific enthalpy [kJ/kg]
i	interest rate [%]
$\dot{m}$	mass flow rate [kg/s]
n	life time of system [year]
P	pressure [bar]

s	specific entropy [kJ/kgK]
$\dot{S}$	entropy rate [kW/K]
T	temperature [$°C$]
$\dot{Z}$	capital investment cost rate [$/h]
μ	salvage value [%]
τ	total annual number of operating hour of system [h]
ϕ	maintenance factor
cnd	condenser
com	compressor
exp	expansion valve
eva	evaporator
H	heat
in	inlet
P	mechanical
S	entropy
T	thermal
W	work
0	dead state
CHE	chemical
P	mechanical
T	thermal
W	work
CRF	capital recovery factor
MOPSA	modified productive structure analysis
PEC	purchased equipment cost
PW	present worth
PWF	present worth factor
SV	salvage value
TCI	total capital investment

REFERENCES

[1] Waheed, M. A., Oni, A. O., Adejuyigbe, S. B., Adewumi, B. A., Fadare, D. A. (2014). Performance enhancement of vapor recompression heat pump. Applied energy, 114, 69-79.

[2] Qureshi, B. A., Zubair, S. M. (2013). Cost optimization of heat exchanger inventory for mechanical subcooling refrigeration cycles. International Journal of Refrigeration, 36(4), 1243-1253.

[3] Kodal, A., Sahin, B., Ekmekci, I., Yilmaz, T. (2003). Thermoeconomic optimization for irreversible absorption refrigerators and heat pumps. Energy Conversion and Management, 44(1), 109-123.

[4] Kodal, A., Sahin, B., Yilmaz, T. (2000). Effects of internal irreversibility and heat leakage on the finite time thermoeconomic performance of refrigerators and heat pumps. Energy Conversion and Management, 41(6), 607-619.

[5] Teyber, R., Trevizoli, P. V., Christiaanse, T. V., Govindappa, P., Niknia, I., Rowe, A. (2017). Permanent magnet design for magnetic heat pumps using total cost minimization. Journal of Magnetism and Magnetic Materials, 442, 87-96.

[6] Esfahani, I. J., Yoo, C. (2014). A highly efficient combined multi-effect evaporation-absorption heat pump and vapor-compression refrigeration part 2: Thermoeconomic and flexibility analysis. Energy, 75, 327-337.

[7] Verda, V., Caccin, M., Kona, A. (2016). Thermoeconomic cost assessment in future district heating networks. Energy, 117, 485-491.

[8] Arat, H., Arslan, O. (2017). Exergoeconomic analysis of district heating system boosted by the geothermal heat pump. Energy, 119, 1159-1170.

[9] Erbay, Z., Hepbasli, A. (2017). Assessment of cost sources and improvement potentials of a ground-source heat pump food drying system through advanced exergoeconomic analysis method. Energy, 127, 502-515.

[10] Sayyadi, H., & Nejatolahi, M. (2011). Thermodynamic and thermoeconomic optimization of a cooling tower-assisted ground source heat pump. Geothermics, 40(3), 221-232.

[11] Erbay, Z., Hepbasli, A. (2017). Exergoeconomic evaluation of a ground-source heat pump food dryer at varying dead state temperatures. Journal of cleaner production, 142, 1425-1435.

[12] Mastrullo, R., Renno, C. (2010). A thermoeconomic model of a photovoltaic heat pump. Applied Thermal Engineering, 30(14-15), 1959-1966.
[13] Akbulut, U., Utlu, Z., Kincay, O. (2016). Exergoenvironmental and exergoeconomic analyses of a vertical type ground source heat pump integrated wall cooling system. Applied Thermal Engineering, 102, 904-921.
[14] Qin N., Hao P. Z. (2017). The operation characteristics of sewage source heat pump system and the analysis of its thermal economic benefits. Applied Thermal Engineering. 124, 1083-1089
[15] Erbay Z., Hepbaşlı A. (2017). Advanced exergoeconomic evaluation of a heat pump food dryer. Biosystems Engineering. 124, 29-39
[16] Kwak, H. Y., You, Y., Oh, S. D., Jang, H. N. (2014). Thermoeconomic analysis of ground- source heat pump systems. International Journal of Energy Research, 38(2), 259-269.
[17] von Spakovsky M. R., Evans R. B. (1993). Engineering functional analysis-part I. ASME J Energy Resour Technol. 155, 86-92.
[18] Rosen, M. A., Dincer, I. (2003). Exergy–cost–energy–mass analysis of thermal systems and processes. Energy Conversion and Management, 44(10), 1633-1651.
[19] Tsatsaronis G., Lin L., Pisa J. (1993). Exergy costing in exergoeconomics. ASME J Energy Resour Technol. 155, 9-16.
[20] Tsatsaronis, G., Moran, M. J. (1997). Exergy-aided cost minimization. Energy Conversion and Management, 38(15-17), 1535-1542.
[21] Tsatsaronis, G., Park, M. H. (2002). On avoidable and unavoidable exergy destructions and investment costs in thermal systems. Energy Conversion and Management, 43(9-12), 1259-1270.
[22] Tsatsaronis G., Lin L. (1990). On exergy costing in exergoeconomics. In: Tsatsaronis G, Bajura RA, Kenney WF, Reistad GM, editors. Computer-aided energy systems analysis. New York: ASME, 1-11.
[23] Kim, S. M., Oh, S. D., Kwon, Y. H., Kwak, H. Y. (1998). Exergoeconomic analysis of thermal systems. Energy, 23(5), 393-406.
[24] Lazzaretto A., Tsatsaronis G. (1997). On the quest for objective equations in exergy costing. In: Ramalingam ML, Lage JG, Mei VC, Chapman JN, editors. Proceedings of the ASME advanced energy systems division. New York: ASME, 413-428.
[25] Lazzaretto A., Tsatsaronis G. (1999). On the calculation of efficiencies and costs in thermal systems. In: Aceves SM, Garimella S, Peterson R, editors. Proceedings of the ASME advanced energy systems division. New York: ASME, 421-430.
[26] Erlach, B., Serra, L., Valero, A. (1999). Structural theory as standard for thermoeconomics. Energy Conversion and Management, 40(15-16), 1627-1649.
[27] Lozano, M. A., Valero, A. (1993). Theory of the exergetic cost. Energy, 18(9), 939-960.
[28] Frangopoulos, C. A. (1987). Thermo-economic functional analysis and optimization. Energy, 12(7), 563-571.
[29] Uysal, C., Kurt, H., Kwak, H. Y. (2017). Exergetic and thermoeconomic analyses of a coal-fired power plant. International Journal of Thermal Sciences, 117, 106-120.
[30] Moran, M. J. (1982). Availability analysis: a guide to efficient energy use. Englewood Cliffs NJ, Prentice Hall,
[31] Enerjisa Başkent Electricity Retail Sales Co.

8

THERMO-ENVIRONMENTAL ANALYSIS AND MULTI-OBJECTIVE OPTIMIZATION OF PERFORMANCE OF ERICSSON ENGINE IMPLEMENTING AN EVOLUTIONARY ALGORITHM

Mohammad H. Ahmadi[1,*], Fathollah Pourfayaz[2], Mohammad Hossein Jahangir[2]

ABSTRACT

This paper makes attempt to optimize a high-temperature differential Ericsson engine with several conditions. A mathematical approach based on the finite-time thermodynamic was proposed with the purpose of gaining thermal efficiency, the output power and the entropy generation rate throughout the Ericsson system with regenerative heat loss, finite rate of heat transfer, finite regeneration process time and conductive thermal bridging loss. In this study, an irreversible Ericsson engine is analyzed thermodynamically in order to optimize its performance. In addition, three Scenarios in multi-objective optimization are presented and the results of them are assessed individually. The first strategy is proposed to maximize the Ecological function, the thermal efficiency and the Exergetic performance criteria. Furthermore, the second strategy is suggested to maximize the Ecological function, the thermal efficiency and Ecological coefficient of performance. The third strategy is proposed to maximize the Ecological function and the thermal efficiency and Dimensionless ecological based thermo-environmental function. Multi-objective evolutionary algorithms based on NSGA-II algorithm was applied to the aforementioned system for calculating the optimum values of decision variables. Decision variables considered in this paper including the regenerator's effectiveness, the high-temperature heat exchanger's effectiveness, the low-temperature heat exchanger's effectiveness, the working fluid temperature in the low-temperature isothermal process and the working fluid temperature in the high-temperature isothermal process. Moreover, Pareto optimal frontier was achieved and an ultimate optimum answer was chosen via three competent decision makers comprising LINMAP, fuzzy Bellman-Zadeh, and TOPSIS approaches. The results from scenarios shown that third scenario is the best scenario.

Keywords: *Evolutionary Algorithms, Decision-Making, Thermodynamic Analysis, Multi-Objective Optimization, Entropy Generation, Ericsson Engine*

INTRODUCTION

One of the simplest types of external-combustion engines is the Ericsson engine which employs a compressible fluid as a working fluid. At Carnot efficiency, the Stirling and Ericsson engines can supposedly be an effective engine to convert heat into mechanical work. Little research has been done on the Ericsson engine and more research is related to the Stirling engine.

The material employed for Stirling engine and Ericsson engine construction effects on the thermal boundary for the operation of this engine. In most cases, the engines work with a cooler and heater temperature of 338 and 923 K, correspondingly [1]. The range of efficiency in Stirling engines vary from 30 to 40% which yielded by normal operating speed varies from 2000 to 4000 rpm, and a usual temperature changes from 923 to 1073 K [2].

Several scholars propose isothermal models like Schmidt's original work (for instance, Urieli and Berchowitz [3], Reader [4] and Hargreaves [5]). Carlson and colleagues [6] improve an ideal approach with non-isothermal heat exchange. These type of approaches propose a development on the isothermal approach as they exclude unfeasibly slow engine speed accompanying with isothermal working spaces and the requirement for infinite heat transfer. Urieli and Kushnir [7] depicted that this analysis can be employed for the purpose of examining the different practical impacts of heat exchangers, non-ideal regenerators, comprising pressure losses and heat transfer. Martaj and colleagues [8] worked on the steady-state operation and illustrated a thermodynamic analysis of a low-temperature Stirling engine, and entropy, energy and exergy balances were reported at each principal component of the engine. The main aims of the Stirling engine inventors can be

This paper was recommended for publication in revised form by Regional Editor Alibakhsh Kasaeian
[1]Faculty of Mechanical Engineering, Shahrood University of Technology, Shahrood, Iran
[2]Department of Renewable Energies and Environmental, Faculty of New Sciences and Tech., University of Tehran, Tehran, Iran
**E-mail address: mohammadhosein.ahmadi@gmail.com, mhosein.ahmadi@shahroodut.ac.ir*

categorized into three main groups: maximum power; maximum efficiency; minimum expenses. Markman and colleagues [9] investigated the mechanical-power losses and thermal-flux of a 200W beta-configuration of the Stirling engine to improve and intensify the efficiency of engine.

A small Stirling engine with 4 watt output power in 900 rpm rotation speed and 0.1 MPa pressure was investigated and constructed by Kagawa and colleagues [10]. Brandhorst and Chapman [11] established a 5 kW engine for usage as a power generator in space usages. Ataer [12] used the Lagrangian approach to analyze the regenerators of Stirling cycle engines. Nakajima and colleagues [13] established a 10 g micro Stirling machine with an approximately 0.05 cm^3 piston swept volume. At 10 Hz, the output power of the engine was 10 mW.

Aramtummaphon [14] evaluated an open cycle Stirling engines by employing steam heated from producer gas. The first engine at a maximum speed of 950 rpm produced a specified power of about 1.36 kW, whereas the second engine, amended from the first one, at a maximum speed of 2200 rpm generated a specified power of about 2.92 kW. Takashi Fukui and colleagues [15] invented and built a micro-engine, and its experimental evaluation was implemented; though, the performance of the micro-engine cannot be scaled to the real one. Iwamoto and colleagues [16] compared the efficiency of high temperature and low-temperature Stirling engines with other types of Stirling engines. They investigations depicted that the LTD Stirling efficiency is about 50% of Carnot efficiency with a similar condition. Wu and colleagues [17] depicted the impacts of regeneration time, heat transfer, and inadequate regeneration on the efficiency of the irreversible Stirling engine cycle. Erbay and Yavuz [18] studied the practical Stirling heat machine for maximum power output circumstances by employing polytropic progressions. They also specified the compression ratio and efficiency at maximum power density and determined the thermal design constraints. Ahmadi and colleagues [19] studied the effects of the Solar Collector Design variables on the Efficiency of Solar Stirling Engine. Ahmadi and colleagues [20-23] proposed a grey-box method to predict the power of Stirling heat engine via machine learning methods.

Investigation of thermodynamic irreversibility in systems acquired significance after the oil crisis in the 1970s to achieve higher efficiencies. The novel approach was entitled as Finite- Time-Thermodynamics (FTT). Primary studies in this field focused on endoreversible power cycle. This engine, named Curzon-Ahlborn-Novikov (CAN), is reversible internally and irreversible externally [24,25]. Compared with Carnot cycle, which works completely irreversible, Can engine yield more realistic results. Moreover, several studies were conducted on the maximum extractable work from irreversible systems [26–28]. For instance, Angulo-Brown established a standard known as ecological function (ECF) [29]. Yan suggested using ambient temperature (T_0) instead of heat sink temperature in Carnot efficiency [30]. Various studies are represented in the literature which has worked on ecological optimization [31–62]. ECOP is another thermo-ecological criterion which is defined and utilized in different thermodynamic cycles [63–72]. A performance coefficient, called exergetic performance criteria (EPC), is another criterion established to find out the relationship between exergy and exergy destruction of a [73–77]. Some studies [78–89] work on obtaining an approach for exergy application in finite time thermodynamic (FTT).

Several studies presented mathematical methods to calculate the overall thermal efficiency of solar powered high-temperature differential dish Stirling engine with regenerator irreversibility and finite heat transfer. Afterwards, the thermal efficiency and absorber working temperature were optimized. [90-92]. Tlili investigated the effects of regenerator effectiveness and internal irreversibility on the thermal efficiency of an endoreversible Stirling heat engine at maximum power condition [93]. Kaushik et al [94-97] studied effects of regeneration and heat transfer of the heat sink and sources on exergy destruction of Stirling and Ericsson engines. Evolutionary algorithms (EA) were originally used throughout the mid-eighties in an effort to unravel the puzzle of this general category [98]. A practical answer to a multi-objective puzzle is to determine a group of answers, each of which fulfills the objectives at a satisfactory degree without being overshadowed by any other answer[99]. Multi-objective optimization issues generally serve a feasibly innumerable group of answers that is to say as Pareto frontier, where examined vectors denote the preeminent probable trade-offs in the objective function area. In this regard, multi-objective optimization of various energy cycles was investigated by numerous nowadays [100-140]. Ahmadi and colleagues [103,105] employed NSGAII to optimize the economic and thermodynamic of a solar-dish Stirling heat engine. They presented another model to evaluate the cooling load of Stirling cryogenic refrigerator cycle as well [113]. Sayyaadi et al. [117] used NSGAII in order to optimize the design elements of a Solar-Driven Heat Engine. Ahmadi and colleagues [121] used MOEA and thermodynamic analysis to optimize an irreversible three-heat-source absorption heat pump. Sadatsakkak et al. [124,125] used thermo-economic analysis and MOEA to optimize an irreversible regenerative closed Brayton cycle and an endoreversible Braysson cycle.

In this study, an irreversible Ericsson engine is thermodynamically investigated in order to optimize its performance. In addition, three scenarios are considered in optimization and obtained results are evaluated. The first strategy is proposed to maximize the Ecological function, the thermal efficiency and the Exergetic performance criteria. Furthermore, the second strategy is suggested to maximize the Ecological function, the thermal efficiency and ECOP. The third strategy is proposed to maximize the Ecological function and the thermal efficiency and Dimensionless ecological based thermo-environmental function. MOEAs jointed with NSGA-II approach was executed in this paper. Decision parameters involved in this paper including the regenerator's effectiveness, the high-temperature heat exchanger's effectiveness, the low-temperature heat exchanger's effectiveness, the working fluid temperature in the low-temperature isothermal process and the working fluid temperature in the high-temperature isothermal process. Moreover, Pareto optimal frontier was achieved and an ultimate optimum answer was chosen via three competent decision makers comprising the LINMAP, fuzzy Bellman-Zadeh, and TOPSIS approaches.

SYSTEM DESCRIPTION

Figure 1 depicts a graphical illustration of an Ericsson heat engine cycle with regenerative heat losses and finite-time heat transfer. As illustrated in Figure 2, ideal Ericsson cycle comprises of 4 progressions containing two isobaric progressions (2–3 and 4–1) in the regenerator and two isothermal (1–2 and 3–4). In a real cycle, it is unfeasible to have an ideal heat transfer in the regenerator, in which the complete amount of absorbed heat (in the process 4–1) is transmitted to the working fluid in the isobaric heating progression (process 2–3). Consequently, a heat transfer loss happens in the regenerator.

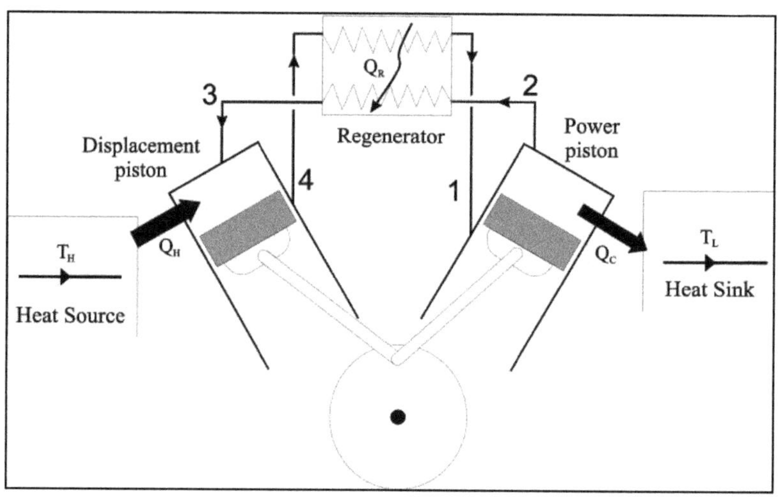

Figure 1. Schematic of proposed Ericsson heat engine

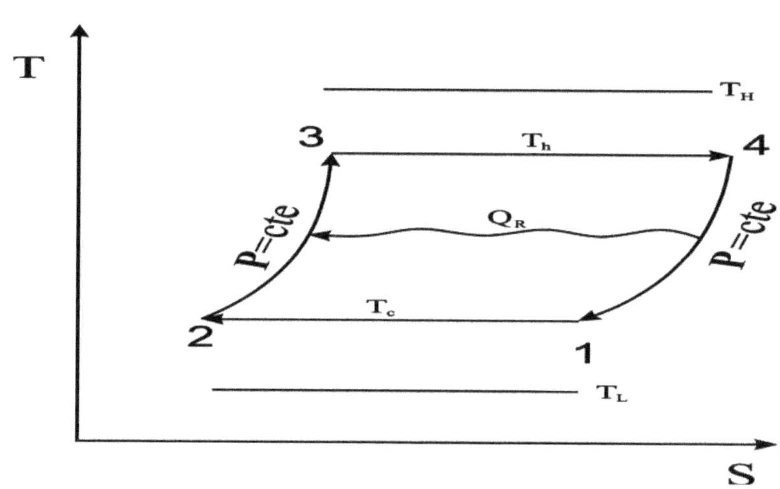

Figure 2. Ericsson engine T-S diagram

THERMODYNAMIC ANALYSIS OF THE SYSTEM

It is worth to stress that the finite heat transfer in the regenerative heat transfer (Q_r) can be calculated by following expression [20,91,92]

$$Q_r = nC_p \varepsilon_R (T_h - T_c) \quad (1)$$

ΔQ_r stands for the heat loss throughout the two regenerative progressions in the cycle and can be determined via the below equation [20, 91,92]:

$$\Delta Q_r = nC_p (1 - \varepsilon_R)(T_h - T_c) \quad (2)$$

C_p represents the working fluid's specific heat capacity in the regenerative progressions (mole), n stands for the mass of the working fluid in terms of mole, ε_r denotes the regenerator's effectiveness, T_h and T_c stand for the working fluid temperatures in the cold space and hot space, correspondingly.

It is not reasonable to pay no attention to the time of two regeneration progressions when compared with two constant temperature progressions included in the suggested approach. So, via the below equation the regeneration time calculated [94-97]:

$$t_R = 2\alpha (T_h - T_c) \quad (3)$$

The heat absorbed between the heat sink and working fluid (Q_L) and the heat released between working fluid and heat source (Q_H), are calculated via the below equations

$$Q_H = nRT_h Ln\lambda + nC_p (1 - \varepsilon_R)(T_h - T_c) \quad (4)$$

$$Q_L = nRT_c Ln\lambda + nC_p (1 - \varepsilon_R)(T_h - T_c) \quad (5)$$

where $\lambda = \frac{p_1}{p_2} = \frac{p_4}{p_3}$ is Ratio of pressure throughout the regenerative progressions. Consequently, we have:

$$Q_H = \left[C_H \varepsilon_H (T_H - T_h) + \xi C_H \varepsilon_H (T_H^4 - T_h^4) \right] t_h \quad (6)$$

$$Q_L = C_L \varepsilon_L (T_c - T_L) t_l \quad (7)$$

in which C_L and C_H denote the external fluids heat capacitance rate in the heat sink and heat source, correspondingly.

$$\varepsilon_H = 1 - e^{-N_H} \quad (8)$$

$$\varepsilon_L = 1 - e^{-N_L} \quad (9)$$

in which ε_H and ε_L stand for the high and low temperature heat exchangers effectiveness, correspondingly and $N_L = U_L A_L / C_L$, $N_H = U_H A_H / C_H$ represent the cyclic period. Using Eqs. (3)- (9), we get that the cyclic period t is:

$$t = \frac{nRT_h Ln\lambda + nC_p(1-\varepsilon_R)(T_h-T_c)}{C_H\varepsilon_H(T_H-T_h)+\xi C_H\varepsilon_H(T_H^4-T_h^4)} + \frac{nRT_c Ln\lambda + nC_p(1-\varepsilon_R)(T_h-T_c)}{C_L\varepsilon_L(T_c-T_L)} + 2\alpha(T_h-T_c) \quad (10)$$

Take into account the cyclic period of the Ericsson engine, the thermal efficiency, the output power, and entropy production of the engine can be determined as following as:

$$P = \frac{W}{t} = \frac{Q_H - Q_L}{t} \quad (11)$$

$$\eta_t = \frac{Q_H - Q_L}{Q_H} \quad (12)$$

$$\sigma = \frac{1}{t}\left(\frac{Q_L}{T_L} - \frac{Q_H}{T_H}\right) \quad (13)$$

Exergy destruction is the measurement of the irreversibilities or lost work in the system and it is equal to environment temperature (T_0, K) multiply entropy generation rate (kW/K). The rate of Exergy destruction (kW) is written as following:

$$\dot{E}_{xD} = T_0\sigma \quad (14)$$

Exergy efficiency:

$$\varphi = \frac{P}{P_R} \quad (15)$$

Ecological function (kW) can be defined as:

$$\dot{E} = P - \dot{E}_{xD} \quad (16)$$

The ECOP and exergetic performance criteria are calculated as following as:

$$ECOP = \frac{P}{T_0\sigma} \quad (17)$$

$$EPC = \frac{Ex}{T_0\sigma} \quad (18)$$

Reversible work per unit time of the system (kW) is the difference of exergy input of the system and exergy output from the system and it is described as following:

$$P_R = \frac{1}{t_{cycl}}\left(Q_H\left(1-\frac{T_0}{T_H}\right) - Q_L\left(1-\frac{T_0}{T_L}\right)\right) \quad (19)$$

where m is the total mass, b_1 is environmental impact of the components (mPts/kg), b_2 is the environmental impact of the fuel (mPts/ MJ), b_3 is the environmental impact of the lost work (mPts/MJ), and b_4 (mPts/MJ), is the environmental impact of the mechanical work. Similarly, the ecological based thermo-environmental function is defined as:

$$B_E = \frac{E}{b_1 m + b_2 Q_H + b_3 E_{xD} + b_4 W} \tag{20}$$

The dimensionless ecological based thermo-environmental function is defined as:

$$b_E = b_4 B_E \tag{21}$$

Substituting Eqs. (3)- (10) intoEqs. (11) and (12) we have,

$$P = \frac{nR(T_h - T_c)Ln\lambda}{\frac{nRT_h Ln\lambda + nC_p\left(1-\varepsilon_R\right)\left(T_h - T_c\right)}{C_H\varepsilon_H\left(T_H - T_h\right) + \xi C_H\varepsilon_H\left(T_H^4 - T_h^4\right)} + \frac{nRT_c Ln\lambda + nC_p\left(1-\varepsilon_R\right)\left(T_h - T_c\right)}{C_L\varepsilon_L\left(T_c - T_L\right)} + 2\alpha\left(T_h - T_c\right)} \tag{22}$$

$$\eta_t = \frac{nR(T_h - T_c)Ln\lambda}{nRT_h Ln\lambda + nC_p\left(1-\varepsilon_R\right)\left(T_h - T_c\right)} \tag{23}$$

MULTI-OBJECTIVE OPTIMIZATION WITH EVOLUTIONARY ALGORITHMS

OPTIMIZATION VIA EA

Genetic Algorithms were firstly proposed by Prof. Holland (1960) by inspiring the concept of natural evolution and Darwinian theorem for optimization purposes [101]. The evolution typically commences from a population of accidentally created individuals and takes place in creations. In each creation, the fitness value of each individual in the population is examined; multiple individuals are stochastically chosen from the present population, and improved to create a fresh population. The fresh population is then employed in the following iteration of the GA. Usually, the GA stops when either an acceptable fitness level was achieved for the population or a maximum number of generations were created. More details of GA can be found in previous works [99, 102].

Also, MOEAs were evolved throughout the past years by frequent examinations on multipart mathematical puzzles and on practical engineering issues and have depicted that they can exclude the complications of conventional approaches [99, 102]. The construction of the MOEA employed in this paper is depicted through Figure 3 [101]. It is worth to highlight that the real values of decision parameters were employed rather than their binary codes.

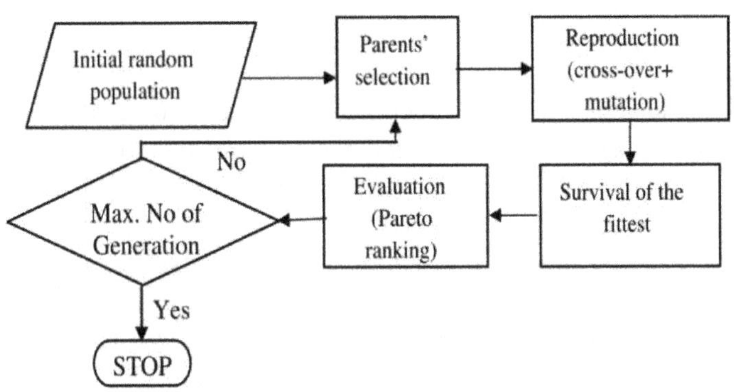

Figure 3. Algorithm steps applied in the study

NSGA-II APPROACH

NSGA-II approach was employed in this paper with the purpose of determining the Pareto frontier by running GA. In this regard, NSGA-II organized the answers based on the Pareto theory and arranging non-dominated answers into non-dominated layers as illustrated in Figure 4. Put it in another way, if N_p stands for the population number, it is classified into N_L layers in which juncture of each two random chosen layers is blank assortment and combination of all layers represents N_p assortment.

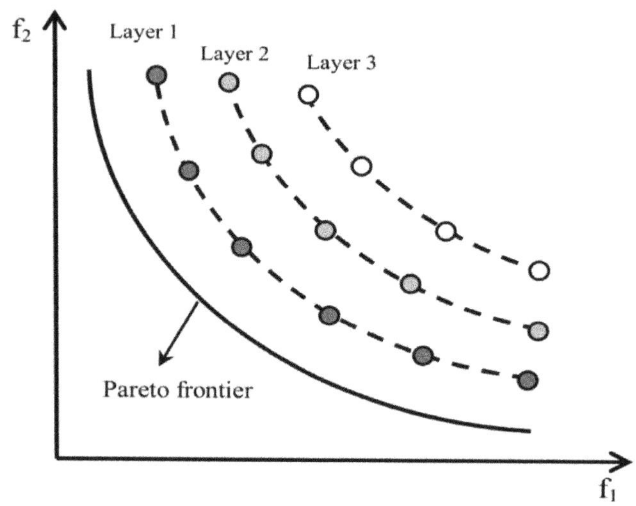

Figure 4. NSGA-II solution layering

The virtual fitness of each answer is equivalent to its layer. Tournament selection was employed for cross over operating in parent choosing between two random chosen layers. So, answer placed on the layer 1, have more opportunity to be chosen for the next creation. Uniform distribution of answers along layers is regulated via an index called "index of crowding distance" for each answer. This criterion is defined as a ratio of detraction of objective functions for two neighbor answers nearby the present answer to the detraction of the minimum and maximum values of that objective. Consequently, for k_{th} objective of j_{th} answer, following expression can be used.

$$i_{dis,j,k} = \frac{f_{k,j-1} - f_{k,j+1}}{f_{k,\max} - f_{k,\min}} \tag{24}$$

For margin answers are allocated an infinite distance index. The summation of individual distance values conforming to each objective stands for the overall crowding distance value as follows:

$$I_{dis,j} = \sum_{k=1}^{M} i_{dis,j,k} \tag{25}$$

in which j represents the individual index and M stands for the number of objectives. Figure 5 depicts a graphical illustration of examination of distance index. In this approach, two variables are determined for each answer:

1) Dominant (Layer) number, N_L, namely the number of answers which control the present answer. Description and definitions of domination were described well in Ref. [98, 100]. Dominant number, for non-dominated answers of the present population is equal to 0, consequently, these answers are located in layer 1. Non-dominated answers for an assortment of the answers not including the layer 1 members are located in layer 2. For M objectives issue with N populations, the number of assessments is equal to MN^2. This process persisted with the purpose of accommodating all answers in their suitable layers. Furthermore, i rank index for each answer is allocated as its layer number, N_L.

2) Crowded comparison operator, n<, defined as following as:

$$\begin{aligned} &\text{A p B if } (\text{rank}_A < \text{rank}_B) \\ &\text{Or:} \\ &\left((\text{rank}_A = \text{rank}_B) \text{ and } \text{I}_{dis,A} > \text{I}_{dis,B}\right) \end{aligned} \tag{26}$$

It reveals that for two answers with dissimilar layers, the answer with the lower layer is desired. Else, for two answers of the same layer, the answer located in the area with a lower concentration of answers is chosen.

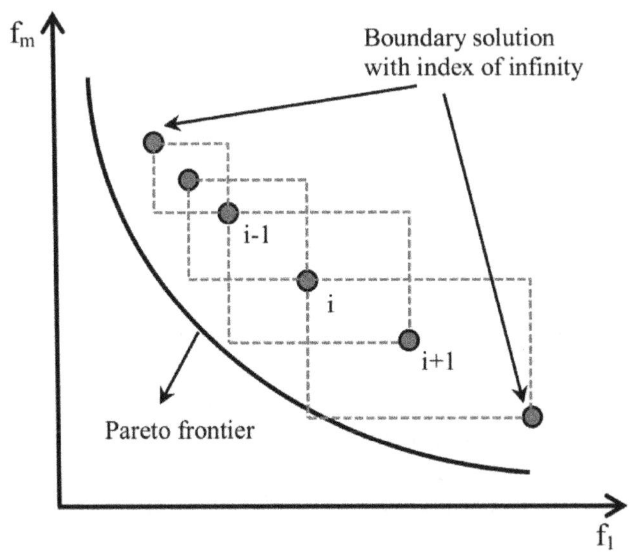

Figure 5. Schematic of distance indexing of individuals in NSGA-II algorithm.Distance indexing of components in NSGA-II algorithms

OBJECTIVE FUNCTION, RESTRAINTS AND DECISION PARAMETERS

The thermal efficiency, the Ecological function and the Exergetic performance criteria three objective functions for the first scenario, which are evaluated via Eqs. (12 and 16 and 18).

The Ecological function, the thermal efficiency, and ECOP three objective functions for the second scenario, which are evaluated via Eqs. (12 and 16 and 17).

The Ecological function, the thermal efficiency, and Dimensionless ecological based thermo-environmental function three objective functions for the third scenario, which are evaluated via Eqs. (12 and 16 and 21). Throughout all scenarios proposed in the present paper the below design parameters were employed:

ε_R : Regenerator's effectiveness

ε_L : The low-temperature heat exchanger effectiveness

ε_H : The high-temperature heat exchanger effectiveness

T_h : The working fluid temperature in the high-temperature isothermal process 3-4

T_c : The working fluid temperature in the low-temperature isothermal process 1-2

Following limitations were included in the optimization process:

$$0.7 \le \varepsilon_R \le 0.95 \tag{27}$$

$$0.7 \le \varepsilon_H \le 0.9 \quad (28)$$

$$0.7 \le \varepsilon_L \le 0.9 \quad (29)$$

$$800 \le T_h \le 1100K \quad (30)$$

$$350 \le T_c \le 500K \quad (31)$$

To determine the optimal design variables of the system, based on genetic algorithm approach a simulation program was coded through Matlab software. Specifications of GA for optimization puzzle are reported in Table 1.

Table 1: Specification of GA for optimization puzzle in this paper

GA Parameters	Value
Population size	400
Population type	Double vector
Tournament size	2
Selection process	Tournament
Maximum number of generations	1000
Mutation	Restriction dependent

Choosing a final optimum answer from Pareto optimal frontier in multi-objective optimization process plays a significant role. In this regard, we should employ decision makers to determine this. Consequently, in this paper three competent decision makers including TOPSIS, Fuzzy and LINMAP were employed as decision makers. Details of these decision makers can be found in previous literature especially references [141, 142].

RESULTS AND DISCUSSION

Results of First Scenario

Via running multi-objective optimization approach the thermal efficiency, the Ecological function and the Exergetic performance criteria are maximized concurrently. The objective functions in the applied optimization, and the restrictions that were employed, are formulated by Eqs. (12 and 16 and 18) and Eqs. (27-31), respectively.

Design variables in optimization process are the low-temperature heat exchanger's effectiveness, regenerator's effectiveness, the working fluid temperature in the high-temperature isothermal process 3-4, the high-temperature heat exchanger's effectiveness, the working fluid temperature in the low-temperature isothermal process 1-2. Following specifications have been considered for Ericson cycle [91,94]:

$$n = 1, C_V = 15J.mol^{-1}.K^{-1}, R = 4.3J.mol^{-1}.K^{-1}, T_H = 1300K, T_L = 300K, T_0 = 285K,$$

$$C_L = C_H = 1000 \ \ WK^{-1}, x = 2\times 10^{-10}, 1 = 2, a = 10^{-5} s/K.$$

Pareto optimal frontier for three objective functions, the objective function associated with the thermal efficiency, the Ecological function and the Exergetic performance criteria of the irreversible Ericsson engine are represented in Figure 6.

$$d_+ = \sqrt{(\eta_t - \eta_{t,n})^2 + (ECF - ECF_n)^2 + (EPC - EPC_n)^2} \quad (32)$$

$$d_- = \sqrt{(\eta_t - \eta_{t,n,non-ideal})^2 + (ECF - ECF_{n,non-ideal})^2 + (EPC - EPC_{n,non-ideal})^2} \quad (33)$$

$$d = \frac{d_+}{(d_+)+(d_-)} \tag{34}$$

$\eta_{t,n}$, ECF_n and EPC_n denote Euclidian the thermal efficiency, the Ecological function and the Exergetic performance criteria. Furthermore, Table 2 comprises the deviation index (*d*) for the outcome of each decision maker.

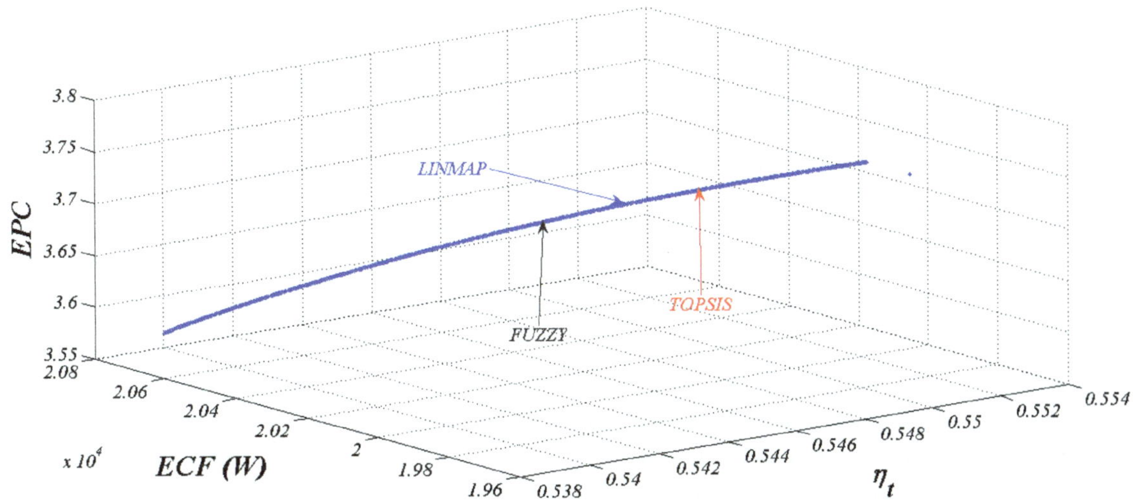

Figure 6. Pareto optimal frontier in the objectives' space of first scenario

Table 2 depicts the optimal outputs achieved for objective functions and decision parameters by executing LINMAP, Fuzzy and TOPSIS approaches for the first scenario. To determine deviations of the results from an ideal and non-ideal solution, following equations were employed.

As reported through Table 2, the deviation indexes for TOPSIS, LINMAP and Fuzzy are 0.036, 0.031 and 0.026, respectively. As clear be seen from this Table, it can conclude that the FUZZY decision-maker has a lower deviation index; consequently the answer which was chosen via the FUZZY decision-maker was selected as a final optimal answer of the multi-objective optimization for the irreversible Ericsson cycle.

Table 2. Outcomes of the decision makers for the first scenario

Decision Making Method	Decision variables					Objectives			Deviation index
	e_R	e_H	e_L	T_h	T_c	h_t	$ECF(W)$	EPC	
TOPSIS	0.950	0.900	0.900	1094.415	364.700	0.548	20104.500	3.721	0.036
LINMAP	0.950	0.900	0.900	1094.390	366.479	0.547	20213.269	3.703	0.031
Fuzzy	0.950	0.900	0.900	1094.386	368.346	0.546	20316.560	3.684	0.026
Ideal point	-	-	-	-	-	0.554	20849.81	3.822	0
Non-ideal point	-	-	-	-	-	0.535	19383.34	3.504	∞

To examine the accuracy of the decision maker's analysis of error was performed. Table 3 demonstrates MAPE (Mean Absolute Percentage Error) and MAAE (Maximum Absolute Percentage Error) for results achieved by the decision makers.

Table 3. Analysis of Error for the results of the first scenario

Decision Making Method	TOPSIS			LINMAP			Fuzzy		
Objectives	h_t	*ECF*	*EPC*	h_t	*ECF*	*EPC*	h_t	*ECF*	*EPC*
Max Error %	2.148	8.404	3.811	1.720	7.400	3.147	0.496	4.849	1.332
Average Error %	1.041	4.670	1.865	0.836	4.277	1.545	0.259	2.467	0.674

Results of Second Scenario

Three objective functions are considered for optimization which contain the thermal efficiency, the Ecological function and ECOP (should be maximized) which formulated via Eqs. (12 and 16 and 17), correspondingly.

Objective functions in this scenario are expressed by Eqs. (12 and 16 and 17) and design variables are formulated with Eqs. (27)-(31).

Design variables in optimization process are the same as the first scenario. Following specifications have been considered for Ericson cycle [91,94]:

$$n = 1, C_V = 15J.mol^{-1}.K^{-1}, R = 4.3J.mol^{-1}.K^{-1}, T_H = 1300K, T_L = 300K, T_0 = 285K,$$

$$C_L = C_H = 1000 \quad WK^{-1}, x = 2' \, 10^{-10}, 1 = 2, a = 10^{-5} s/K.$$

Figure 7 depicts the Pareto frontier in the suggested objectives' space achieved in the optimization scenario. Three ultimate answers were chosen by the LINMAP, Fuzzy Bellman-Zadeh, and TOPSIS decision makers which are highlighted in this figure. According to Figure 7, the obtained points by LINMAP and TOPSIS are approached towards each other. Also, it was shown that the optimal value of the thermal efficiency varied from 53.467% to 54.844% and the optimal value of the Ecological function was between 19.972 (kW) and 20.850 (kW) and the optimal value of the ECOP was between 2.400 and 2.615.

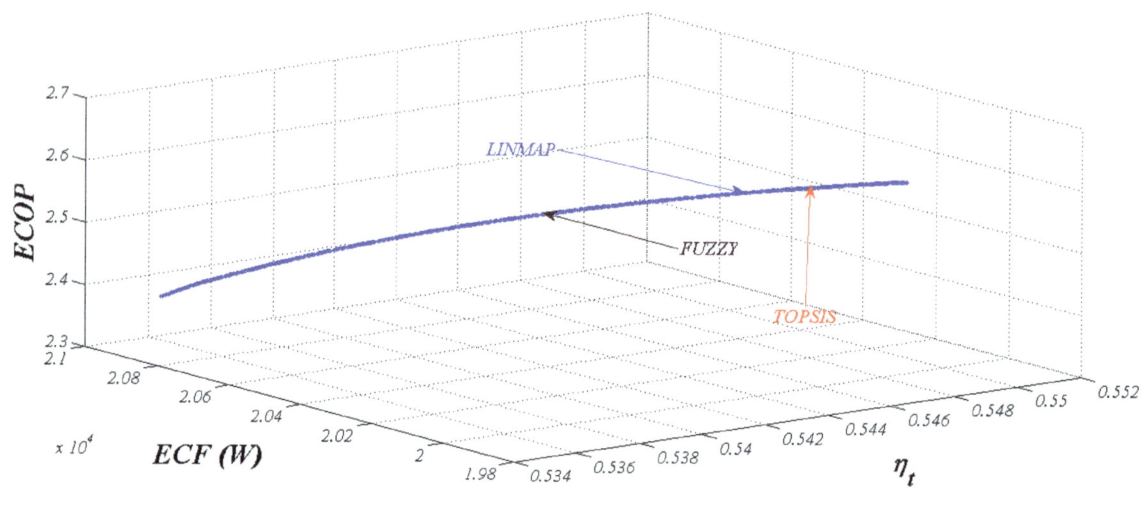

Figure 7. Pareto optimal frontier in the objectives' space of second scenario

Table 4 reports the optimal outputs achieved for objective functions and decision parameters via running TOPSIS, Fuzzy and LINMAP approaches for second scenario. To determine deviations of the results from an ideal and non-ideal solution, following equations were employed.

$$d_+ = \sqrt{(\eta_t - \eta_{t,n})^2 + (ECF - ECF_n)^2 + (ECOP - ECOP_n)^2} \qquad (35)$$

$$d_- = \sqrt{(\eta_t - \eta_{t,n,non-ideal})^2 + (ECF - ECF_{n,non-ideal})^2 + (ECOP - ECOP_{n,non-ideal})^2} \quad (36)$$

$$d = \frac{d_+}{(d_+)+(d_-)} \quad (37)$$

$\eta_{t,n}$, ECF_n and $ECOP_n$ denote Euclidian the thermal efficiency, the Ecological function and the ECOP. Furthermore, Table 4 comprises the deviation index (*d*) for the outcome of each decision maker.

As reported in Table 4, the deviation indexes for TOPSIS, LINMAP and Fuzzy are 0.035, 0.030 and 0.017, respectively. As clear be seen from this Table, it can conclude that the FUZZY decision-maker has a lower deviation index; consequently the answer which was chosen via the FUZZY decision-maker was selected as a final optimal answer of the multi-objective optimization for the irreversible Ericsson cycle.

Table 4. Outcomes of the decision makers for the second scenario

Decision Making Method	Decision variables					Objectives			Deviation index
	e_R	e_H	e_L	T_h	T_c	h_t	$ECF(W)$	$ECOP$	
TOPSIS	0.950	0.900	0.900	1088.063	364.556	0.547	20120.226	2.594	0.035
LINMAP	0.950	0.900	0.900	1088.062	366.269	0.546	20226.707	2.576	0.030
Fuzzy	0.950	0.900	0.900	1088.039	371.284	0.543	20487.154	2.526	0.017
Ideal point	-	-	-	-	-	0.554	20849.810	2.713	0
Non-ideal point	-	-	-	-	-	0.535	19383.340	2.399	∞

Finally, deviations of the final answers gained by each decision maker and ideal answer are assessed. Table 5 demonstrates MAPE and MAAE of results obtained via the aforesaid decision makers.

Table 5. Analysis of error for the results of the second scenario

Decision Making Method	TOPSIS			LINMAP			Fuzzy		
Objectives	h_t	ECF	$ECOP$	h_t	ECF	$ECOP$	h_t	ECF	$ECOP$
Max Error %	2.148	8.404	3.811	1.720	7.400	3.147	0.496	4.849	1.332
Average Error %	1.041	4.670	1.865	0.836	4.277	1.545	0.259	2.467	0.674

Results of Third Scenario

Throughout this scenario we attempted to maximize the thermal efficiency, the Ecological function and Dimensionless ecological based thermo-environmental function at the same time. The objective functions, and the limitations which were employed, are expressed by Eqs. (12 and 16 and 21) and Eqs. (27-31), respectively. Design variables in optimization process are the same as the first scenario. Following specifications have been considered for Ericson cycle [91, 94]:

$$n = 1, C_V = 15J.mol^{-1}.K^{-1}, R = 4.3J.mol^{-1}.K^{-1}, T_H = 1300K, T_L = 300K, T_0 = 285K,$$

$$C_L = C_H = 1000\ WK^{-1}, x = 2\times 10^{-10}, l = 2, a = 10^{-5} s/K.$$

Pareto optimal frontier for three objective functions, the objective function associated with the Ecological function, the thermal efficiency, and Dimensionless ecological based thermo-environmental function of the irreversible Ericsson engine are represented in Figure 8.

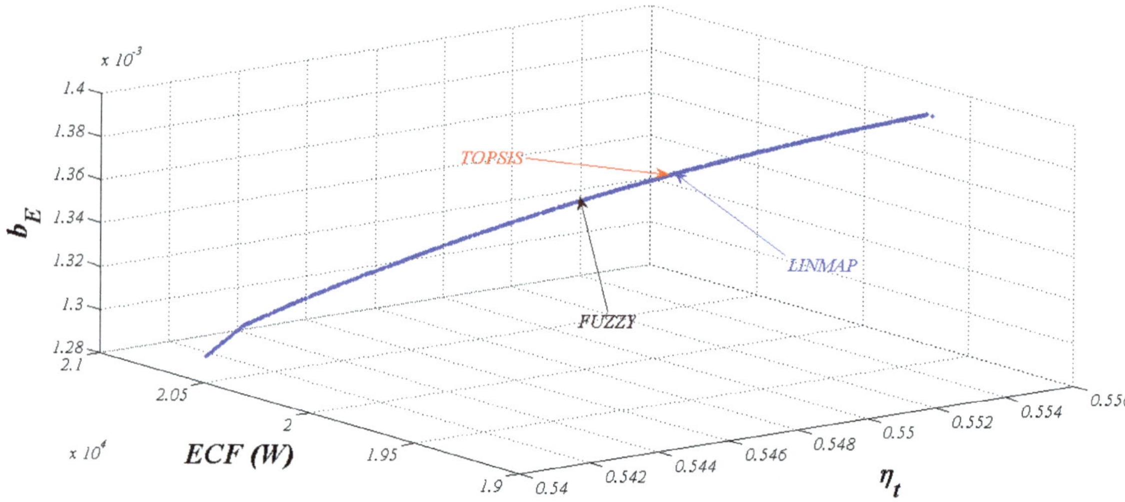

Figure 8. Pareto optimal frontier in the objectives' space of third scenario

Table 6 reports the optimal outputs achieved for objective functions and decision parameters via running TOPSIS, Fuzzy and LINMAP approaches for third scenario. To determine deviations of the results from an ideal and non-ideal solution, following equations were employed.

$$d_+ = \sqrt{(\eta_t - \eta_{t,n})^2 + (ECF - ECF_n)^2 + (b_E - b_{E,n})^2} \tag{38}$$

$$d_- = \sqrt{(\eta_t - \eta_{t,n,non-ideal})^2 + (ECF - ECF_{n,non-ideal})^2 + (b_E - b_{E,n,non-ideal})^2} \tag{39}$$

$$d = \frac{d_+}{(d_+)+(d_-)} \tag{40}$$

$\eta_{t,n}$, ECF_n and $b_{E,n}$ denote Euclidian the thermal efficiency, the Ecological function and the Dimensionless ecological based thermo-environmental function. Furthermore, Table 6 comprises the deviation index (*d*) for the outcome of each decision maker.

As reported through Table 6, the deviation indexes for TOPSIS, LINMAP and Fuzzy are 0.052, 0.043 and 0.035, respectively. As clear be seen from this Table, it can conclude that the FUZZY decision-maker has a lower deviation index; consequently the answer which was chosen via the FUZZY decision-maker was selected as a final optimal answer of the multi-objective optimization for the irreversible Ericsson cycle.

Table 6. Outcomes of the decision makers for the third scenario

Decision Making Method	Decision variables					Objectives			Deviation index
	e_R	e_H	e_L	T_h	T_c	h_t	$ECF(W)$	b_E ' 10^3	
TOPSIS	0.950	0.900	0.900	1096.679	360.197	0.552	19772.681	1.378	0.052
LINMAP	0.950	0.900	0.900	1096.679	362.556	0.550	19948.829	1.366	0.043
Fuzzy	0.950	0.900	0.900	1096.668	365.063	0.549	20115.565	1.353	0.035
Ideal point	-	-	-	-	-	0.554	20849.810	1.400	0
Non-ideal point	-	-	-	-	-	0.535	19383.340	1.200	∞

Finally, deviations of the final answers gained by each decision maker and ideal answer are assessed. Table 7 explicates MAAE and MAPE of results obtained via the aforesaid decision makers.

Table 7. Analysis of error analysis for the results of the third scenario

Decision Making Method	TOPSIS			LINMAP			Fuzzy		
Objectives	h_t	*ECF*	b_E	h_t	*ECF*	b_E	h_t	*ECF*	b_E
Max Error %	2.148	8.404	3.811	1.720	7.400	3.147	0.496	4.849	1.332
Average Error %	1.041	4.670	1.865	0.836	4.277	1.545	0.259	2.467	0.674

Figure 9 depicts the comparison between thermal efficiency gained from three scenarios. As clear be seen from Figure 9, the third scenario has the highest value of thermal efficiency compared to other scenarios. Moreover, the lowest value of thermal efficiency is for the second scenario. It should be noted that thermal efficiency was an objective function for all the scenarios.

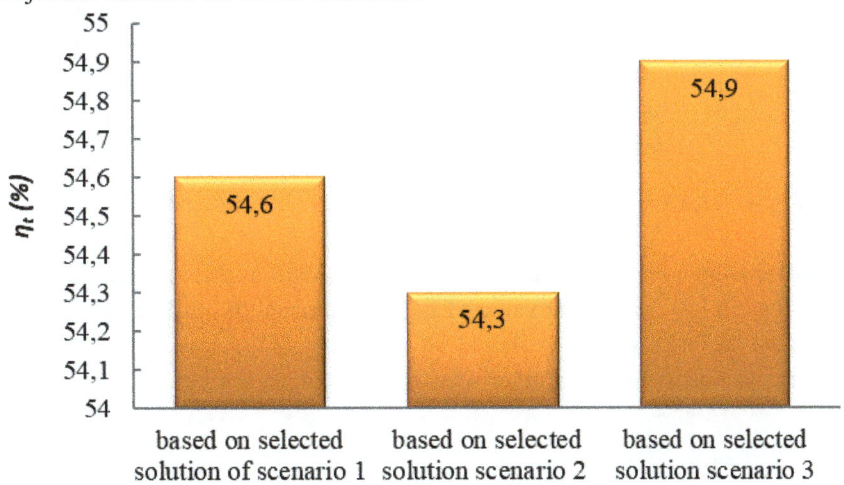

Figure 9. Comparison of thermal efficiency between different optimization scenarios

Figure 10 depicts the comparison between ECF gained from three scenarios. As clear be seen from Figure 10, the second scenario has the highest value of ECF compared to other scenarios. Moreover, the lowest value of ECF is for the third scenario. It should be noted that ECF was an objective function for all the scenarios.

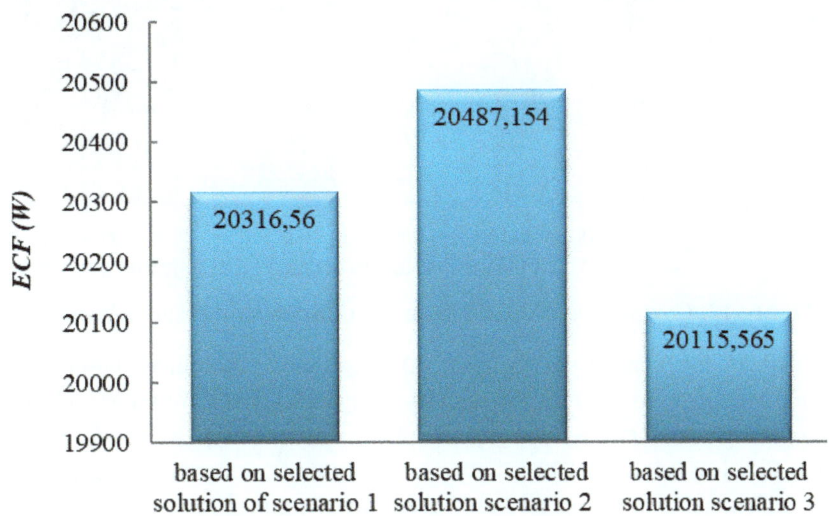

Figure 10. Comparison of ECF between different optimization scenarios

Figure 11 depicts the comparison between EPC gained from three scenarios. As clear be seen from Figure 11, the third scenario has the highest value of EPC compared to other scenarios. Moreover, the lowest value of EPC is for the second scenario. It should be noted that EPC was an objective function just for the first scenario and the values of EPC in other scenarios were calculated at optimum conditions gained from optimization process.

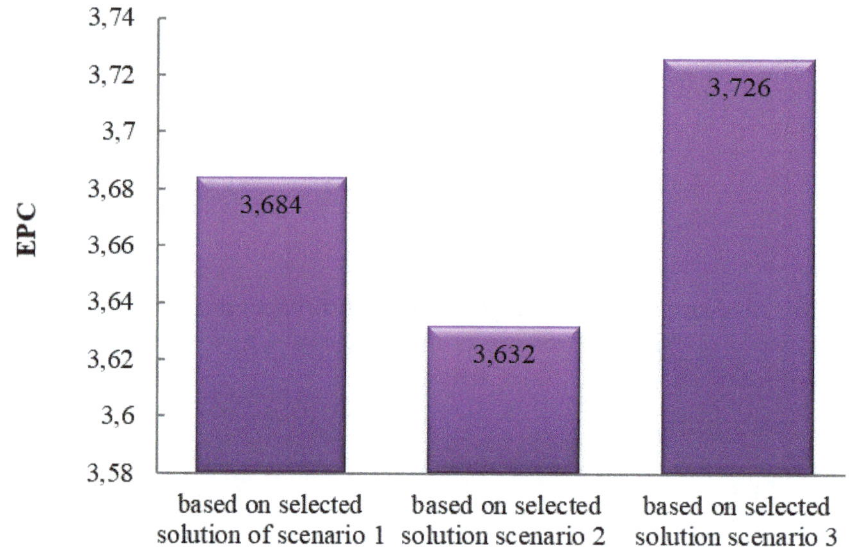

Figure 11. Comparison of EPC between different optimization scenarios

Figure 12 depicts the comparison between ECOP gained from three scenarios. As clear be seen from Figure 12, the third scenario has the highest value of ECOP compared to other scenarios. Moreover, the lowest value of ECOP is for the second scenario. It should be noted that ECOP was an objective function just for the second scenario and the values of ECOP in other scenarios were calculated at optimum conditions gained from optimization process.

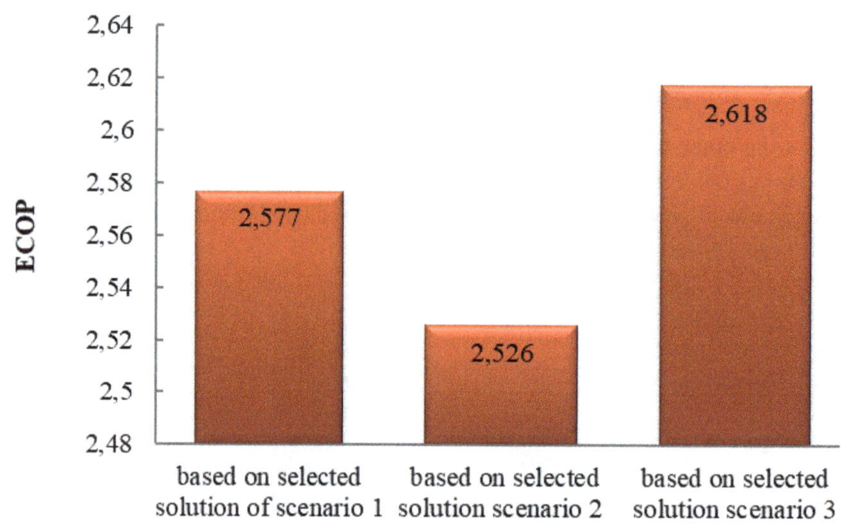

Figure 12. Comparison of ECOP between different optimization scenarios

Figure 13 depicts the comparison between Dimensionless ecological based thermo-environmental function gained from three scenarios. As clear be seen from Figure 13, the third scenario has the highest value of Dimensionless ecological based thermo-environmental function compared to other scenarios. Moreover, the lowest value of dimensionless ecological based thermo-environmental function is for the second scenario. It should be noted that dimensionless ecological based thermo-environmental function was an objective function just for the third scenario and the values of dimensionless ecological based thermo-environmental function in other scenarios were calculated at optimum conditions gained from optimization process.

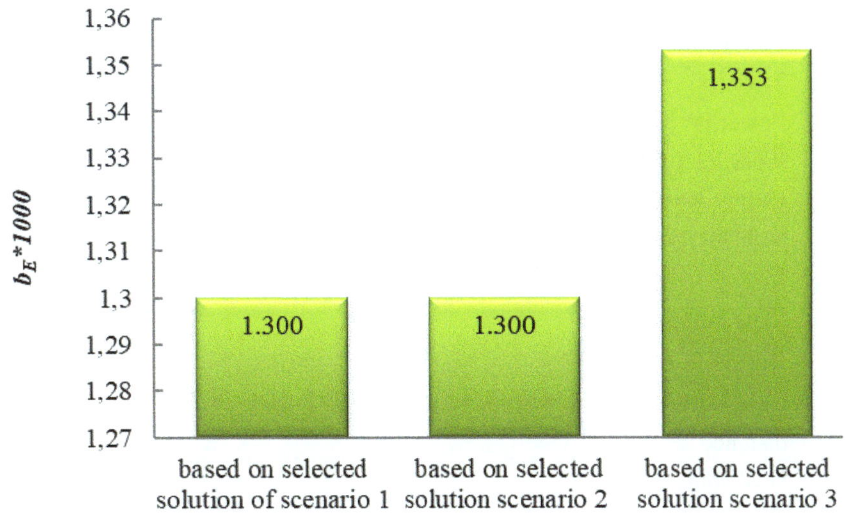

Figure13. Comparison of dimensionless ecological based thermo-environmental function between different optimization scenarios

Finally, as it is clearly seen from Figures 9 through 13, it can conclude that the third scenario was the best scenario in comparison other proposed scenarios. This is main due to the condition when b_E is maximum the values of ECOP, EPC and thermal efficiency are maximum. In other words, maximizing b_E results in gaining maximum values of ECOP, EPC and thermal efficiency.

CONCLUSIONS

This paper made attempt to illustrate multi-objective optimization of Ericson system based on finite-time thermodynamics analysis. In this regard, the optimum values of the Ecological function, the thermal efficiency, Exergetic performance criteria, the ECOP, and dimensionless ecological based thermo-environmental function of the Ericsson engine have been determined. The thermal efficiency, output power, and entropy generation rate throughout the engine have been chosen as parallel objective functions in the optimization process. Furthermore, the low temperature heat exchanger's effectiveness (ε_L),the regenerator's effectiveness (ε_R), the working fluid temperature in the high temperature isothermal process(T_h), the high temperature heat exchanger's effectiveness (ε_H), and working fluid temperature in the low temperature isothermal process(T_c) have been chosen as design variables with definite limitations in optimization process. MOEA based on NSGA-II approach was applied to the aforementioned system for calculating the optimum values of decision variables. Moreover, Pareto optimal frontier was achieved and an ultimate optimum answer was chosen via three competent decision makers comprising the LINMAP, fuzzy Bellman-Zadeh, and TOPSIS approaches. If the main goal is ECF, the results of the second scenario are the best. Also, If the main goals are thermal efficiency, ECOP, EPC and Dimensionless ecological based thermo-environmental function the results of the third scenario are the best.

NOMECLATURE

A	Area,[m^2]
C	Heat capacitance rate,[W/K]
C_p	Specific heat capacity,[$Jmol^{-1}K^{-1}$]
h	heat transfer coefficient[$Wm^{-2}K^{-1}$]
n	Number of mole[-]
N	Number of heat transfer units[-],
P	Power output ,[W]

Q	Heat, [J]
R	The gas constant,[$Jmol^{-1}K^{-1}$]
S	Entropy,[J/K]
T	Temperature,[K]
t	Time ,[s]
P	Power,[W]
p	Pressure of the working fluid
W	Output work, [J]
0	ambient
1	Inlet
1,2,3,4	State points
2	Outlet
c	Cold side
dis	distance
H	Heat source
h	Hot side
L	Cold side/Heat sink
R	Regenerator
rec	absorber
η	Thermal efficiency[-]
ε	Effectiveness and emissivity factor[-]
λ	Ratio of pressure during the regenerative processes [-]
δ	Stefan's constant [$Wm^{-2}K^{-4}$]
σ	Entropy production [W/K]

REFERENCES

[1] Tlili, I., and Sa'ed, A. (2013). Thermodynamic evaluation of a second order simulation for Yoke Ross Stirling engine. Energy conversion and management, 68, 149-160.

[2] Formosa, F., and Despesse, G. (2010). Analytical model for Stirling cycle machine design. Energy Conversion and Management, 51(10), 1855-1863.

[3] Berchowitz, D. M., and Urieli, I. (1984). Stirling cycle engine analysis. Adam Hilger Ltd, Bristol.

[4] Reader, G. T. (1983). Hooper C. Stirling engines. London and New York: EandFN Spon.

[5] Hargreaves, C. M. (1991). The Phillips stirling engine.

[6] Carlson, H., Commisso, M. B., and Lorentzen, B. (1990, August). Maximum obtainable efficiency for engines and refrigerators based on the Stirling cycle. In Proceedings of the 25th Intersociety Energy Conversion Engineering Conference (Vol. 5, pp. 366-371). IEEE.

[7] Urieli, I., and Kushnir, M. (1982). The ideal adiabatic cycle-A rational basis for Stirling engine analysis. In IECEC'82; Proceedings of the Seventeenth Intersociety Energy Conversion Engineering Conference (pp. 1662-1668).

[8] Martaj, N., Grosu, L., and Rochelle, P. (2007). Thermodynamic study of a low temperature difference Stirling engine at steady state operation. International Journal of Thermodynamics, 10(4), 165.

[9] Markman, M. A., SHMATOK, I., and KRASOVSKII, V. (1983). An experimental investigation of a low-power Stirling engine. Geliotekhnika, (3), 19-24.

[10] Kagawa, N., Araoka, K., Sakuma, T., and Ichikawa, S. (1990, August). Design and development of a miniature Stirling engine. In Proceedings of the 25th Intersociety Energy Conversion Engineering Conference (Vol. 5, pp. 442-447). IEEE.

[11] Brandhorst Jr, H. W., and Chapman Jr, P. A. (2008). New 5 kW free-piston Stirling space convertor developments. Acta Astronautica, 63(1-4), 342-347.

[12] Ataer, Ö. E. (2002). Numerical analysis of regenerators of free-piston type Stirling engines using Lagrangian formulation. International journal of refrigeration, 25(5), 640-652.

[13] Nakajima, N., Ogawa, K., and Fujimasa, I. (1989). Study on microengines: miniaturing Stirling engines for actuators. Sensors and Actuators, 20(1-2), 75-82.

[14] Aramtummaphon, D. (1996). A study of the feasibility of using heat energy from producer gas for running Stirling engine by steam as working Fluid (Doctoral dissertation, Master thesis, King Mongkut's University of technology Thonburi).
[15] Fukui, T., Shiraishi, T., Murakami, T., and Nakajima, N. (1999). Study on high specific power micro-Stirling engine. JSME International Journal Series B Fluids and Thermal Engineering, 42(4), 776-782.
[16] Iwamoto, S., Toda, F., Hirata, K., Takeuchi, M., and Yamamoto, T. (1997, May). Comparison of low-and high-temperature differential Stirling engines. In Proceedings of eighth International Stirling engine conference (pp. 29-38).
[17] Wu, F., Chen, L., Wu, C., and Sun, F. (1998). Optimum performance of irreversible Stirling engine with imperfect regeneration. Energy Conversion and Management, 39(8), 727-732.
[18] Erbay, L. B., and Yavuz, H. (1997). Analysis of the Stirling heat engine at maximum power conditions. Energy, 22(7), 645-650.
[19] Ahmadi, M. H., and Hosseinzade, H. (2012). Investigation of solar collector design parameters effect onto solar stirling engine efficiency. J Applied Mechanic Engg, 1(102), 2.
[20] Ahmadi, M. H., Aghaj, S. S. G., and Nazeri, A. (2013). Prediction of power in solar stirling heat engine by using neural network based on hybrid genetic algorithm and particle swarm optimization. Neural Computing and Applications, 22(6), 1141-1150.
[21] Ahmadi, M. H., Ahmadi, M. A., Sadatsakkak, S. A., and Feidt, M. (2015). Connectionist intelligent model estimates output power and torque of stirling engine. Renewable and Sustainable Energy Reviews, 50, 871-883.
[22] Ahmadi, M., Ahmadi, M. A., Mehrpooya, M., and Rosen, M. (2015). Using GMDH neural networks to model the power and torque of a stirling engine. Sustainability, 7(2), 2243-2255.
[23] Toghyani, S., Ahmadi, M. H., Kasaeian, A., and Mohammadi, A. H. (2016). Artificial neural network, ANN-PSO and ANN-ICA for modelling the Stirling engine. International Journal of Ambient Energy, 37(5), 456-468.
[24] Curzon, F. L., and Ahlborn, B. (1975). Efficiency of a Carnot engine at maximum power output. American Journal of Physics, 43(1), 22-24.
[25] Novikov, I. I. (1958). The efficiency of atomic power stations (a review). Journal of Nuclear Energy (1954), 7(1-2), 125-128.
[26] Wu, C. (1988). Power optimization of a finite-time Carnot heat engine. Energy, 13(9), 681-687.
[27] Yan, Z., and Chen, L. (1995). The fundamental optimal relation and the bounds of power output and efficiency for an irreversible Carnot engine. Journal of Physics A: Mathematical and General, 28(21), 6167.
[28] Chen, J. (1994). The maximum power output and maximum efficiency of an irreversible Carnot heat engine. Journal of Physics D: Applied Physics, 27(6), 1144.
[29] Angulo-Brown, F. (1991). An ecological optimization criterion for finite-time heat engines. Journal of Applied Physics, 69(11), 7465-7469.
[30] Chen, L., Xiaoqin, Z., Sun, F., and Wu, C. (2004). Ecological optimization for generalized irreversible Carnot refrigerators. Journal of Physics D: Applied Physics, 38(1), 113.
[31] Chen, L., Zhang, W., and Sun, F. (2007). Power, efficiency, entropy-generation rate and ecological optimization for a class of generalized irreversible universal heat-engine cycles. Applied Energy, 84(5), 512-525.
[32] Chen, L., Zhou, J., Sun, F., and Wu, C. (2004). Ecological optimization for generalized irreversible Carnot engines. Applied Energy, 77(3), 327-338.
[33] Chen, L., Xiaoqin, Z., Sun, F., and Wu, C. (2007). Exergy-based ecological optimization for a generalized irreversible Carnot heat-pump. Applied Energy, 84(1), 78-88.
[34] Ge, Y., Chen, L., and Sun, F. (2016). Progress in finite time thermodynamic studies for internal combustion engine cycles. Entropy, 18(4), 139.
[35] Chen, L., Meng, F., and Sun, F. (2016). Thermodynamic analyses and optimization for thermoelectric devices: The state of the arts. Science China Technological Sciences, 59(3), 442-455.
[36] Cheng, C. Y. (1997). The ecological optimization of an irreversible Carnot heat engine. Journal of physics D: Applied physics, 30(11), 1602.
[37] Xia, D., Chen, L., Sun, F., and Wu, C. (2006). Universal ecological performance for endo-reversible heat engine cycles. International journal of ambient energy, 27(1), 15-20.
[38] Zhang, W., Chen, L., Sun, F., and Wu, C. (2007). Exergy-based ecological optimal performance for a universal endoreversible thermodynamic cycle. International Journal of Ambient Energy, 28(1), 51-56.
[39] Chen, L., Zhu, X., Sun, F., and Wu, C. (2006). Exergy-based ecological optimization of linear phenomenological heat-transfer law irreversible Carnot-engines. Applied Energy, 83(6), 573-582.
[40] Zhu, X., Chen, L., Sun, F., and Wu, C. (2003). The ecological optimisation of a generalised irreversible Carnot engine for a generalised heat transfer law. International journal of ambient energy, 24(4), 189-194.
[41] Zhu, X., Chen, L., Sun, F., and Wu, C. (2005). Effect of heat transfer law on the ecological optimization of a generalized irreversible Carnot engine. Open Systems and Information Dynamics, 12(03), 249-260.
[42] Li, J., Chen, L., and Sun, F. (2011). Ecological performance of an endoreversible Carnot heat engine with complex heat transfer law. International Journal of Sustainable Energy, 30(1), 55-64.

[43] Li, J., Chen, L., and Sun, F. (2011). Ecological performance of a generalized irreversible Carnot heat engine with complex heat transfer law. Int. J. Energy and Environment, 2(1), 57-70.
[44] Tu, Y., Chen, L., Sun, F., and Wu, C. (2006). Exergy-based ecological optimisation for an endoreversible Brayton refrigeration cycle. International Journal of Exergy, 3(2), 191-201.
[45] Chen, L., Xiaoqin, Z., Sun, F., and Wu, C. (2004). Ecological optimization for generalized irreversible Carnot refrigerators. Journal of Physics D: Applied Physics, 38(1), 113.
[46] Chen, L., Xiaoqin, Z., Sun, F., and Wu, C. (2007). Exergy-based ecological optimization for a generalized irreversible Carnot heat-pump. Applied Energy, 84(1), 78-88.
[47] Chen, L., Xiaoqin, Z., Sun, F., and Wu, C. (2007). Exergy-based ecological optimization for a generalized irreversible Carnot heat-pump. Applied Energy, 84(1), 78-88.
[48] Chen, L., Xiaoqin, Z., Sun, F., and Wu, C. (2007). Ecological optimisation of a generalised irreversible Carnot refrigerator for a generalised heat transfer law. International Journal of Ambient Energy, 28(4), 213-219.
[49] Li, J., Chen, L., Sun, F., and Wu, C. (2011). Ecological performance of an endoreversible Carnot refrigerator with complex heat transfer law. International Journal of Ambient Energy, 32(1), 31-36.
[50] Chen, L., Li, J., and Sun, F. (2012). Ecological optimization of a generalized irreversible Carnot refrigerator in the case of Q∝(Δ T n) m. International Journal of Sustainable Energy, 31(1), 59-72.
[51] Tyagi, S. K., Kaushik, S. C., and Salohtra, R. (2002). Ecological optimization and parametric study of irreversible Stirling and Ericsson heat pumps. Journal of Physics D: Applied Physics, 35(16), 2058.
[52] Zhu, X., Chen, L., Sun, F., and Wu, C. (2005). Effect of heat transfer law on the ecological optimization of a generalized irreversible Carnot engine. Open Systems and Information Dynamics, 12(03), 249-260.
[53] Zhu, X., Chen, L., Sun, F., and Wu, C. (2005). The ecological optimization of a generalized irreversible Carnot heat pump for a generalized heat transfer law. Journal of the Energy Institute, 78(1), 5-10.
[54] Chen L, Li J, Sun F, and Wu C (2009). Effect of a complex generalized heat transfer law on ecological performance of an endoreversible Carnot heat pump. Int J Ambient Energy, 30, 102–8.
[55] Zhu, X., Chen, L., Sun, F., and Wu, C. (2001). Optimum performance of a generalized irreversible Carnot heat pump with a generalized heat transfer law. Physica Scripta, 64(6), 584.
[56] Liu, X., Chen, L., Wu, F., and Sun, F. (2009). Ecological optimization of an irreversible harmonic oscillators Carnot heat engine. Science in China Series G: Physics, Mechanics and Astronomy, 52(12), 1976-1988.
[57] Wang, W. H., Chen, L. G., Sun, F. R., and Wu, C. (2006). Optimal heat conductance distribution and optimal intercooling pressure ratio for power optimisation of irreversible closed intercooled regenerated Brayton cycle. Journal of the Energy Institute, 79(2), 116-119.
[58] Wang, W., Chen, L., and Sun, F. (2011). Ecological optimisation of an irreversible-closed ICR gas turbine cycle. International Journal of Exergy, 9(1), 66-79.
[59] Tyagi, S. K., Kaushik, S. C., and Salhotra, R. (2002). Ecological optimization and performance study of irreversible Stirling and Ericsson heat engines. Journal of Physics D: Applied Physics, 35(20), 2668.
[60] Zhu, X., Chen, L., Sun, F., and Wu, C. (2006). Exergy-based ecological optimization for a generalized irreversible Carnot refrigerator. Journal of the Energy Institute, 79(1), 42-46.
[61] Wu C, Chen L, and Sun F (2006). Ecological optimization performance of an irreversible quantum SI engine powering with an ideal Fermi gas. Open Sys Inform Dynam, 13, 55–66.
[62] Acikkalp, E. (2013). Models for optimum thermo-ecological criteria of actual thermal cycles. Thermal Science, 17(3).
[63] Ust, Y., Sahin, B., and Sogut, O. S. (2005). Performance analysis and optimization of an irreversible dual-cycle based on an ecological coefficient of performance criterion. Applied energy, 82(1), 23-39.
[64] Ust, Y., and Sahin, B. (2007). Performance optimization of irreversible refrigerators based on a new thermo-ecological criterion. International Journal of Refrigeration, 30(3), 527-534.
[65] Ust, Y., Sahin, B., Kodal, A., and Akcay, I. H. (2006). Ecological coefficient of performance analysis and optimization of an irreversible regenerative-Brayton heat engine. Applied Energy, 83(6), 558-572.
[66] Sogut, O. S., Ust, Y., and Sahin, B. (2006). The effects of intercooling and regeneration on the thermo-ecological performance analysis of an irreversible-closed Brayton heat engine with variable-temperature thermal reservoirs. Journal of Physics D: Applied Physics, 39(21), 4713.
[67] Ust, Y. (2010). Effect of regeneration on the thermo-ecological performance analysis and optimization of irreversible air refrigerators. Heat and mass transfer, 46(4), 469-478.
[68] Ust, Y. (2009). Performance analysis and optimization of irreversible air refrigeration cycles based on ecological coefficient of performance criterion. Applied Thermal Engineering, 29(1), 47-55.
[69] Ust, Y., Sogut, O. S., Sahin, B., and Durmayaz, A. (2006). Ecological coefficient of performance (ECOP) optimization for an irreversible Brayton heat engine with variable-temperature thermal reservoirs. Journal of the Energy Institute, 79(1), 47-52.
[70] Ust, Y., Sahin, B., and Kodal, A. (2006). Performance analysis of an irreversible Brayton heat engine based on ecological coefficient of performance criterion. International Journal of Thermal Sciences, 45(1), 94-101.

[71] Üst, Y., Sahin, B., and Kodal, A. (2005). Ecological coefficient of performance (ECOP) optimization for generalized irreversible Carnot heat engines. Journal of the Energy Institute, 78(3), 145-151.
[72] Ust, Y., Safa, A., and Sahin, B. (2005). Ecological performance analysis of an endoreversible regenerative Brayton heat-engine. Applied Energy, 80(3), 247-260.
[73] Ust, Y., Akkaya, A. V., and Safa, A. (2011). Analysis of a vapour compression refrigeration system via exergetic performance coefficient criterion. Journal of the Energy Institute, 84(2), 66-72.
[74] Akkaya, A. V., Sahin, B., and Erdem, H. H. (2008). An analysis of SOFC/GT CHP system based on exergetic performance criteria. International Journal of Hydrogen Energy, 33(10), 2566-2577.
[75] Akkaya, A. V., Sahin, B., and Erdem, H. H. (2007). Exergetic performance coefficient analysis of a simple fuel cell system. International Journal of Hydrogen Energy, 32(17), 4600-4609.
[76] Ust, Y., Sahin, B., and Kodal, A. (2007). Optimization of a dual cycle cogeneration system based on a new exergetic performance criterion. Applied Energy, 84(11), 1079-1091.
[77] Ust, Y., Sahin, B., and Yilmaz, T. (2007). Optimization of a regenerative gas-turbine cogeneration system based on a new exergetic performance criterion: exergetic performance coefficient. Proceedings of the Institution of Mechanical Engineers, Part A: Journal of Power and Energy, 221(4), 447-456.
[78] Gonca, G. (2017). Application of a novel thermo-ecological performance criterion: effective ecological power density (EFECPOD) to a Joule-Brayton cycle (JBC) turbine. Journal of Thermal Engineering, 3(5), 1478-1488.
[79] Mironova, V. A., Tsirlin, A. M., Kazakov, V. A., and Berry, R. S. (1994). Finite-time thermodynamics: Exergy and optimization of time-constrained processes. Journal of applied physics, 76(2), 629-636.
[80] Sieniutycz, S., and von Spakovsky, M. R. (1998). Finite time generalization of thermal exergy. Energy conversion and management, 39(14), 1423-1447.
[81] Sieniutycz, S. (1998). Generalized Carnot problem of maximum work in finite time via Hamilton–Jacobi–Bellman theory. Energy conversion and management, 39(16-18), 1735-1743.
[82] Sieniutycz, S. (1999). Carnot problem of maximum work from a finite resource interacting with environment in a finite time. Physica A: Statistical Mechanics and its Applications, 264(1-2), 234-263.
[83] Sieniutycz, S. (1997). Hamilton-Jacobi-Bellman theory of dissipative thermal availability. Physical Review E, 56(5), 5051.
[84] Li, J., Chen, L., and Sun, F. (2009). Optimum work in real systems with a class of finite thermal capacity reservoirs. Mathematical and Computer Modelling, 49(3-4), 542-547.
[85] Li, J., Chen, L., and Sun, F. (2010). Maximum work output of multistage continuous Carnot heat engine system with finite reservoirs of thermal capacity and radiation between heat source and working fluid. Thermal Sci, 14(1), 1-9.
[86] Li, J., Chen, L. G., and Sun, F. R. (2009). Extremal work of an endoreversible system with two finite thermal capacity reservoirs.
[87] Xia, S. J., Chen, L. G., and Sun, F. R. (2010). Effects of mass transfer laws on finite time exergy. Journal of the Energy Institute, 83(4), 210-216.
[88] Xia, S., Chen, L., and Sun, F. (2011). Power-optimization of non-ideal energy converters under generalized convective heat transfer law via Hamilton-Jacobi-Bellman theory. Energy, 36(1), 633-646.
[89] Xia, S. J., Chen, L. G., and Sun, F. R. (2012). Finite time exergy with generalised heat transfer law. Journal of the Energy Institute, 85(2), 70-77.
[90] Sharma, A., Shukla, S. K., and Rai, A. K. (2011). Finite time thermodynamic analysis and optimization of solar-dish stirling heat engine with regenerative losses. Thermal Science, 15(4).
[91] Yaqi, L., Yaling, H., and Weiwei, W. (2011). Optimization of solar-powered Stirling heat engine with finite-time thermodynamics. Renewable energy, 36(1), 421-427.
[92] Ahmadi, M. H., Sayyaadi, H., and Hosseinzadeh, H. (2015). Optimization of Output Power and Thermal Efficiency of Solar-Dish Stirling Engine Using Finite Time Thermodynamic Analysis. Heat Transfer—Asian Research, 44(4), 347-376.
[93] Tlili, I. (2012). Finite time thermodynamic evaluation of endoreversible Stirling heat engine at maximum power conditions. Renewable and Sustainable Energy Reviews, 16(4), 2234-2241.
[94] Kaushik, S. C., and Kumar, S. (2001). Finite time thermodynamic evaluation of irreversible Ericsson and Stirling heat engines. Energy Conversion and Management, 42(3), 295-312.
[95] Kaushik, S. C., and Kumar, S. (2000). Finite time thermodynamic analysis of endoreversible Stirling heat engine with regenerative losses. Energy, 25(10), 989-1003.
[96] Kaushik, S. C., Tyagi, S. K., and Mohan, S. (2003). Performance evaluation of an irreversible Stirling heat engine cycle. International journal of ambient energy, 24(3), 149-156.
[97] Tyagi, S. K., Chen, J., Lin, G., and Kaushik, S. C. (2004). Thermoeconomic optimization and parametric study of an irreversible Ericsson heat engine cycle. Int. J. Appl. Thermodynamics, 7(4), 189-198.
[98] Veldhuizen, D.A.V., Lamont, G.B. (2000). Multiobjective evolutionary algorithms: analyzing the state-of-the-art, Evolutionary computation, 8, 125-147

[99] Konak, A., Coit, D. W., and Smith, A. E. (2006). Multi-objective optimization using genetic algorithms: A tutorial. Reliability Engineering and System Safety, 91(9), 992-1007.
[100] Bäck, T., Fogel, D. B., and Michalewicz, Z. (1997). Handbook of evolutionary computation. CRC Press.
[101] Ahmadi, M. H., Hosseinzade, H., Sayyaadi, H., Mohammadi, A. H., and Kimiaghalam, F. (2013). Application of the multi-objective optimization method for designing a powered Stirling heat engine: design with maximized power, thermal efficiency and minimized pressure loss. Renewable Energy, 60, 313-322.
[102] Zhang, J., Zhu, H., Yang, C., Li, Y., and Wei, H. (2011). Multi-objective shape optimization of helico-axial multiphase pump impeller based on NSGA-II and ANN. Energy Conversion and Management, 52(1), 538-546.
[103] Ahmadi, M. H., Sayyaadi, H., Dehghani, S., and Hosseinzade, H. (2013). Designing a solar powered Stirling heat engine based on multiple criteria: maximized thermal efficiency and power. Energy Conversion and Management, 75, 282-291.
[104] Lazzaretto, A., and Toffolo, A. (2004). Energy, economy and environment as objectives in multi-criterion optimization of thermal systems design. Energy, 29(8), 1139-1157.
[105] Ahmadi, M. H., Sayyaadi, H., Mohammadi, A. H., and Barranco-Jimenez, M. A. (2013). Thermo-economic multi-objective optimization of solar dish-Stirling engine by implementing evolutionary algorithm. Energy Conversion and Management, 73, 370-380.
[106] Ahmadi, M. H., Ahmadi, M. A., Maleki, A., Pourfayaz, F., Bidi, M., and Açıkkalp, E. (2017). Exergetic sustainability evaluation and multi-objective optimization of performance of an irreversible nanoscale Stirling refrigeration cycle operating with Maxwell–Boltzmann gas. Renewable and Sustainable Energy Reviews, 78, 80-92.
[107] Atashkari, K., Nariman-Zadeh, N., Gölcü, M., Khalkhali, A., and Jamali, A. J. E. C. (2007). Modelling and multi-objective optimization of a variable valve-timing spark-ignition engine using polynomial neural networks and evolutionary algorithms. Energy Conversion and Management, 48(3), 1029-1041.
[108] Ahmadi, M. H., Ahmadi, M. A., Bayat, R., Ashouri, M., and Feidt, M. (2015). Thermo-economic optimization of Stirling heat pump by using non-dominated sorting genetic algorithm. Energy Conversion and Management, 91, 315-322.
[109] Ahmadi, M. H., Ahmadi, M. A., Mellit, A., Pourfayaz, F., and Feidt, M. (2016). Thermodynamic analysis and multi objective optimization of performance of solar dish Stirling engine by the centrality of entransy and entropy generation. International Journal of Electrical Power and Energy Systems, 78, 88-95.
[110] Toghyani, S., Kasaeian, A., and Ahmadi, M. H. (2014). Multi-objective optimization of Stirling engine using non-ideal adiabatic method. Energy Conversion and Management, 80, 54-62.
[111] Ahmadi, M. H., Ahmadi, M. A., Pourfayaz, F., and Bidi, M. (2016). Thermodynamic analysis and optimization for an irreversible heat pump working on reversed Brayton cycle. Energy conversion and management, 110, 260-267.
[112] Ahmadi, M. H., Ahmadi, M. A., Mehrpooya, M., Feidt, M., and Rosen, M. A. (2016). Optimal design of an Otto cycle based on thermal criteria. Mechanics and Industry, 17(1), 111.
[113] Ahmadi, M. H., Ahmadi, M. A., Mohammadi, A. H., Feidt, M., and Pourkiaei, S. M. (2014). Multi-objective optimization of an irreversible Stirling cryogenic refrigerator cycle. Energy Conversion and Management, 82, 351-360.
[114] Ahmadi, M. H., Ahmadi, M. A., Mohammadi, A. H., Mehrpooya, M., and Feidt, M. (2014). Thermodynamic optimization of Stirling heat pump based on multiple criteria. Energy Conversion and Management, 80, 319-328.
[115] Ahmadi, M. H., Mohammadi, A. H., Dehghani, S., and Barranco-Jimenez, M. A. (2013). Multi-objective thermodynamic-based optimization of output power of Solar Dish-Stirling engine by implementing an evolutionary algorithm. Energy conversion and Management, 75, 438-445.
[116] Ahmadi, M. H., Mohammadi, A. H., and Pourkiaei, S. M. (2016). Optimisation of the thermodynamic performance of the Stirling engine. International Journal of Ambient Energy, 37(2), 149-161.
[117] Sayyaadi, H., Ahmadi, M. H., and Dehghani, S. (2014). Optimal design of a solar-driven heat engine based on thermal and ecological criteria. Journal of Energy Engineering, 141(3), 04014012.
[118] Sahraie, H., Mirani, M. R., Ahmadi, M. H., and Ashouri, M. (2015). Thermo-economic and thermodynamic analysis and optimization of a two-stage irreversible heat pump. Energy conversion and management, 99, 81-91.
[119] Ahmadi, M. H., Ahmadi, M. A., Mehrpooya, M., Hosseinzade, H., and Feidt, M. (2014). Thermodynamic and thermo-economic analysis and optimization of performance of irreversible four-temperature-level absorption refrigeration. Energy conversion and management, 88, 1051-1059.
[120] Ahmadi, M. H., and Ahmadi, M. A. (2015). Thermodynamic analysis and optimization of an irreversible Ericsson cryogenic refrigerator cycle. Energy Conversion and Management, 89, 147-155.
[121] Ahmadi, M. H., Ahmadi, M. A., Mehrpooya, M., and Sameti, M. (2015). Thermo-ecological analysis and optimization performance of an irreversible three-heat-source absorption heat pump. Energy Conversion and Management, 90, 175-183.

[122] Ahmadi, M. H., Ahmadi, M. A., and Feidt, M. (2016). Performance optimization of a solar-driven multi-step irreversible brayton cycle based on a multi-objective genetic algorithm. Oil and Gas Science and Technology–Revue d'IFP Energies nouvelles, 71(1), 16.
[123] Ahmadi, M. H., Ahmadi, M. A., and Feidt, M. (2015). Thermodynamic analysis and evolutionary algorithm based on multi-objective optimization of performance for irreversible four-temperature-level refrigeration. Mechanics and Industry, 16(2), 207.
[124] Sadatsakkak, S. A., Ahmadi, M. H., and Ahmadi, M. A. (2015). Thermodynamic and thermo-economic analysis and optimization of an irreversible regenerative closed Brayton cycle. Energy conversion and management, 94, 124-129.
[125] Sadatsakkak, S. A., Ahmadi, M. H., Bayat, R., Pourkiaei, S. M., and Feidt, M. (2015). Optimization density power and thermal efficiency of an endoreversible Braysson cycle by using non-dominated sorting genetic algorithm. Energy Conversion and Management, 93, 31-39.
[126] Ahmadi, M. H., Ahmadi, M. A., and Sadatsakkak, S. A. (2015). Thermodynamic analysis and performance optimization of irreversible Carnot refrigerator by using multi-objective evolutionary algorithms (MOEAs). Renewable and Sustainable Energy Reviews, 51, 1055-1070.
[127] Ahmadi, M. H., and Mehrpooya, M. (2015). Thermo-economic modeling and optimization of an irreversible solar-driven heat engine. Energy conversion and management, 103, 616-622.
[128] Ahmadi, M. H., Ahmadi, M. A., and Pourfayaz, F. (2015). Performance assessment and optimization of an irreversible nano-scale Stirling engine cycle operating with Maxwell-Boltzmann gas. The European Physical Journal Plus, 130(9), 190.
[129] Ahmadi, M. H., Ahmadi, M. A., Shafaei, A., Ashouri, M., and Toghyani, S. (2016). Thermodynamic analysis and optimization of the Atkinson engine by using NSGA-II. International Journal of Low-Carbon Technologies, 11(3), 317-324.
[130] Ahmadi, M. H., Dehghani, S., Mohammadi, A. H., Feidt, M., and Barranco-Jimenez, M. A. (2013). Optimal design of a solar driven heat engine based on thermal and thermo-economic criteria. Energy Conversion and Management, 75, 635-642.
[131] Ahmadi, M. H., Ahmadi, M. A., and Pourfayaz, F. (2016). Thermodynamic analysis and evolutionary algorithm based on multi-objective optimization performance of actual power generating thermal cycles. Applied Thermal Engineering, 99, 996-1005.
[132] Ahmadi, M. H., and Ahmadi, M. A. (2016). Multi objective optimization of performance of three-heat-source irreversible refrigerators based algorithm NSGAII. Renewable and Sustainable Energy Reviews, 60, 784-794.
[133] Hussein, A. K. (2015). Multi-objective optimization of a three-dimensional internally finned tube based on Response Surface Methodology (RSM). Journal of Thermal Engineering, 1(2), 131-142.
[134] Ahmadi, M. H., Nabakhteh, M. A., Ahmadi, M. A., Pourfayaz, F., and Bidi, M. (2017). Investigation and optimization of performance of nano-scale Stirling refrigerator using working fluid as Maxwell–Boltzmann gases. Physica A: Statistical Mechanics and its Applications, 483, 337-350.
[135] Wong, J. Y., Sharma, S., and Rangaiah, G. P. (2016). Design of shell-and-tube heat exchangers for multiple objectives using elitist non-dominated sorting genetic algorithm with termination criteria. Applied Thermal Engineering, 93, 888-899.
[136] Rahdar, M. H., Heidari, M., Ataei, A., and Choi, J. K. (2016). Modeling and optimization of R-717 and R-134a ice thermal energy storage air conditioning systems using NSGA-II and MOPSO algorithms. Applied Thermal Engineering, 96, 217-227.
[137] Aminyavari, M., Najafi, B., Shirazi, A., and Rinaldi, F. (2014). Exergetic, economic and environmental (3E) analyses, and multi-objective optimization of a CO2/NH3 cascade refrigeration system. Applied Thermal Engineering, 65(1-2), 42-50.
[138] Mamaghani, A. H., Najafi, B., Shirazi, A., and Rinaldi, F. (2015). Exergetic, economic, and environmental evaluations and multi-objective optimization of a combined molten carbonate fuel cell-gas turbine system. Applied Thermal Engineering, 77, 1-11.
[139] Ahmadi, M. H., Jokar, M. A., Ming, T., Feidt, M., Pourfayaz, F., and Astaraei, F. R. (2018). Multi-objective performance optimization of irreversible molten carbonate fuel cell–Braysson heat engine and thermodynamic analysis with ecological objective approach. Energy, 144, 707-722.
[140] Ahmadi, M. H., Ahmadi, M. A., Açıkkalp, E., Alhuyi Nazari, M., Arab Pour Yazdi, M., and Kumar, R. (2018). New thermodynamic analysis and optimization of performance of an irreversible diesel cycle. Environmental Progress and Sustainable Energy, 37(4), 1475-1490.
[141] Mazur, V. (2007). Fuzzy thermoeconomic optimization of energy-transforming systems. Applied Energy, 84(7-8), 749-762.
[142] Olson, D. L. (1996). Decision aids for selection problems. Springer Science and Business Media.

9

NUMERICAL INVESTIGATION OF EFFECTS OF WORKING CONDITIONS ON PERFORMANCE OF PEM FUEL CELL

M. Ozdogan[1,*], L. Namli[1], A. Durmus[1]

ABSTRACT

In this study, the effects of the working pressure and temperature on the performance of the PEM fuel cell were investigated numerically. Non-isothermal, steady-state and single-phase model was used to examine the behaviour of the proton exchange membrane (PEM) fuel cells in the three-dimensional condition. The three-dimensional single-cell model has been developed within FLUENT 6.3 software by utilizing the PEMFC module. The results of polarization (voltage) variation curves and current density distribution were given and compared with each other. According to the results obtained, by keeping humidification and cell temperatures in equilibrium, the performance of the cell improves with the increasing cell temperature. In addition, the current density of the cell increases with the increasing operating pressure.

Keywords: *PEM Fuel Cell, Working Conditions, Three-Dimensional Modelling*

INTRODUCTION

Energy consumption in the world is increasing day by day. The share of renewable energy in the world's energy consumption is 9% (7% hydraulic and 2% other renewable) [1]. Dissemination of fuel cells which is another renewable energy source is important for the security of energy supply and for a healthy environment. Fuel cells are electrochemical energy converters which convert the fuel directly into electricity without any burning process. There are many types of fuel cells based on types of electrode used. The proton exchange membrane fuel cells (PEMFC) are widely used in fuel cells due to its low working temperature, high power density, low pollution, quick start up and shutdown capability [2]. The performance of the PEMFC depends on many factors such as working condition, design and material parameters [3].

Güvenlioğlu and Stenger [4] modelled a PEMFC which is steady-state and two-dimensional. They assumed that distribution of temperature was homogenous. They investigated the effect of the channel geometry and the relative humidity of the reactants on the performance of the cell. They obtained higher current densities at the smaller flow length of channel and thickness of the bipolar plate. Kim et al. [5] compared two- and three-dimensional models to examine the effect of two-dimensional modelling PEMFC. They found out similar result in each model but they got more accurate results from the three-dimensional model. Dadda et al. [6] developed a two-dimensional unsteady model to investigate heat and mass transfer at the PEMFC. They gave changes of distribution of temperature and water concentration with time as a result. Berning and Djilali [7] investigated the effects of temperature, pressure, the concentration of reactants, the thickness and the pore structure of the gas diffusion electrode and the thickness of the flow channel on the performance of PEMFC. Carcadea et al. [8] developed three-dimensional, steady-state and single-phase model in order to investigate the mass and charge transfer in a PEMFC. They modelled two types of PEM fuel cells which were conventional and interdigitated flow fields and compared fuel cell types each other. Lobato et al. [9] modelled PEMFC with three different types of flow channel design: 4-step serpentine, parallel and pin-type to investigate the effects of channel geometries. The results of the model in this study showed a similar performance at serpentine and pin-type flow channels. On the other hand, at the parallel type flow channel, the model results have shown low performance. Reffeira et al. [10] modelled the one-dimensional PEMFC which has two-phase flow. They compared the results obtained by the two-phase flow model with those found by the single-phase flow model. Ahmadi et al. [11] investigated numerically and experimentally the influence of the operating pressure and gas diffusion layer geometrical configuration on the PEM fuel cell performance. Heidary et al. [12] investigated numerically the effect of temperature, pressure and saturation level on the cathode catalyst layer. According to their results, the cell performance increases with the

This paper was recommended for publication in revised form by Tolga Taner
[1]*Department of Mechanical Engineering, Ondokuz Mayıs University, Samsun, TURKEY*
**E-mail address: muhammet.ozdogan@omu.edu.tr*

increasing temperature and pressure. They found out that saturation level the most effective parameter on cell performance and the increasing saturation reduced the cell performance.

The present study, the effects of the working pressure and temperature on the performance of the PEM fuel cell were investigated numerically. To investigate the effects of the working condition on the performance of the PEM fuel cell, a three-dimensional model was developed. Three-dimensional single-cell was modelled with FLUENT 6.3 software by utilizing a PEMFC module. After creating a model, the convergence curves were checked to examine the accuracy of the model. Also, the independence of the solution from the number of mesh elements was examined by using five different mesh structures. The model was confirmed with the experimental data that the Wang et al. [13] had published. After verifying the model, analyses were performed for the various temperature and operating pressure values. Consequently, to assess the effects of the working pressure and temperature on the performance of the PEM fuel cell, the polarization (voltage) and power curves, the distribution of the current density that is in the form of the counter line graphics, the current density curves according to pressure and temperature were given and evaluated systematically.

NUMERICAL MODEL

To investigate the working conditions on the performance of PEMFC, three-dimensional single-cells were modelled. A cross-sectional view of the PEMFC is shown in Figure 1 and dimensions of the cell are given Table 1. In this study, three-dimensional single-cell was modelled with FLUENT 6.3 software by utilizing a PEMFC module. The assumptions made in improving the model are as follows:

- The fuel cell operates under steady state condition.
- Flow is laminar.
- Membrane, gas diffusion layer and catalyst layer are treated as isotropic porous media.
- Reactant gases are regarded as ideal gases.
- Water is only in the vapour phase.

The governing equations of mass conservation, momentum conservation, energy conservation, species concentration, proton transport and electron transport are as follows:

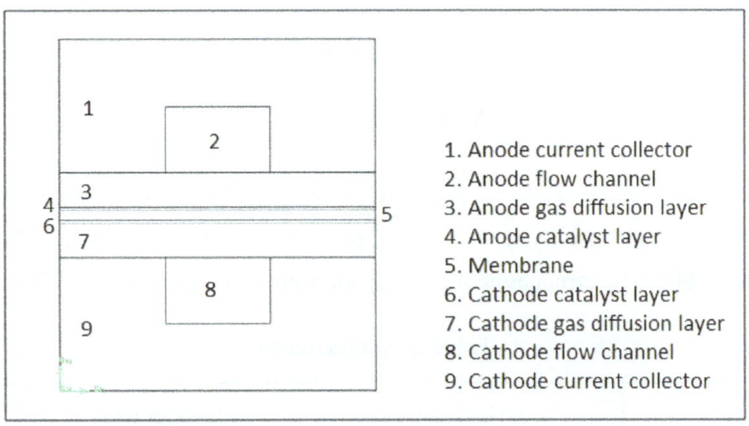

Figure 1. A cross-sectional view of the PEMFC

Table 1. Geometrical parameters of the fuel cell model

Parameter	**MEC-1**
Channel length (mm)	50
Channel height (mm)	1
Channel width (mm)	1
Current collector width (mm)	3
Current collector height (mm)	2
Gas diffusion layers thickness (mm)	0.3
Catalyst layers thickness (mm)	0.01
Membrane thickness (mm)	0.05

- Mass conservation equation

$$\nabla.(\varepsilon\rho\vec{u}) = 0 \tag{1}$$

- Momentum conservation equation

$$\nabla.(\varepsilon\rho\vec{u}\vec{u}) = -\varepsilon\nabla p + \nabla.(\varepsilon\mu\nabla\vec{u}) + S_u \tag{2}$$

In the momentum conservation equation, S_u are external body forces and presented in Table 2.

- Energy conservation equation

$$\nabla.(\varepsilon\rho\vec{u}h) = \nabla.(k\nabla T) + S_h \tag{3}$$

Where, h is the enthalpy, k is the thermal conductivity and S_h is the energy source term and is defined in Table 2.

- Species concentration equation

$$\nabla.(\varepsilon\vec{u}C_k) = \nabla.(D_k^{eff}\nabla C_k) + S_k \tag{4}$$

In Equation 4, S_k is the source term and is given Table 2, D_k^{eff} is the effective gas species diffusivity and can be expressed as:

$$D_k^{eff} = \varepsilon^{1.5} D_k \tag{5}$$

- Proton transport and electron transport equation

$$\nabla.(\sigma_m \nabla\phi_m) = S_i \tag{6}$$

$$\nabla.(\sigma_s \nabla\phi_s) = S_e \tag{7}$$

In the proton transport and electron transport equation, σ_m is the ionic conductivity, ϕ_m is the membrane potential, σ_s is the electrical conductivity, ϕ_s is the solid phase potential. S_i and S_e are the source terms of these equations and are given Table 2. Electrochemical and transport properties are listed in Table 3.

Table 2. Source terms

Component / Source terms	S_u (Momentum)	S_h (Energy)	S_k (Species)	S_e (Potential)	S_i (Potential)
Current collectors	0	0	0	0	0
Gas channels	0	0	0	0	0
Gas diffusion layers	$-\frac{\mu}{K}\varepsilon^2\vec{u}$	0	0	0	0
Catalyst layer	$-\frac{\mu}{K}\varepsilon^2\vec{u}$	$h_{reaction} + j_{a.c}.\eta_{a.c} + \frac{i_s^2}{\sigma_s^{eff}} + \frac{i_m^2}{\sigma_m^{eff}} + h_{phase}$	$H_2: -\frac{M_{H_2}}{2F} j_a$ $O_2: -\frac{M_{O_2}}{4F} j_c$ $H_2O: \frac{M_{H_2O}}{2F} j_c + \frac{n_d M_{H_2O}}{2F} j_c$	j	j
Membrane	$-\frac{\mu}{K}\varepsilon^2\vec{u}$	0	0	0	0

Table 3. Electro chemical and transport properties

Parameter [Units]	**Value**
Anode reference current density [A.m^{-3}]	1.5x10^{9}
Anode reference molar concentration[kmol.m^{-}	1
Anode concentration exponent	0.5
Anode transfer coefficient	2
Anode backing layer porosity	0.4
Cathode reference current density [A.m^{-3}]	4x10^{6}
Cathode reference molar concentration	1
Cathode concentration exponent	1
Cathode transfer coefficient	2
Cathode backing layer porosity	0.4
Open-circuit voltage [V]	0.95
Reference diffusivity of hydrogen [m^{2}.s^{-1}]	9.15x10^{-5}
Reference diffusivity of oxygen [m^{2}.s^{-1}]	2.2x10^{-5}
Reference diffusivity of water vapour [m^{2}.s^{-1}]	2.56x10^{-5}

Boundary and Operating Condition

Boundary conditions have to be defined to solve the discretized equation. At the inlet of the gas flow channels, the mass flow inlet boundary conditions are defined as a boundary condition. Flow rate, temperature, and spices concentrations are described as the inlet values at the flow channels. Well-humidified hydrogen enters the anode side flow channel and well-humidified air enters the cathode side flow channel. The values of mass flow rates of air and hydrogen are 4.68x10^{-5} kg.s^{-1} and 1.762x10^{-6} kg.s^{-1}, respectively. At the outlet of the gas flow channels, pressure outlet boundary conditions are used. Solution zone, except the current collector plates areas, are defined as fluid, the current collector plates are defined as a solid. Top and bottom surfaces of the cell are assumed as walls which have a constant temperature.

Solution Method

The equations were solved in the FLUENT 6.3 software which uses finite volume method to discretize equations. The solution zones were created and meshed in the GAMBIT software as shown in Figure 2. To investigate independent of grid size and balancing solution time, the solution zones were meshed in five different mesh sizes. In all calculations, the SIMPLE algorithm was used to solve the discretized equations of pressure-velocity coupling. To discretize pressure, the standard scheme was used and the first-order upwind scheme was used to discretize other equations. For all characteristics, iterations were continued until solutions converge up to 10^{-6}.

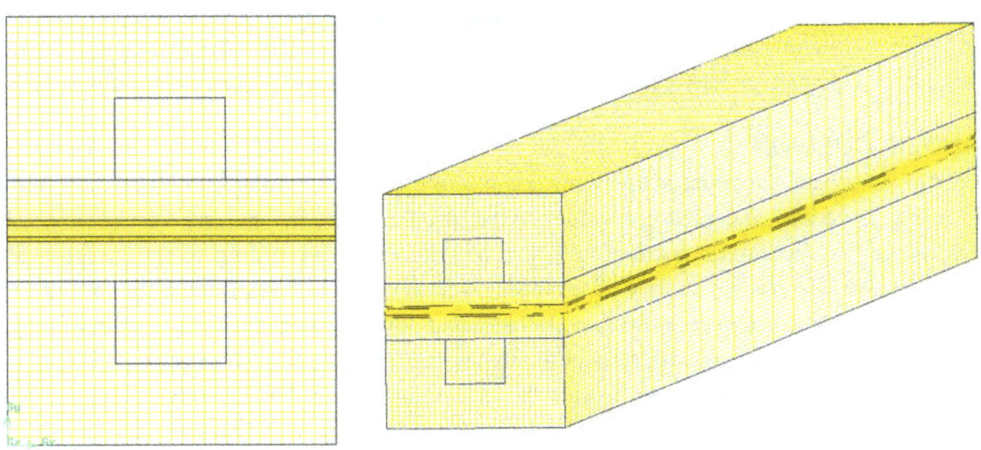

Figure 2. Mesh structure of PEMFC.

Model Validation

In order to validate the model, the results obtained from the computations were compared with the experimental data which was taken from the study of Wang et al. [13]. Geometrical parameters of the reference study were given Table 4. To compare the model predictions and experimental results, the polarization curves were shown in Figure 3. As can be seen in Figure 3, the model results are overwhelmingly in good agreement with the experimental data from the literature [13] except for the values of current density greater than about 1.1 A/cm^2. The difference in this interval is probably due to the poor performance of the PEM fuel cell because of the adverse effects of water flooding which was ignored at the model.

Table 4. Geometrical parameters of the reference study [13]

Parameter	MEC-1
Channel length (mm)	70
Channel height (mm)	1
Channel width (mm)	1
Current collector width (mm)	3
Current collector height (mm)	2
Gas diffusion layers thickness (mm)	0.3
Catalyst layers thickness (mm)	0.0129
Membrane thickness (mm)	0.108

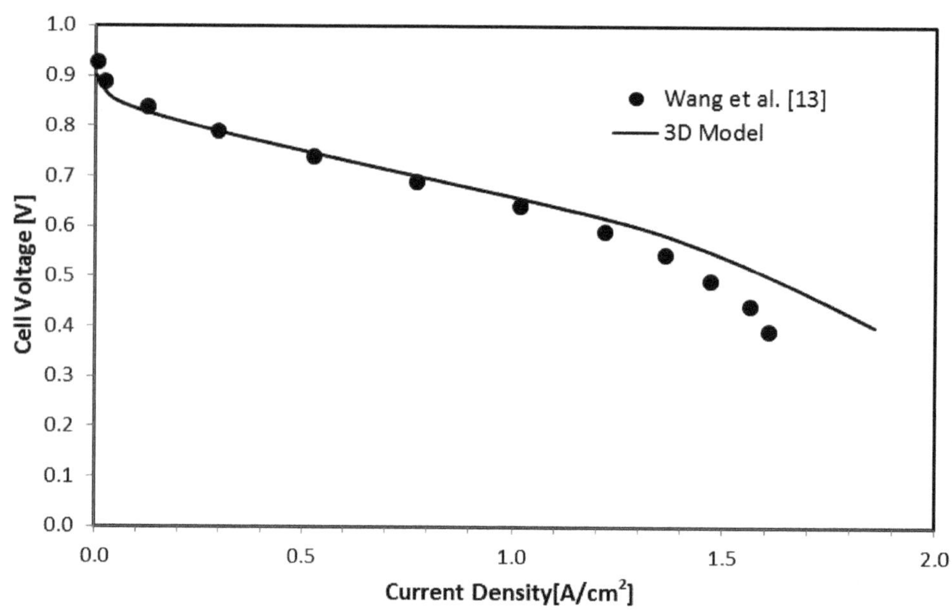

Figure 3. Comparison between the model predictions and the experimental results [13]

RESULTS AND DISCUSSION

In order to investigate the effect of operating conditions on cell performance, the analyses were performed for three different temperature and pressure values. After the analyses, the polarization (voltage) and power variations with respect to the current density are given in Figures 4-6 according to the various values of the temperature and pressure. The current density distributions were given in Figures 7 and 8 for various temperatures (313 K, 333 K and 353 K) and pressures (100 kPa, 200 kPa and 300 kPa) in the form of contour line graphics. In addition, the current density variations with respect to temperature and pressure were given in Figures 9 and 10.

Figure 4, Figure 5 and Figure 6 demonstrate the variations of the polarization (voltage) and power curves in the conditions of three different temperatures and operating pressures. As it can be seen in Figures 4-6, the effect of temperature is extremely clear at the values of the high current densities and it can be said that the increase of temperature increases cell voltage and cell power. Especially after the 1.1 A/cm^2 of the current density, these differences have become more obvious for both the polarization (voltage) and power. The results are similar for the three operating pressures but the effect of temperature is more pronounced at high pressures. The cell voltage and cell power increase with increasing operating pressure. At all conditions, the cell potential decreases with increasing current density; also with increasing current density, firstly, power reaches its maximum value and then decreases.

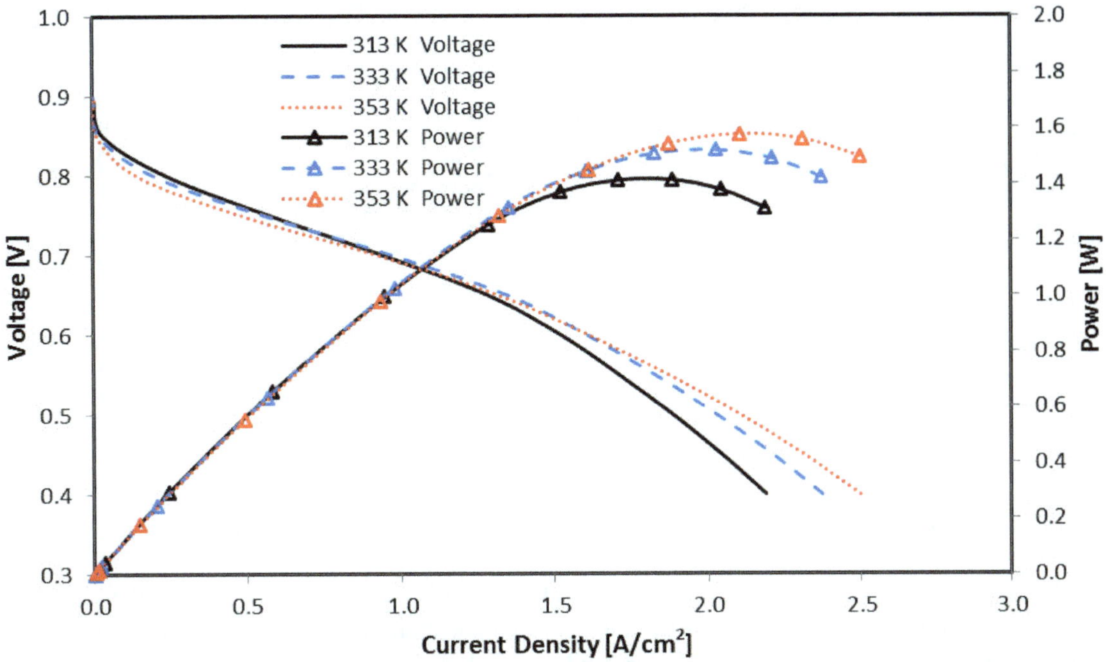

Figure 4. Effect of temperature on the cell performance for 100 kPa operating pressure

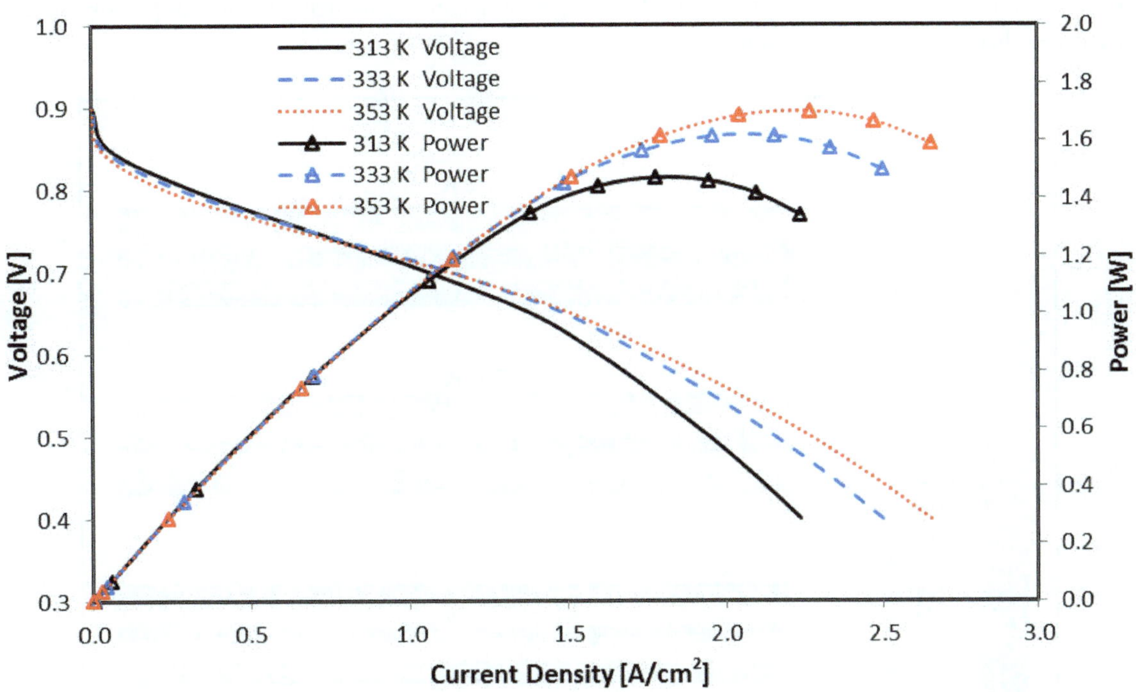

Figure 5. Effect of temperature on the cell performance for 200 kPa operating pressure

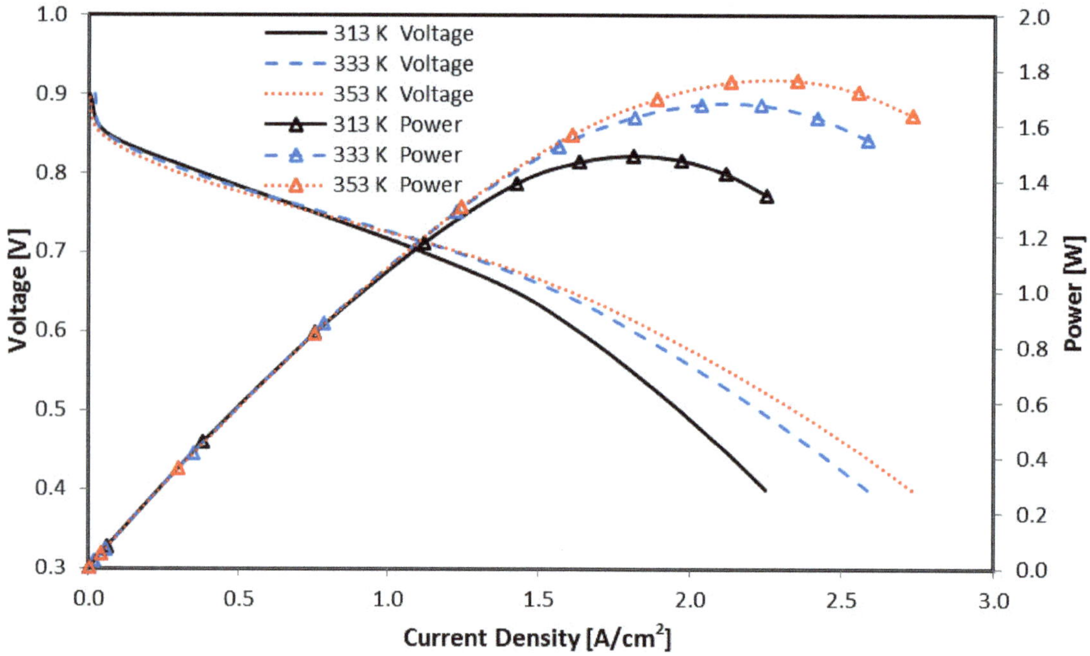

Figure 6. Effect of temperature on the cell performance for 300 kPa operating pressure

For different temperatures, the current density distribution is given as counter graphic in Figure 7. The comparison was carried out at 300 kPa pressure and 0.5 V cell potential. It is seen that increasing the temperature increases the amount of maximum current density and the nonhomogeneous distribution of the current density. The maximum current densities occur near the flow channel and the amount of current density decreases towards the outlet cross-section. Figure 8 shows the current density distribution in the form of counter line graphics for three different operating pressures. The graphics were drawn by using data obtained for a temperature of 353 K and a cell potential of 0.5 V. Referring to Figure 7 and Figure 8, it can be said that the effect of pressure on current density is less than the effect of temperature.

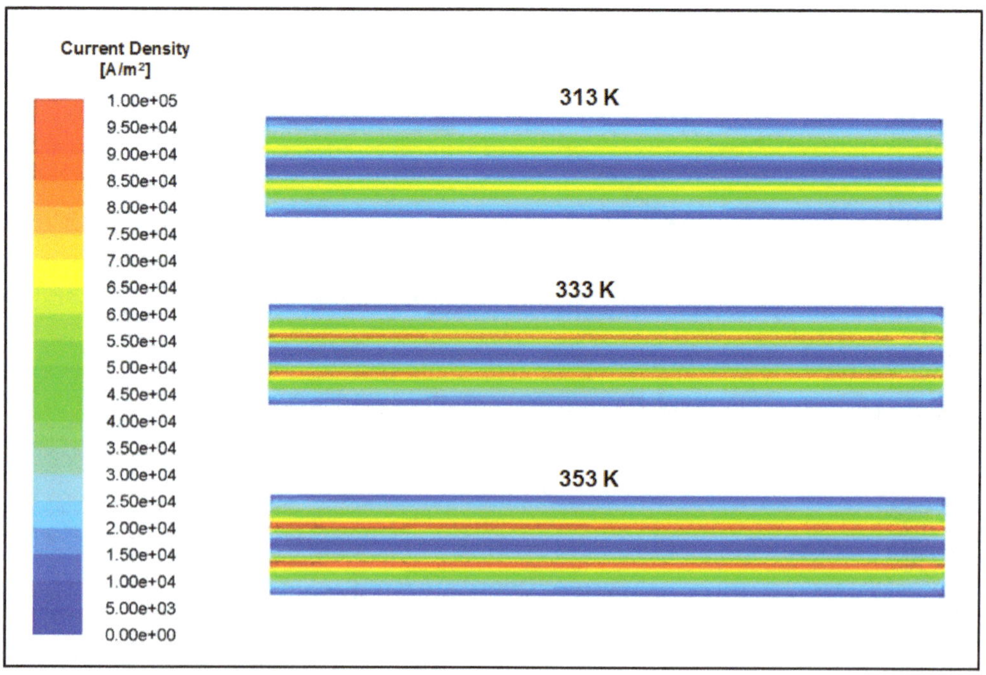

Figure 7. Current density distribution on the cross section z=2.67 mm (300 kPa operating pressure and 0.5 V cell voltage)

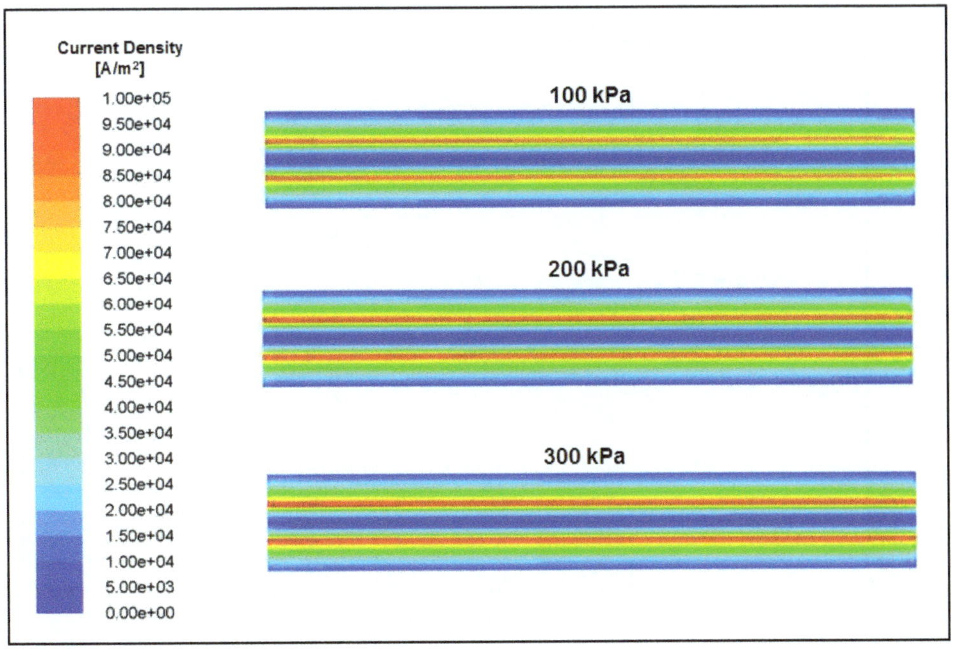

Figure 8. Current density distribution on the cross section z=2.67 mm (353 K temperature and 0.5 V cell voltage)

For 0.5 V, 0.6 V and 0.7 V cell voltages and 300 kPa operating pressure, the current density variation with respect to the temperature were given in Figure 9. As seen in Figure 9, it can be said that the current density value increases with increasing temperature and the effect of temperature variation at low temperatures and low voltages on the current density is much more apparent. Figure 10 shows the current density change curves with respect to the pressure of 0.5, 0.6 and 0.7 V cell voltages for the temperature of 333K. As shown in Figure 10, the current density magnitude increases with increasing pressure and the effect of pressure variation at low pressure on current density is much more apparent. Referring to Figure 9 and Figure 10, it can be said that the effect of the temperature variation on the current density is greater than the variation of the pressure.

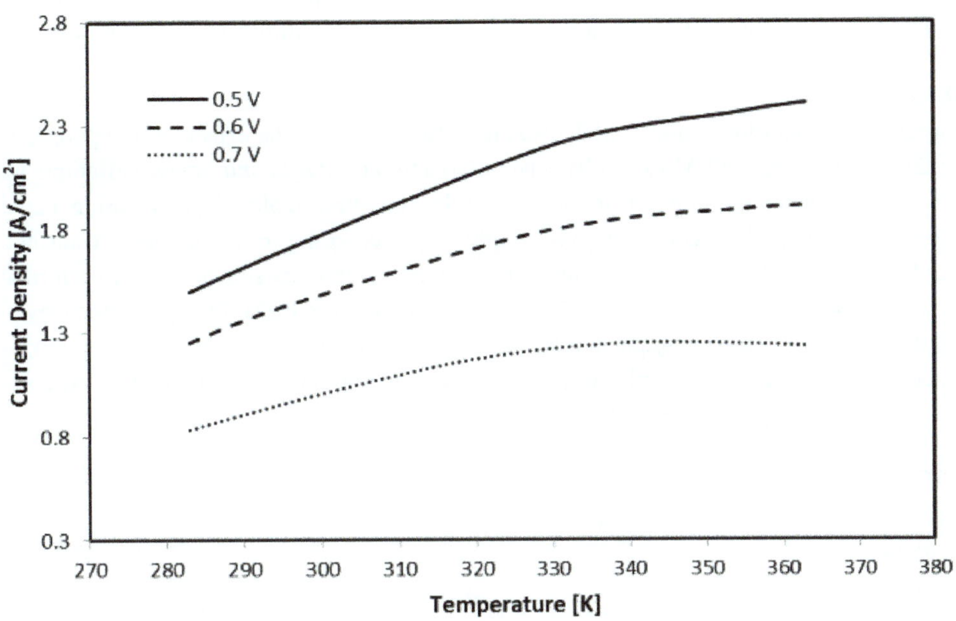

Figure 9. Effect of temperature on the PEM cell current density for different cell voltage (300 kPa operating pressure)

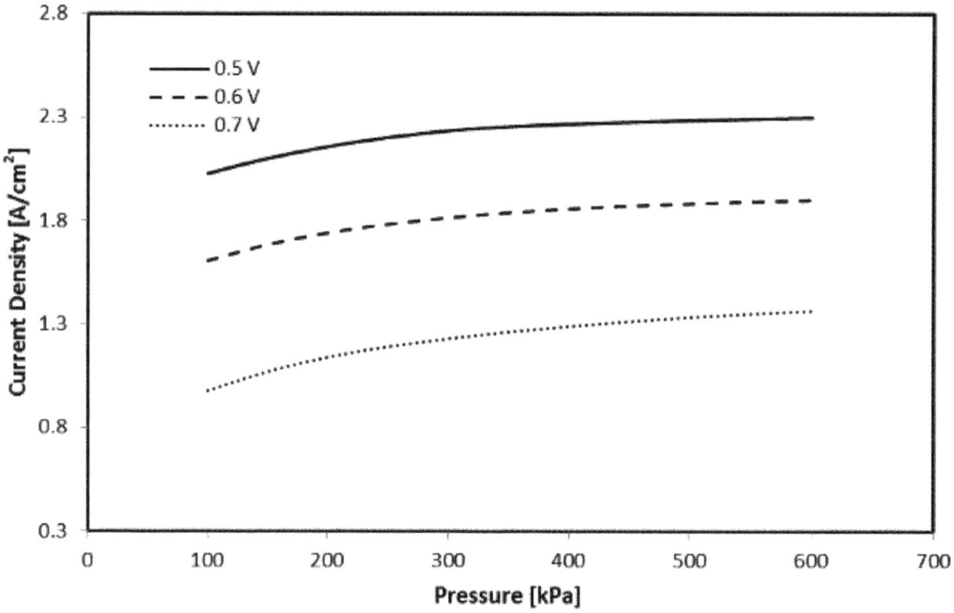

Figure 10. Effect of operating pressure on the PEM cell current density for different cell voltage (333 K)

Compared with similar studies in the literature, it can be concluded that the results are compatible with the literature. Yuan et al. [14], studied the effect of operating parameters on the cell performance for the PEM fuel cell with parallel flow channel structure. Their results show that, as in the current study, the increase in working pressure and working temperature increase cell performance. Besides, while increasing the current density, the effects of heat and pressure increase. With the increasing current density, the cell potential decreases and the power values increase after reaching the maximum value power values decrease. Additionally, their results showed that the effect of temperature on cell performance is more pronounced than the effect of pressure. In this study, the current density is about 2.5 A/cm^2 at 0.4V cell potential. Yuan et al. [14], found the current density about 1.1 A/cm^2 at 0.4V cell potential. On the other hand, Obayopo et al. [15] who have modelled the single PEM fuel cell, found the current density about 3 A/cm^2 at 0.4V cell potential. The main reason for the different results is that the reactant flow rates were different. Also, it can be said that the design parameters are effective in different results. In general, it can be said that the results obtained in the current study and in the literature show a similar tendency.

CONCLUSIONS

In this study, a three-dimensional model was used to investigate the effects of temperature and pressure on the performance of the single PEM fuel cells. The polarization (voltage) and power variations with respect to current density, current density distributions in the form of the contour line plots and the current density variations with respect to temperature and pressure are given graphically. According to the obtained result, the cell voltage and cell power have increased with increasing temperature and the cell has shown similar performance for three different pressure values. Nevertheless, the effect of temperature on PEM fuel performance is much more pronounced at the high pressures. Consequently, by keeping humidification and cell temperatures in equilibrium, the performance of the cell improves with the increasing cell temperature and the cell voltage and cell power increase with the increasing operating pressure.

NOMENCLATURE

ε Porosity
ρ Density [kg/m^3]
$\vec{u}$ Velocity vector [m/s]
p Pressure [Pa]
μ Viscosity [kg m/s]
h Enthalpy [kJ/kg]

k	Thermal conductivity [W/m K]
T	Temperature [K]
C_k	Molar concentration of species [mol/m^3]
D_k	Species diffusivity [m^2/s]
σ_m	Ionic conductivity [S/m]
σ_s	Electrical conductivity [S/m]
ϕ_m	Membrane potential [V]
ϕ_s	Solid phase potential [V]
S	Source terms
K	Permeability [m^2]
$h_{reaction}$	Enthalpy of reaction [kJ/kg]
h_{phase}	Enthalpy of water vaporization [kJ/kg]
j	Transfer current density [A/m^3]
η	Overpotantial [V]
i_s	Current density [A/m^2]
i_m	Ionic current density [A/m^2]
σ_s^{eff}	Effective electric conductivity [S/m]
σ_m^{eff}	Effective ionic conductivity [S/m]
M	Molecular weight of species [kg/mol]
F	Faraday constant [96.487 C/mol]
n_d	Electro-osmosis drag coefficient

REFERENCES

[1] Sungur, B., Ozdogan, M., Topaloglu, B., Namli, L. (2017). Technical and economic evaluation of micro cogeneration systems in the context of global energy consumption (in Turkish), Engineer and Machinery, 58(686), 1-20.

[2] Khazaee, I., Ghazikhani, M., (2012). Numerical simulation and experimental comparison of channel geometry on performance of a PEM fuel cell, Arabian Journal for Science and Engineering, 37(8), 1-13.

[3] Wang, X. D., Duan Y. Y., Yan W. M., Peng X. F. (2008). Effects of flow channel geometry on cell performance for PEM fuel cells with parallel and interdigitated flow fields, Electrochimica Acta, 53(16), 5334-5343.

[4] Güvelioğlu, G. H., Stenger, H. G. (2005). Computational fluid dynamics modeling of polymer electrolyte membrane fuel cells, Journal of Power Sources, 147, 95-106. 275-281.

[5] Kim, G., Sui, P. C., Shah, A. A., Djilali, N. (2010) Reduced-dimensional models for straight-channel proton exchange membrane fuel cells, Journal of Power Sources, 195, 3240-3249.

[6] Dadda, B., Abboudi, S., Ghezal, A. (2013) Transient two-dimensional model of heat and mass transfer in a PEM fuel cell membrane, International Journal of Hydrogen Energy, 38, 7092-7101.

[7] Berning, T., Djilali, N. (2003). Three-dimensional computational analysis of transport phenomena in a PEM fuel cell—a parametric study, Journal of Power Sources, 124, 440-452.

[8] Carcadea, E., Ene, H., Ingham, D. B., Lazar, R., Ma, L., Pourkashanian M., Stefanescu I. (2005) Numerical simulation of mass and charge transfer for a PEM fuel cell, International Communications in Heat and Mass Transfer, 32, 1273-1280.

[9] Lobato, J., Canizares P., Rodrigo, A. M., Pinar, F. J., Mena, E., Ubeda, D. (2010). Three-dimensional model of a 50 cm2 high-temperature PEM fuel cell. Study of the flow channel geometry influence, International Journal of Hydrogen Energy, 35, 5510-5520.

[10] Ferreira, R. B., Falcão, D. S., Oliveira, V. B., Pinto, A. M. F. (2015). A one-dimensional and two-phase flow model of a proton exchange membrane fuel cell, Journal of Chemical Technology and Biotechnology, 90(9), 1547-1551.

[11] Ahmadi, N., Dadvand, A., Rezazadeh, S., Mirzaee, I. (2016) Analysis of the operating pressure and GDL geometrical configuration effect on PEM fuel cell performance, Journal of the Brazilian Society of Mechanical Sciences and Engineering, 38(8), 2311-2325.

[12] Heidary, H., Kermani, M. J., Khajeh-Hosseini-Dalasm, N. (2016). Performance analysis of PEM fuel cells cathode catalyst layer at various operating conditions. International Journal of Hydrogen Energy, 41(47), 22274-22284.

[13] Wang, L., Husar, A., Zhou T., Liu, H. (2003). A parametric study of PEM fuel cell performances, International Journal of Hydrogen Energy, 28, 1263-1277.

[14] Yuan, W., Tang, Y., Pan, M., Li Z., Tang, B. (2010). Model prediction of effects of operating parameters on proton exchange membrane fuel cell performance, Renewable Energy, 35(3), 656-666.
[15] Obayopo, S. O., Bello-Ochende, T., Meyer, J. P. (2013). Three-dimensional optimization of a fuel gas channel of a proton exchange membrane fuel cell for maximum current density. International Journal of Energy Research, 37(3), 228-241.

10

NUMERICAL SIMULATION OF NATURAL CONVECTION MELTING IN 2D AND 3D ENCLOSURES

N.S. Bondareva[1], M.A. Sheremet[1,*]

ABSTRACT

Natural convection melting in 2D and 3D enclosures with a local heater is studied numerically. The present research is related to a development of effective cooling system for the electronic devices using the phase change material that is essentially important nowadays. The domain of interest includes vertical cold walls, adiabatic horizontal walls and a discrete heater of constant high temperature located on the bottom adiabatic wall. The cavity is filled with a phase change material (PCM) in solid state at the beginning of the process. During the heating from the heat source PCM is melting. Numerical solution of the present problem has been conducted using the dimensionless transformed variables such as stream function and vorticity for 2D cavity and vector potential functions and vorticity vector for 3D cavity with appropriate initial and boundary conditions. The developed numerical technique has been verified comprehensively. Obtained results have shown a potential of the used methods for 2D and 3D problems. It has been found that, melting process is more intensive in 3D case and the heat transfer rate at the heater is greater for 2D in comparison with 3D data.

Keywords: *Natural Convection, Melting, Numerical Analysis, Square and Cubical Cavities*

INTRODUCTION

Nowadays, one of effective cooling methods for modern electronic devices is a usage of phase change materials (PCM). These materials have a moderate melting temperature, high latent melting energy and high heat capacity. Heat generation or absorption inside these materials due to the phase transformation with constant temperature is widely used not only for electronic cooling but also in solar energy storages, building temperature control systems and in other systems [1–14].

Thus, Fan et al. [2] experimentally have analyzed the effects of melting temperature and the presence of internal fins on behavior of PCM for thermal management of electronics. It has been found that application of a PCM with a higher melting temperature can extend a longer time of protection of the devices from over-heating. Gharbi et al. [3] have studied experimentally the heat transfer behavior of PCM in a cavity with a local heater having a radiator system. The authors have revealed that an addition of PCM extends twice the critical time of the heat sink and the radiator system consisting of solid fins essentially changes the thermal control. Numerical simulation of PCM melting within an enclosure with internal horizontal rectangular fins has been performed by Sharifi et al. [4]. It has been found that utilization of horizontal rectangular fins initially promotes rapid melting, followed by slower melting once PCM in the inter-fin regions has been liquefied. Gharbi et al. [5] have studied experimentally the thermal performance of PCM inside vertical cavity with three discrete heaters on vertical wall. The flow structures and heat transfer patterns have been analyzed for different heat flux cycles. Taghilou and Talati have examined numerically [6] a melting process within a finned rectangular cavity filled with PCM using the lattice Boltzmann method. The obtained results have shown that using of heat-conducting solid fins can lead to more essential melting of PCM. Prieto et al. [7] have investigated numerically the thermal behavior of PCM heat exchangers for heating systems. The RT60 paraffin has been used for analysis. The authors used a full CFD model in Ansys Fluent 14.5 and found that melting and solidification in the vertical arrangement are slower than in the horizontal one. Madruga and Mendoza [8] have studied numerically the melting of n-octadecane under the thermocapillary effect at a free surface inside the circular sections. It has been shown that thermocapillary effect on the free surface has a powerful influence on the heat transfer rate, decreasing a total melting time. Wang et al. [9] have experimentally analyzed the thermal behavior of the vertical cylindrical battery with composite paraffin and fin structure. Ren and Chan [10] have studied numerically the PCM melting process in a cavity with internal fins under the lateral heating using the lattice Boltzmann method. They have found that using internal fins could enhance the heat transfer process in the PCM cavity, namely, the PCM

This paper was recommended for publication in revised form by Regional Editor Bekir Yilbas
[1]Department of Theoretical Mechanics, Tomsk State University, Tomsk, Russia
**E-mail address: Michael-sher@yandex.ru*

melting speed can be increased with the length of the internal fins. Moreover, the melting speed is enhanced when the PCM cavity is inclined towards the counter clockwise direction. Wang et al. [11] have investigated numerically an opportunity to optimize the latent thermal energy storage system using the PCM and internal fins. Obtained results have shown that fin-length, fin-ratio and angle between neighbor fins have affected the heat transfer enhancement inside the PCM storage system. Tabassum et al. [12] have examined the melting process of PCM in the annular gap using the finite volume method. Mathematical model has been formulated on the basis of the enthalpy-porosity approach using the primitive variables. Comprehensive analysis has revealed that the eccentric position of the inner tube in the inverse triangular annulus is an excellent configuration for the storage system among the studied heat exchangers. Yang et al. [15] have explored the thermal performance of the low melting point metal (LMPM) based internal finned heat sink using the experimental setup and commercial Fluent package. It has been found that LMPMs can be a very good candidate as a PCM for thermal energy storage and thermal management.

Abovementioned analysis of several works on PCM melting shows that in the case of numerical study the primitive variables are the main variables for investigation and 2D study is a wide-spread approach, while in the case of 3D simulation the commercial packages or open access codes are used.

The objective of the present study is to numerically analyze unsteady melting driven by natural convection in 2D and 3D cavities having local heaters using finite difference method and transformed variables such as stream function and vorticity for 2D cavity and vector potential functions and vorticity vector for 3D cavity. It should be noted that, it the first time for usage of vector potential functions and vorticity vector for numerical analysis of natural convection melting within cubical cavity with a local heater. 2D analysis of the natural convection melting in a square cavity has been performed earlier by Bondareva and Sheremet [16, 17] using also the uniform magnetic field. 3D analysis has been conducted also in the case of uniform magnetic field by Bondareva and Sheremet [18, 19]. The present work is a more detailed comparison of the obtained data for 2D and 3D cases with highlighting an opportunity to use non-primitive variables for simulation of PCM melting process in 3D domains.

GOVERNING EQUATIONS AND NUMERICAL TECHNIQUE

We consider 2D and 3D cavities (Figure 1) with a heat source of constant temperature mounted on the bottom wall. At the beginning of the process the domain of interest is filled with PCM in solid state having fusion temperature. Two opposite vertical walls of the enclosure have a low constant temperature, while other walls are adiabatic. Melt is considered to be incompressible, radiatively non-participating and the Boussinesq approximation is valid. The melt motion and heat transfer in the cavity are assumed to be two-dimensional for the square cavity (Figure 1*a*) and three-dimensional for the cubical enclosure (Figure 1*b*). It should be noted that 2D cavity can be considered as a middle cross-section of 3D cavity. The melt flow and heat transfer are described using Navier-Stokes equations, including the energy equation for the melt formulated on the basis of enthalpy approach.

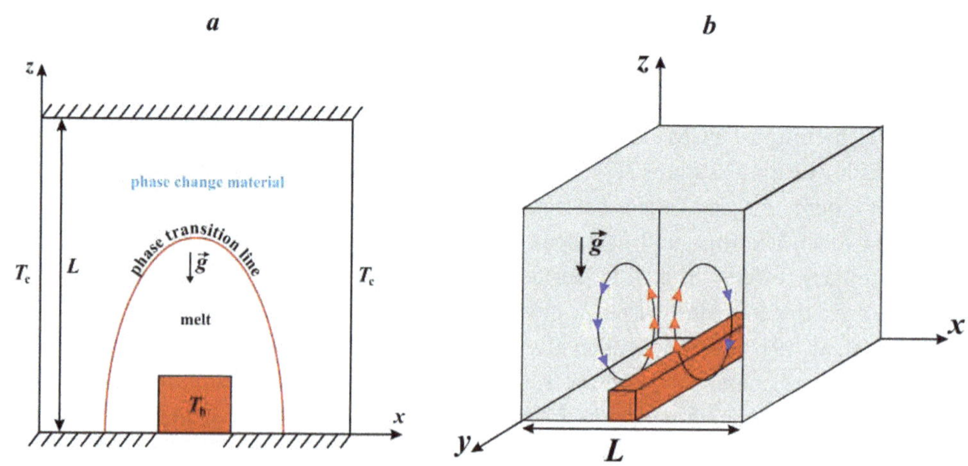

Figure 1. Schematic diagram of the physical system

The governing partial differential equations can be written as follows in tensor form for the liquid state of material (inside the melt zone) [16–19]:

$$div\mathbf{V} = 0 \tag{1}$$

$$\rho_l\left(\frac{\partial \mathbf{V}}{\partial t} + \left(\mathbf{V},\bar{\nabla}\right)\mathbf{V}\right) = -\bar{\nabla}p + \mu_l\nabla^2\mathbf{V} - \rho_l\beta\left(T - T_m\right)\mathbf{g} \tag{2}$$

$$\frac{\partial h}{\partial t} + \left(\mathbf{V},\bar{\nabla}h\right) = k_l\nabla^2 T \tag{3}$$

It should be noted that here we used the enthalpy formulation for the energy equation.

For the solid state of PCM we utilized the heat conduction equation in following form

$$\partial h/\partial t = k_s\nabla^2 T \tag{4}$$

Here $h = \begin{cases} \rho_s c_s T, & T < T_m, \\ \rho_s c_s T_m + \rho_l L_m + \rho_l c_l (T - T_m), & T \geq T_m \end{cases}$ is the enthalpy.

To combine the energy equations for melt and solid material and to delete the miscoordinations for enthalpy function at solid-liquid interface the following smoothing function was utilized:

$$\varphi = \begin{cases} 0, & T < T_m - \eta \\ \left[T - \left(T_m - \eta\right)\right]/2\eta, & T_m - \eta \leq T \leq T_m + \eta \\ 1, & T > T_m + \eta \end{cases} \tag{5}$$

It is worth noting that solution of the fundamental problems is more useful to perform in dimensionless variables. Therefore, we introduce the following dimensionless variables in Cartesian coordinates:

$$X = x/L,\ Y = y/L,\ Z = z/L,\ U = u\big/\sqrt{g\beta(T_h - T_m)L},\ V = v\big/\sqrt{g\beta(T_h - T_m)L},$$
$$W = w\big/\sqrt{g\beta(T_h - T_m)L},\ \tau = t\sqrt{g\beta(T_h - T_m)/L},\ \Theta = (T - T_m)/(T_h - T_m)$$

Further, in the case of 2D cavity (Figure 1*a*) we define the dimensionless stream function Ψ ($U = \partial\Psi/\partial Y$, $V = -\partial\Psi/\partial X$) and vorticity $\Omega = \partial V/\partial X - \partial U/\partial Y$ and the governing equations for 2D natural convection melting can be written as follows taking into account the definitions of enthalpy and smoothing function Eq. (5) [16, 17]:

$$\frac{\partial^2\Psi}{\partial X^2} + \frac{\partial^2\Psi}{\partial Z^2} = -\Omega \tag{6}$$

$$\frac{\partial\Omega}{\partial\tau} + U\frac{\partial\Omega}{\partial X} + W\frac{\partial\Omega}{\partial Z} = \sqrt{\frac{Pr}{Ra}}\left(\frac{\partial^2\Omega}{\partial X^2} + \frac{\partial^2\Omega}{\partial Z^2}\right) + \frac{\partial\Theta}{\partial X} \tag{7}$$

$$\zeta(\varphi)\left(\frac{\partial\Theta}{\partial\tau} + U\frac{\partial\Theta}{\partial X} + W\frac{\partial\Theta}{\partial Z}\right) + Ste\cdot\left(\frac{\partial\varphi}{\partial\tau} + +U\frac{\partial\varphi}{\partial X} + W\frac{\partial\varphi}{\partial Z}\right) = \frac{\xi(\varphi)}{\sqrt{Ra\cdot Pr}}\left(\frac{\partial^2\Theta}{\partial X^2} + \frac{\partial^2\Theta}{\partial Z^2}\right) \tag{8}$$

In the case of 3D cavity (Figure 1*b*) we define the dimensionless vector potential functions in the following form [18–20]:

$$\mathbf{V} = \mathrm{rot}(\boldsymbol{\Psi}) \Rightarrow \begin{cases} U = \dfrac{\partial \Psi_z}{\partial Y} - \dfrac{\partial \Psi_y}{\partial Z}, \\ V = \dfrac{\partial \Psi_x}{\partial Z} - \dfrac{\partial \Psi_z}{\partial X}, \quad W = \dfrac{\partial \Psi_y}{\partial X} - \dfrac{\partial \Psi_x}{\partial Y} \end{cases}$$

and vorticity vector in the form [18–20]:

$$\boldsymbol{\Omega} = \mathrm{rot}(\mathbf{V}) \Rightarrow \begin{cases} \Omega_x = \dfrac{\partial W}{\partial Y} - \dfrac{\partial V}{\partial Z}, \\ \Omega_y = \dfrac{\partial U}{\partial Z} - \dfrac{\partial W}{\partial X}, \quad \Omega_z = \dfrac{\partial V}{\partial X} - \dfrac{\partial U}{\partial Y} \end{cases}$$

Taking into account these new functions and dimensionless variables, the governing equations (1)–(4) for 3D cavity can be written as follows [18–20]:

$$\frac{\partial \Omega_x}{\partial \tau} + U\frac{\partial \Omega_x}{\partial X} + V\frac{\partial \Omega_x}{\partial Y} + W\frac{\partial \Omega_x}{\partial Z} - \Omega_x\frac{\partial U}{\partial X} - \Omega_y\frac{\partial U}{\partial Y} - \Omega_z\frac{\partial U}{\partial Z} = \sqrt{\frac{Pr}{Ra}}\left(\frac{\partial^2 \Omega_x}{\partial X^2} + \frac{\partial^2 \Omega_x}{\partial Y^2} + \frac{\partial^2 \Omega_x}{\partial Z^2}\right) + \frac{\partial \Theta}{\partial Y}, \quad (9)$$

$$\frac{\partial \Omega_y}{\partial \tau} + U\frac{\partial \Omega_y}{\partial X} + V\frac{\partial \Omega_y}{\partial Y} + W\frac{\partial \Omega_y}{\partial Z} - \Omega_x\frac{\partial V}{\partial X} - \Omega_y\frac{\partial V}{\partial Y} - \Omega_z\frac{\partial V}{\partial Z} = \sqrt{\frac{Pr}{Ra}}\left(\frac{\partial^2 \Omega_y}{\partial X^2} + \frac{\partial^2 \Omega_y}{\partial Y^2} + \frac{\partial^2 \Omega_y}{\partial Z^2}\right) - \frac{\partial \Theta}{\partial X}, \quad (10)$$

$$\frac{\partial \Omega_z}{\partial \tau} + U\frac{\partial \Omega_z}{\partial X} + V\frac{\partial \Omega_z}{\partial Y} + W\frac{\partial \Omega_z}{\partial Z} - \Omega_x\frac{\partial W}{\partial X} - \Omega_y\frac{\partial W}{\partial Y} - \Omega_z\frac{\partial W}{\partial Z} = \sqrt{\frac{Pr}{Ra}}\left(\frac{\partial^2 \Omega_z}{\partial X^2} + \frac{\partial^2 \Omega_z}{\partial Y^2} + \frac{\partial^2 \Omega_z}{\partial Z^2}\right), \quad (11)$$

$$\frac{\partial^2 \Psi_x}{\partial X^2} + \frac{\partial^2 \Psi_x}{\partial Y^2} + \frac{\partial^2 \Psi_x}{\partial Z^2} = -\Omega_x, \quad (12)$$

$$\frac{\partial^2 \Psi_y}{\partial X^2} + \frac{\partial^2 \Psi_y}{\partial Y^2} + \frac{\partial^2 \Psi_y}{\partial Z^2} = -\Omega_y, \quad (13)$$

$$\frac{\partial^2 \Psi_z}{\partial X^2} + \frac{\partial^2 \Psi_z}{\partial Y^2} + \frac{\partial^2 \Psi_z}{\partial Z^2} = -\Omega_z, \quad (14)$$

$$\zeta(\varphi)\left(\frac{\partial \Theta}{\partial \tau} + U\frac{\partial \Theta}{\partial X} + V\frac{\partial \Theta}{\partial Y} + W\frac{\partial \Theta}{\partial Z}\right) + Ste\left(\frac{\partial \varphi}{\partial \tau} + U\frac{\partial \varphi}{\partial X} + +V\frac{\partial \varphi}{\partial Y} + W\frac{\partial \varphi}{\partial Z}\right) = \frac{\xi(\varphi)}{\sqrt{Ra \cdot Pr}}\left(\frac{\partial^2 \Theta}{\partial X^2} + \frac{\partial^2 \Theta}{\partial Y^2} + \frac{\partial^2 \Theta}{\partial Z^2}\right). \quad (15)$$

where $\zeta(\varphi) = \frac{\rho_s c_s}{\rho_l c_l} + \varphi\left(1 - \frac{\rho_s c_s}{\rho_l c_l}\right)$; $\xi(\varphi) = \frac{k_s}{k_l} + \varphi\left(1 - \frac{k_s}{k_l}\right)$; $Ra = g(\rho\beta)_l (T_h - T_m)(\rho c)_l L^3 / (\mu_l k_l)$ is the Rayleigh number; $Pr = \mu_l c_l / k_l$ is the Prandtl number; $Ste = L_m / \left[c_l (T_h - T_m)\right]$ is the Stefan number.

Two-dimensional governing equations (6)–(8) are subject to the following initial and boundary conditions [16, 17]:

$$\left.\begin{aligned}
&\tau = 0:\ \ \Psi = \Omega = 0,\ \Theta = -\eta;\\
&\tau > 0:\ \ \Psi = 0,\ \ \ \Omega = -\frac{\partial^2\Psi}{\partial X^2} - \frac{\partial^2\Psi}{\partial Z^2}\ \ \text{at melting front;}\\
&\Psi = 0,\ \ \ \Omega = -\frac{\partial^2\Psi}{\partial X^2},\ \ \ \Theta = \Theta_c\ \ \text{at } X = 0 \text{ and } X = 1;\\
&\Psi = 0,\ \ \ \Omega = -\frac{\partial^2\Psi}{\partial Z^2},\ \ \ \frac{\partial\Theta}{\partial Z} = 0\ \ \text{at } Z = 0 \text{ and } Z = 1;\\
&\Theta = 1\ \ \text{at heat source surface.}
\end{aligned}\right\} \tag{16}$$

In the case of 3D cavity we have the following initial and boundary conditions [18, 19]:

$$\begin{aligned}
&\tau = 0:\ \Psi_x = \Psi_y = \Psi_z = 0,\ \Omega_x = \Omega_y = \Omega_z = 0,\ \Theta = -\eta;\\
&\tau > 0:\ \Psi_x = \Psi_y = \Psi_z = 0,\ \ \ \mathbf{\Omega} = -\nabla^2\mathbf{\Psi}\ \ \text{at melting front;}\\
&\frac{\partial\Psi_x}{\partial X} = \Psi_y = \Psi_z = 0,\ \mathbf{\Omega} = -\nabla^2\mathbf{\Psi}, \Theta = \Theta_c\ \ \text{at } X = 0 \text{ and } X = 1;\\
&\Psi_x = \frac{\partial\Psi_y}{\partial Y} = \Psi_z = 0,\ \mathbf{\Omega} = -\nabla^2\mathbf{\Psi},\ \frac{\partial\Theta}{\partial Y} = 0 \text{ at } Y = 0 \text{ and } Y = 1;\\
&\Psi_x = \Psi_y = \frac{\partial\Psi_z}{\partial Z} = 0,\ \mathbf{\Omega} = -\nabla^2\mathbf{\Psi},\ \frac{\partial\Theta}{\partial Z} = 0 \text{ at } Z = 0 \text{ and } Z = 1;\\
&\Theta = 1\ \ \text{at heat source surface.}
\end{aligned} \tag{17}$$

Governing equations with corresponding initial and boundary conditions for 2D cavity Eqs. (6)–(8) and (16) and for 3D cavity Eqs. (9)–(15) and (17) have been solved by finite difference method of the second order accuracy [16–22]. The diffusive terms have been approximated by central differences. The convective terms have been discretized by the second order Samarsky monotonic difference scheme. The parabolic equations have been solved on the basis of Samarsky locally one-dimensional scheme. The obtained systems of algebraic equations have been solved by Thomas algorithm. The partial differential equations for the stream function or vector potential functions have been discretized by means of the five/seven-points difference scheme on the basis of central differences for the partial derivatives. The obtained linear discretized equations have been solved by the successive over relaxation method. It should be noted that some difficulties have been met in definition of vorticity vector components at the phase line. This definition has been performed using the additional nodes and more complicated approximation of the Poisson equations for vector potential functions.

The comprehensive verification has been carried out including grid independence test and some natural convection problems. Detailed description of this validation part can be found in [16–20].

In the case of 3D grid independency test, we analyzed the position of solid-liquid interface at the middle cross-section $Y = 0.5$ for $Ra = 3.2\cdot10^6$ with three time moments such as $\tau = 520.2$, $\tau = 867$, $\tau = 1387.2$ and using two different grids of 60×60×60 points (solid lines) and 100×100×100 points (dashed lines). Figure 2 shows a location of solid-liquid interface for the considered grids.

Taking into account the non-significant changes in the obtained locations, a uniform grid of 60×60×60 points has been used for analysis.

Figure 3 shows a good comparison for the position of phase line with experimental data of Gau and Viskanta [23] in the case of natural convection melting of pure gallium in a rectangular cavity. The dimension of the cavity used in the experiments is 6.35 cm and 8.89 cm. The left hot wall and the right cold wall are maintained at temperatures 38.35°C and 28.3°C, respectively, and the cavity was initially filled by solid PCM at temperature 28.3°C. The horizontal walls are insulated. The physical properties of pure gallium were chosen at 32°C, a temperature that is representative of the temperature range of the experiment. One can find in Figure 3 a location of phase line at different time moments. At the beginning of the melting process (2 min) the main heat transfer mechanism is a heat conduction, therefore the phase line is parallel to the vertical isothermal wall. Further, convection develops with a formation of convective cell in the upper part of the cavity, where phase line bends. Comparing the location of the calculated phase line with experimental data it is possible to conclude that main differences are for initial time level (for some points at 6 min the relative discrepancy is about 20%). While for more intensive melting this discrepancy is low (at 17 min the relative discrepancy is less than 10%). It should be noted that the main reason for such differences is an anisotropic gallium used in the experiments.

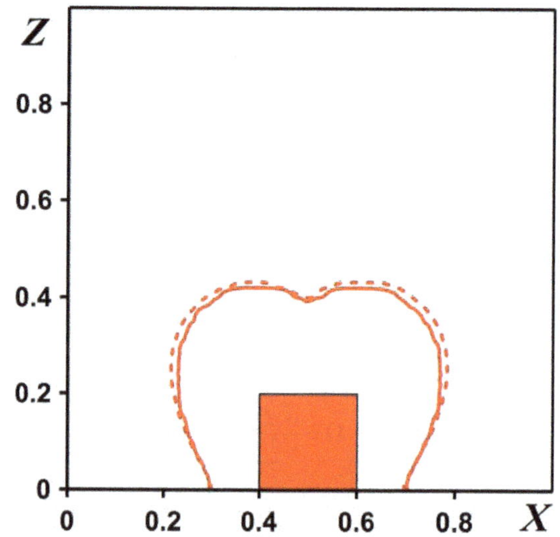

Figure 2. Locations of phase line at middle cross-section $Y = 0.5$ for $Ra = 3.2{\cdot}10^6$ and $\tau = 520.2$ for two grids of $60{\times}60{\times}60$ points (solid lines) and $100{\times}100{\times}100$ points (dashed lines)

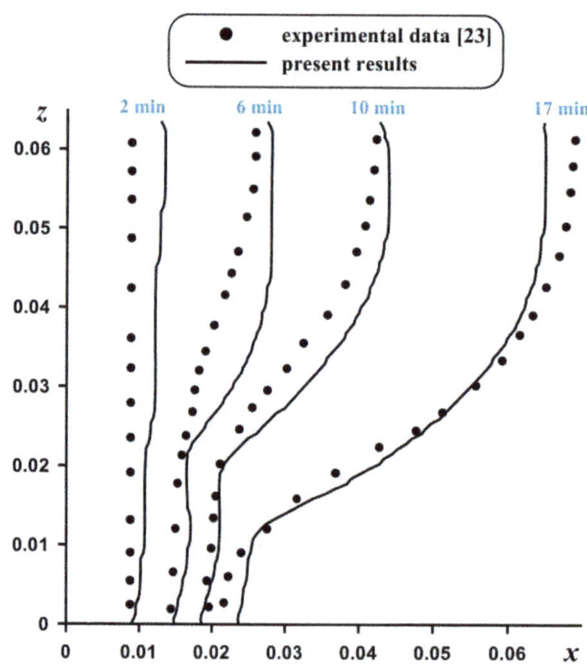

Figure 3. Comparison of phase line position

RESULTS AND DISCUSSION

Numerical investigation of the boundary-value problems has been carried out at the following values of dimensionless complexes: Rayleigh number ($Ra = 5{\cdot}10^4$, $5{\cdot}10^7$), Prandtl number ($Pr = 48.36$), Stefan number ($Ste = 5.53$). Particular efforts have been focused on the effects of Rayleigh number and dimensionless time on the fluid flow, heat and mass transfer characteristics for 2D and 3D cavities.

Figure 4 demonstrates an evolution of streamlines and isotherms in 2D cavity and in the middle cross-section $Y = 0.5$ of 3D cavity for $Ra = 5{\cdot}10^4$. It should be noted that, small value of buoyancy force does not lead to differences between 2D and 3D formulations. One can find only some differences in a location of isotherm $\Theta = -0.2$ near the upper wall. At the same time an increase in time leads to a diminution of this difference.

Regardless of the dimensionless time value, two convective cells are formed inside the melt zone, illustrating an ascending flow in central part and descending flows near the phase transition line. An increase in time leads to a growth of the melting zone and a development of convective thermal plume over the heater. More essential melting occurs in vertical direction in comparison with horizontal direction.

Figure 5 demonstrates distributions of streamlines and isotherms for 2D and 3D cases at $Ra = 5{\cdot}10^7$, $\tau = 219$. It is an initial time level for high Rayleigh number value. One can find a presence of some differences in melting flow structure and as a result in temperature profiles. Main differences are in convective cells over the heater due to different heat fluxes for 2D and 3D surfaces.

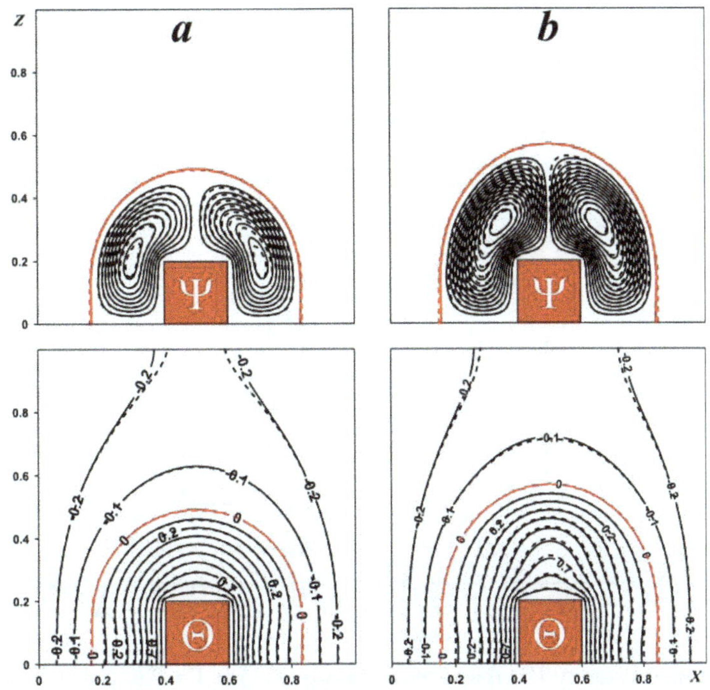

Figure 4. Isolines of Ψ and Θ for $Ra = 5{\cdot}10^4$: $\tau = 344.8$ – *a*, $\tau = 517.2$ – *b* (solid lines – 2D and dashed lines – 3D)

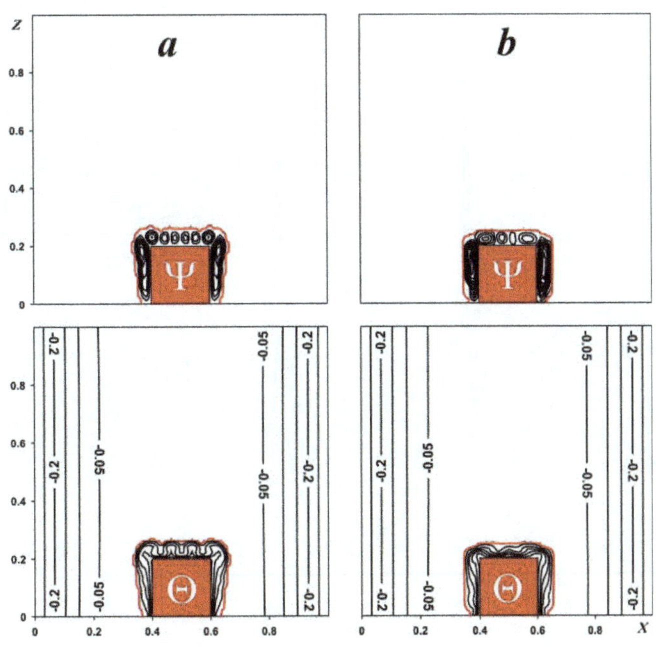

Figure 5. Isolines of Ψ and Θ for $Ra = 5{\cdot}10^7$, $\tau = 219$: 2D – *a*, 3D – *b*

More detailed comparison between 2D and 3D results can be performed on the basis of vertical velocity and temperature profiles presented in Figure 6 for $Ra = 4{\cdot}10^5$ and different values of the dimensionless time. First of all it is possible to analyze an evolution of the melting process in the following form. Heating from the local heat source reflects a formation of ascending convective flow in central part and two descending flows near the phase line. Motion velocity in these flows increases with time and melting distributes also in horizontal direction taking into account the zero velocity values. At initial time, melt circulation in 2D case is more essential in comparison with 3D case, but this time these differences vanish. A displacement of the central ascending flow in 3D case from the symmetry line can be explained by a formation of the transient 3D regime with unstable thermal plume.

Temperature profiles (see Figure 6*b*) also reflect more essential heating in 2D case at initial time, while further on temperature in cubical cavity is greater then temperature in a square cavity.

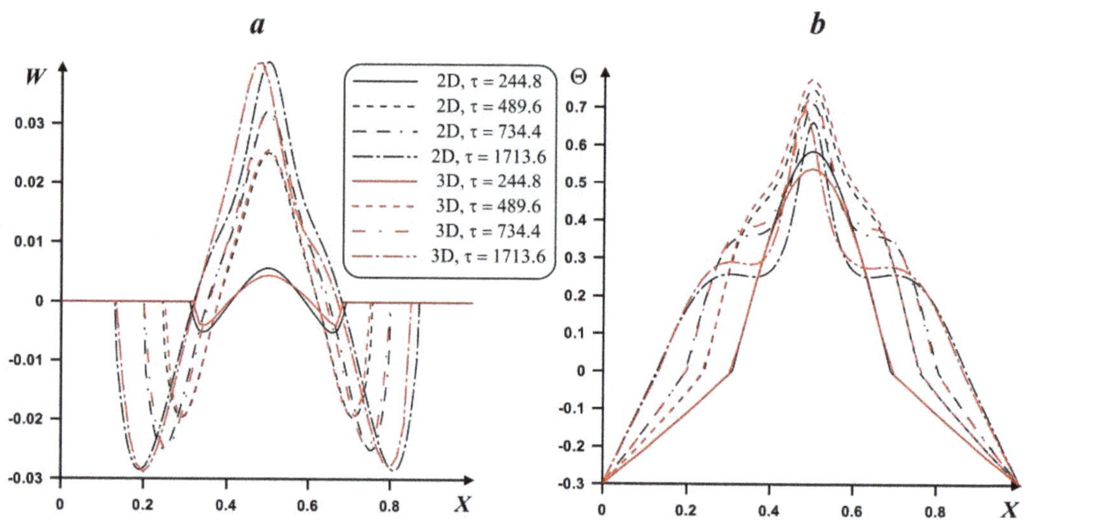

Figure 6. Profiles of vertical velocity W (*a*) and temperature Θ (*b*) at $Y = 0.5$, $Z = 0.3$ for 2D and 3D cavities at $Ra = 4{\cdot}10^5$ and different values of the dimensionless time

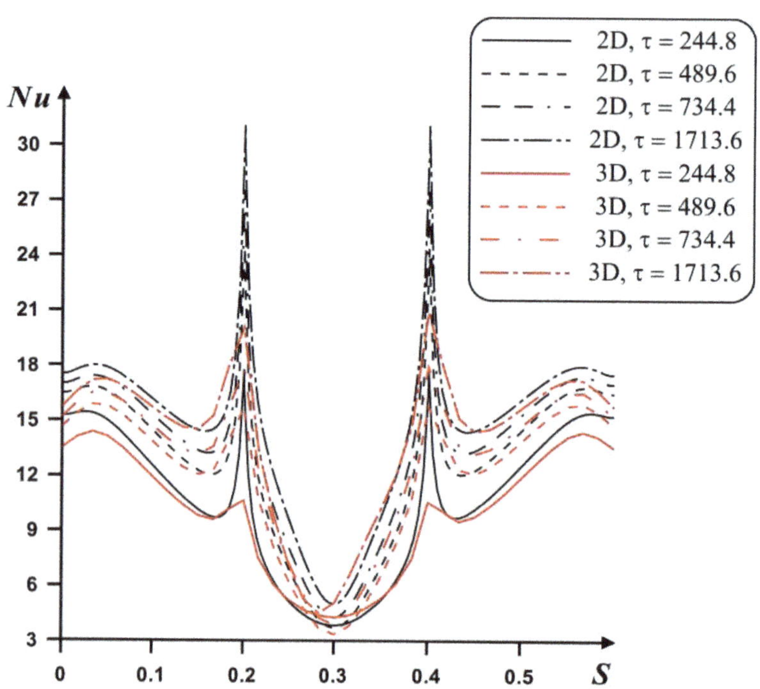

Figure 7. Evolution of the local Nusselt number along the heat source edge for $Ra = 4{\cdot}10^5$

Evolution of the local Nusselt number at $Ra = 4 \cdot 10^5$ for 2D and 3D cases is shown in Figure 7. Behavior of *Nu* along the heater surface is similar for 2D and 3D cases. Along the vertical surfaces the local Nusselt number decreases from the bottom part till upper surface due to a formation of expanding thermal boundary layer near these surfaces. Along the upper heater wall one can find a maximum *Nu* values near the edges of the upper surface and a minimum value in the central part. The main reason for such behavior is a formation of a thermal plume in the central part where the temperature gradient decreases. The local Nusselt number increases with time due to more essential heating. 2D case illustrates great values of *Nu* in comparison with 3D case.

CONCLUDING REMARKS

Natural convection melting of paraffin in a cubical cavity and square cavity with a local heater along the bottom wall has been analyzed numerically. Governing equations formulated in dimensionless "vector potential functions - vorticity vector" for 3D case and "stream function – vorticity" for 2D case have been solved by finite difference method of the second order accuracy. Comparison between 2D and 3D data illustrates an efficiency of the developed mathematical model and numerical technique. The presented approach to numerical simulation of 2D and 3D natural convection melting problem can be used for analysis of effective cooling system for electronic devices and new versions of the thermal energy storages. Comparison between 2D and 3D cases has shown that at initial time level melting in 2D occurs intensively, while with time this behavior changes. Also local Nusselt number along the heater surface is greater for 2D case in comparison with 3D data.

ACKNOWLEDGMENTS

This work was conducted as a government task of the Ministry of Education and Science of the Russian Federation (Project Number 13.9724.2017/8.9).

NOMECLATURE

c	specific heat
g	gravitational acceleration
h	enthalpy
k	thermal conductivity
L	length of the cavity
L_m	latent heat
Nu	local Nusselt number
p	dimensional pressure
Pr	Prandtl number
Ra	Rayleigh number
Ste	Stephan number
T	dimensional temperature
t	dimensional time
T_c	low constant temperature
T_h	heat source temperature
T_m	melting temperature
u, v, w	dimensional velocity components
U, V, W	dimensionless velocity components
$\mathbf{V}$	velocity vector
x, y, z	dimensional Cartesian coordinates
X, Y, Z	dimensionless Cartesian coordinates
β	melt thermal expansion coefficient
η	part of the smoothing function domain
Θ	dimensionless temperature
μ	dynamic viscosity
$\xi(\varphi)$, $\zeta(\varphi)$	additional functions
ρ	density
τ	dimensionless time
φ	smoothing function
Ψ_x, Ψ_y, Ψ_z	dimensionless vector potential functions
$\Omega_x, \Omega_y, \Omega_z$	dimensionless vorticity vector components

c	cold
h	hot
l	liquid
s	solid material

REFERENCES

[1] Raoux, S., Wuttig, M. (2009). Phase change materials: Science and Applications. Berlin: Springer.

[2] Fan, L. W., Xiao, Y. Q., Zeng, Y., Fang, X., Wang, X., Xu, X., ... & Cen, K. F. (2013). Effects of melting temperature and the presence of internal fins on the performance of a phase change material (PCM)-based heat sink. International Journal of Thermal Sciences, 70, 114-126.

[3] Gharbi, S., Harmand, S., Jabrallah, S. B. (2015). Experimental comparison between different configurations of PCM based heat sinks for cooling electronic components. Applied Thermal Engineering, 87, 454-462.

[4] Sharifi, N., Bergman, T. L., Faghri, A. (2011). Enhancement of PCM melting in enclosures with horizontally-finned internal surfaces. International journal of heat and mass transfer, 54(19-20), 4182-4192.

[5] Gharbi, S., Harmand, S., Jabrallah, S. B. (2017). Experimental study of the cooling performance of phase change material with discrete heat sources – Continuous and intermittent regimes, Applied Thermal Engineering, 111, 103–111.

[6] Numerical investigation on the natural convection effects in the melting process of PCM in a finned container using lattice Boltzmann method

[7] Prieto, M. M., Suarez, I., Gonzalez, B. (2017). Analysis of the thermal performance of flat plate PCM heat exchangers for heating systems, Applied Thermal Engineering, 116, 11–23.

[8] Madruga, S., Mendoza, C. (2017). Heat transfer performance and melting dynamic of a phase change material subjected to thermocapillary effects. International Journal of Heat and Mass Transfer, 109, 501-510.

[9] Wang, Z., Zhang, H., Xia, X. (2017). Experimental investigation on the thermal behavior of cylindrical battery with composite paraffin and fin structure, International Journal of Heat Mass Transfer, 109, 958–970.

[10] Ren, Q., Chan, C. L. (2016). GPU accelerated numerical study of PCM melting process in an enclosure with internal fins using lattice Boltzmann method, International Journal of Heat Mass Transfer, 100, 522–535.

[11] Wang, P., Yao, H., Lan, Z., Peng, Z., Huang, Y., Ding, Y. (2016). Numerical investigation of PCM melting process in sleeve tube with internal fins, Energy Conversion and Management, 110, 428–435.

[12] Tabassum, T., Hasan, M., Begum, L. (2017). 2-D numerical investigation of melting of an impure PCM in the arbitrary-shaped annuli. International Journal of Thermal Sciences, 114, 296-319.

[13] Korti, A. I. N. (2016). Numerical simulation on the effect of latent heat thermal energy storage unit, Journal of Thermal Engineering, 2, 598–606.

[14] Kladisios, P., Stegou-Sagia, A. (2016).Using phase change materials in photovoltaic systems for cell temperature reduction: A finite difference simulation approach, Journal of Thermal Engineering, 2(4), 897–906.

[15] Yang, X. H., Tan, S. C., Ding, Y. J., Wang, L., Liu, J., Zhou, Y. X. (2017). Experimental and numerical investigation of low melting point metal based PCM heat sink with internal fins. International Communications in Heat and Mass Transfer, 87, 118-124.

[16] Bondareva, N. S., Sheremet, M. A. (2016). Mathematical simulation of melting inside a square cavity with a local heat source, Thermophysics and Aeromechanics, 23, 553–565.

[17] Bondareva, N. S., Sheremet, M. A. (2016). Effect of inclined magnetic field on natural convection melting in a square cavity with a local heat source. Journal of Magnetism and Magnetic Materials, 419, 476-484.

[18] Bondareva, N. S., Sheremet, M. A. (2017). 3D natural convection melting in a cubical cavity with a heat source. International Journal of Thermal Sciences, 115, 43-53.

[19] Bondareva, N. S., Sheremet, M. A. (2017). Natural convection heat transfer combined with melting process in a cubical cavity under the effects of uniform inclined magnetic field and local heat source. International Journal of Heat and Mass Transfer, 108, 1057-1067.

[20] Martyushev, S. G., Sheremet, M. A. (2014). Conjugate natural convection combined with surface thermal radiation in a three-dimensional enclosure with a heat source. International Journal of Heat and Mass Transfer, 73, 340-353.

[21] Shenoy, A., Sheremet, M., Pop, I. (2016). Convective flow and heat transfer from wavy surfaces: viscous fluids, porous media and nanofluids. Boca Raton: CRC Press.

[22] Öztop, H., Selimefendigil, F., Abu-Nada, E., Al-Salem, K. (2016). Recent developments of computational methods on natural convection in curvilinear shaped enclosures, Journal of Thermal Engineering, 2(2), 693–698.

[23] Gau, C., Viskanta, R. (1986). Melting and solidification of a pure metal on a vertical wall. Journal of Heat Transfer, 108(1), 174-181.

EXERGETIC EFFICIENCY ANALYSIS OF A COMBINED POWER PLANT OF A CONTAINER SHIP

Y. Durmusoglu[1,*] , G.Kocak[1]

ABSTRACT

Energy problems in the world require more efficient use of energy. Thus, in recent years there has been a significant increase in scientific studies on energy efficiency. Approximately 80% of world trade is carried by sea. Therefore, transportation of such a big volume requires large-scale energy consumption. For this reason, energy efficiency is very important in the maritime sector. The International Maritime Organization (IMO) which is the authority on maritime transport in the world executes obligations and recommendations for energy efficiency and environmental pollution issues. An important method of scientifically measuring energy efficiency is exergy analysis. It is observed that the performance analysis by exergy method is not applied to the shipping sector sufficiently, so this paper is tried to fill this gap in maritime area. In this article, the performance of the combined power system of a container ship is analyzed by the exergy method. According to the obtained results, the total exergy efficiency of the system is observed to be 49%.

Keywords: *Shipping, Energy Efficiency, Exergy, Combined Power System*

INTRODUCTION

Ships are floating power plants. They produce energy and exploit this energy mainly for powering the propulsion systems during their port-to-port voyages. Moreover, ships require electricity for maneuvering, hotels and operation of all electrical and electronic equipment. A main engine and small size generators that are using fossil fuels supply all energy requirements of a ship. Besides there are also some systems producing electricity from turbo generator powered using the waste heat and/or a shaft generator. There are various auxiliary machinery on board ship powered by electrical energy. Taking into consideration all these facts, cost of the energy production on ships may reach high levels annually. For this reason, usage of energy efficiently on ships should be considered seriously. The International Maritime Organization (IMO), which is the authority on maritime transport in the world, executes obligations and recommendations for energy efficiency and environmental pollution issues. The energy efficiency requirements were adopted as amendments to MARPOL Annex VI in 2011 and entered into force on 2013 [1]. Therefore, the shipping companies should obey the rules and regulations declared by IMO.

One of the popular methods for evaluating energy efficiency is exergy analysis which is based on second law of thermodynamics. The most prominent feature of the exergy analysis is its capability to analyze individual components of the system separately. Thus, it is possible to monitor the location and level of the irreversibility of the components. Combined power systems are one of the methods used to increase energy efficiency. These systems consist of two different power cycles, the primary and secondary cycles. On board ships, these combinations are often confronted as Diesel and Rankine power cycles. In our study a turbo-charged (T/C) diesel engine combined with a turbo generator (T/G) is inspected which is explained elaborately in next section.

Feng et al. (1998) defined a new parameter to evaluate the efficiency of a cogeneration power plant which is called cogeneration efficiency. In this study the authors also carried out an exergy efficiency and make comparison with their parameter. They show that the compared results are very similar [2]. Franco and Russo (2002), studied exergetic efficiency of a combined cycle (CC) plant which consists of a Brayton cycle and a Rankine cycle. They could increase the overall plant efficiency to 60% which is mentioned as the goal of the sector by authors [3]. Yilmaz, T. (2004) made an optimization of waste heat recovery system and carried out the exergetic performance analysis based on alternative performance criteria [4].

Danov and Gupta (2004), developed computational tools in their studies to investigate the mutual operation between a turbocharged diesel engine and a heat recovery steam generator (HRSG) in combined- cycle power plants [5-6]. Tien et al. (2005) studied a waste heat recovery for generating electricity through a steam turbine on a ship. An iterative mathematical model is proposed utilizing heat transfer laws. They show that the proposed mathematical model is comparable with the experimental results [7]. Ebadi and Bandpy (2005) applied an exergy analysis method to a 116 MW gas turbine plant. They evaluated the effect of change in the turbine inlet

This paper was recommended for publication in revised form by Regional Editor Mustafa Kılıç
[1]Istanbul Technical University, Department of Marine Engineering, 34940, Tuzla, Istanbul, Turkey
**E-mail: ydurmusoglu@itu.edu.tr*

temperature on the exergetic efficiency and exergy destruction in the plant and confirmed that inlet temperature is crucial for exergy efficiency and exergy destruction [8]. Kamate and Gangavati (2008), analyzed a 20.70 MW cogeneration plant using the exergetic, thermal and fuel saving methods. They achieved 69% energy utilization factor and 28% exergetic efficiency [9]. Abusoglu and Kanoglu (2009a, 2009b) studied exergy analysis of a diesel engine powered cogeneration system (DEPC). The authors applied their exergy methodology on a reel plant which has a total installed electricity capacity of 25.3MW, installed in Gaziantep, Turkey. They concluded that the exergy efficiency of the plant is determined to be 40.6% [10,11].

In the report of MAN Diesel and Turbo (MAN, 2012), it is shown that the possibility to generate an electrical output from the main engine power by utilising exhaust gas energy in a waste heat recovery system is up to 11% [12]. Saidur et al. (2012) made a review of studies for waste heat recovery from the internal combustion engines (ICE). In this study mainly thermoelectric generators (TEG), organic rankine cycle (ORC), six-stroke cycle IC engines and developments of turbocharger (T/C) technologies are inspected. [13]. Shu et al. (2013) studied a review of waste heat recovery on two-stroke IC engine aboard ships. In their study, they considered basic principles, novel methods, existing designs, theoretical and experimental analyses, economics and feasibility [14]. Grljušić et. al. (2014), inspected a combined heat and power (CHP) production of a Suezmax-size oil tanker that utilizes the low-temperature waste heat energy for all heating and electricity requirements during navigation. They analyzed various configurations for waste heat recovery and observed that a supercritical organic Rankine cycle (ORC) performs maximum efficiency [15]. Modern large diesel engines are about 50% efficient in utilizing the fuel heat energy and the remainder is lost to the environment as waste heat [16]. Baldi and Gabrielli predicted, based on exergy analysis, an achievable fuel savings of 4–16% for a medium range tanker by the use of watse heat recovery system (WHRS) [17].

M. H. Seyyedvalilu and et al. (2015) carried out an exergy and exergoeconomic analysis and parametric study of a diesel engine based combined heat and power (CHP) system. In the results, they showed that how influences the heating power, exergetic efficiency, and exergy destruction cost and exergoeconomic factor of the CHP system in all environment temperatures with the cycle design parameters [18]. Benvenuto et al. (2014) studied a system to increase the efficiency of a waste heat recovery system and compared it with a traditional waste heat recovery system which consists of a steam turbine (ST) and exhaust gas turbine (EGT). In this study, the authors designed a system which utilizes heat energy of scavenging air and jacket cooling water to increase feed water temperature. They could increase the efficiency of overall system from 56% to 57.5% [19]. Marty et. al. (2016), analyzed a ship power plant on a simulation. They carried out exergy analysis and proposed some solutions to improve efficiency of the system. They mainly propose to replace fresh water generator with a reverse osmoses system and utilize waste heat in an ORC system [20]. Baldi et. al.(2015a and 2015b) studied exergy analysis of a chemical tanker and a cruise ship. They find out that the exergetic value of the exhaust gasses represent the 18% of the engine power output. Therefore, it is a significant point for energy recovery [21, 22]. A WHRS can utilize the remaining wasted heat for producing mechanical/-electrical power which can then feed the demand for propulsion and auxiliary services at no additional fuel costs and zero associated CO_2 emissions [23].

There are many studies of exergy analysis method which are generally applied for shore power plants. However, for marine power plants there are very few studies which are mentioned above. In this study, thermodynamic performance analysis of a 4200 TEU container ship combined power system is carried out based on exergetic approach which is not studied before. The data used for the analysis is obtained from a realistic engine room simulator (ERS) of Kongsberg maritime company. This data is used in the mathematical model of the power system which is modelled in Engineering Equation Solver (EES). By this study, the main power generation of a container ship is analyzed and it is possible to evaluate the exergy efficiency of each component to carry out necessary improvements.

MATERIAL AND METHODS

Ship Combined Systems

In marine power systems, the energy efficiency increase from waste heat can be realized in various ways. Most confronted methods are cogeneration and combined power systems. The combined power system utilizes supplementary waste heat recovery sources for preheating of the feed water that will increase the obtainable steam. Therefore, superheated steam can be produced which is used for electric power production of a marine power plant [16]. Today's power systems overall efficiency can be increased up to 55% in combined power systems which use waste heat recovery. A sample combined marine power system is shown in Figure 1. In this system, a diesel engine is used for primary power production. The waste heat of diesel engine's exhaust gas is recovered by exhaust gas boiler and exhaust gas turbine. Besides, a steam turbine is powered by the superheated steam from the boiler which drives a turbo generator (T/G). Even more the T/G is supported by an exhaust gas turbine.

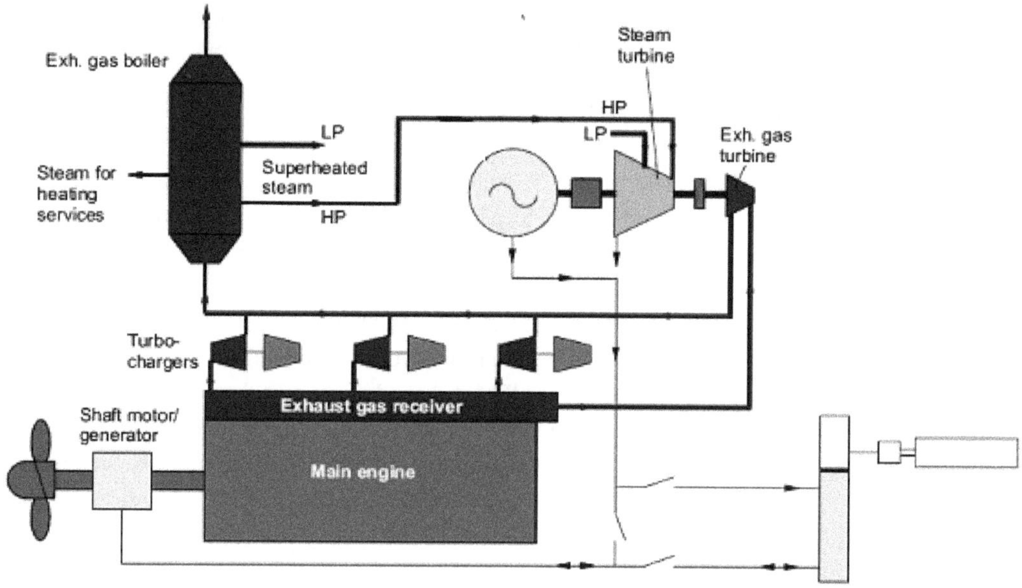

Figure 1. A ship energy efficiency application by MAN B&W in 2007 [16].

Exergy Analysis

Exergy is defined as the maximum useful work that can be obtained from a reversible thermodynamic system while the system comes into equilibrium with its environment. The environmental condition is called as dead state condition in which pressure is P_0 and temperature is T_0. Exergy is a significant tool for measurement of energy efficiency due to possibility of component-based analysis of the system efficiency and irreversibility. The loss of exergy, or irreversibility, provides a generally applicable quantitative measure for inefficiencies [24].

Exergy balances of the power plant should be determined in order to perform the exergy analysis. The exergy balance equation is formed as:

$$\sum \dot{E}_Q - \sum \dot{E}_{\dot{W}} + \sum \dot{E}_{in} - \sum \dot{E}_{out} = \dot{E}_D \tag{1}$$

where $\dot{E}_D$ is the exergy destruction. The exergy destruction is the irreversible energy called also as anergy. The total exergy ($\dot{E}$) of a system can be shown as follows:

$$\dot{E} = \dot{m} \cdot e \tag{2}$$

where $\dot{m}$ is the mass flow rate in kg/s and e is the specific exergy in kJ/kg. The exergies of heat and work are expressed as follows,

$$\dot{E}_{\dot{Q}} = \dot{Q}\left(1 - \frac{T_0}{T}\right) \tag{3}$$

$$\dot{E}_{\dot{W}} = \dot{W} \tag{4}$$

where T is the temperature of heat source in K and W is the total work done by the system in kW. Substituting Eq.(2-4) in Eq.(1), we obtain:

$$\sum \dot{Q}\left(1-\frac{T_0}{T}\right)-\dot{W}+\sum_i \dot{m}_i e_i-\sum_o \dot{m}_o e_o=\dot{E}_D \tag{5}$$

where e_i is input specific exergy and e_o is output specific exergy. The specific exergy (e) consists of physical exergy (e^{PH}), chemical exergy (e^{CH}), potential exergy (e^{PE}) and kinetic exergy (e^{KE}). As a general approach, the kinetic and potential exergy are negligible [24]. The physical exergy and the chemical exergy are defined as follows, respectively:

$$e^{PH}=\left(h-h_0\right)-T_0\left(s-s_0\right) \tag{6}$$

$$e^{CH}=\sum_{i=1}^{n} X_i e_i^{CH}+RT_0\sum_{i=1}^{n} X_i \ln(X_i) \tag{7}$$

where X_i is the moles of ith gases and e_i^{CH} is the standard chemical exergy of gases and gas mixtures and selected from a table of standard chemical exergies [25]. Therefore, the specific exergy can be expressed as,

$$e=e^{PH}+e^{CH} \quad . \tag{8}$$

Eq.(8) is not sufficient for calculation of specific exergy of the fuel. Thus, the specific exergy of the fuel is calculated by:

$$e_f=LHV\cdot\phi_i \tag{9}$$

where ϕ_i is a parameter which depends on different fuel types (i=solid, fluid, gas) and LHV is the lower heat value of fuel in kJ/kg. The ϕ_i is calculated for solid, fluid and gas fuel types by the following experimental equations proposed by Bejan (1996) [26], respectively.

$$\phi_s=1.0437+0.1882\frac{H}{C}+0.0610\frac{O}{C}+0.0404\frac{N}{C} \tag{10}$$

$$\phi_f=1.0401+0.1728\frac{H}{C}+0.0432\frac{O}{C}+0.2169\frac{S}{C}\left(1-2.0628\frac{H}{C}\right) \tag{11}$$

$$\phi_g=1.033+0.0169\frac{H}{C}-\frac{0.0698}{C} \tag{12}$$

To analyze the exergetic efficiency of a thermodynamic system it is required to identify the terms product ($\dot{E}_P$) and fuel ($\dot{E}_F$). The product is the exergy that contains the benefits obtained while the fuel is the exergy provided for the subsystem through the resources [27]. In the combined power system analyzed in this study, the exergy of propulsion power and electrical power of T/G is product and the exergy of heavy fuel oil is the fuel. Therefore, the exergetic efficiencies of subsystems are calculated by:

$$\varepsilon=\frac{\dot{E}_P}{\dot{E}_F} \tag{13}$$

Eventually, the overall exergetic efficiency of the combined power plant is determined as:

$$\varepsilon_{os} = \frac{\dot{E}_P}{\dot{E}_F} = \frac{\dot{W}_{ME} + \dot{W}_{TG}}{\dot{m}_f LHV} \tag{14}$$

where, m_f is the mass flow rate of the fuel. There are two exergy destruction ratios which are useful for comparing various components of the same system. These are the rate of exergy destruction ratios (y_D) and (y_D^*). First exergy destruction ratio (y_D) is comparison of the exergy destruction in a system component to the fuel provided to the overall system which is formulized as:

$$y_D = \frac{\dot{E}_D}{\dot{E}_{F,total}} \tag{15}$$

The latter is (y_D^*) the comparison of the exergy destruction in a system component to the total exergy destruction rate within the system which is expressed as:

$$y_D^* = \frac{\dot{E}_D}{\dot{E}_{D,total}} \tag{16}$$

Exergy Analysis of the Marine Power Plant

The ship's engine room is a very complex power system. There are various types of marine power plants. Some ships may have shaft generator system whereas another ship may have turbo generator etc. The plant analyzed in this study is a combined power plant of a container ship with a diesel engine as primary power system and a turbo generator as a secondary power system. The specifications of the container ship are listed in Table 1.

Table 1. Specifications of the container ship

Length overall	295	m
TEU	4200	-
Breadth	32	m
Draught	12.6	m
DWT	55000	tons
Service speed	25	knots

The data used for the analysis is obtained from a realistic engine room simulator (ERS) of Kongsberg maritime company while the plant is running at the condition of full speed ahead and full loaded. Then, analyzes is performed by building the mathematical model of the power system with Engineering Equation Solver (EES). The specifications of the marine power plant are listed in Table 2.

Table 2. Specifications of the power plant.

Type	Sulzer 12RTA 84 C	-
Bore	84	cm
Stroke	240	cm
Cylinders	12	-
Engine speed	102	rpm (full load)
Power	42.85	MW (full load)
sfoc	168	g/kW.h (full load)
FOC	7.7	ton/day

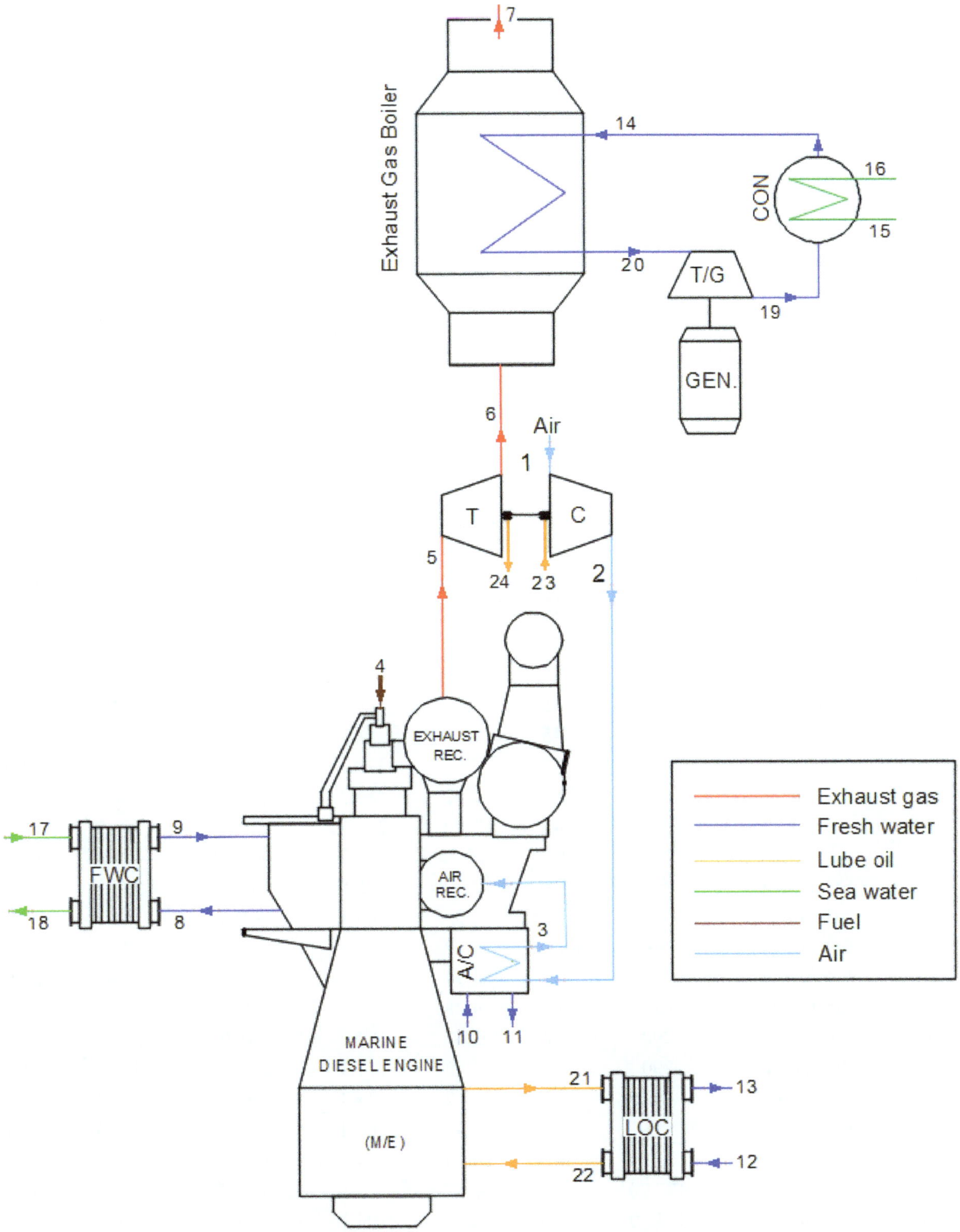

Figure 2. Combined marine power plant and data points.

The data is obtained from 24 state points for exergetic analysis of the combined marine power plant which are shown in Fig.2. This system uses a diesel engine as a primary power system for propulsion of the ship. There are two components for waste heat recovery in the system. These are waste heat boiler (WHR) and three sets of turbo chargers (T/C). The superheated steam produced in the boiler is feeding the turbo generator for electrical power generation. Besides there are auxiliary systems such as lubricating oil cooler (LOC), air cooler (A/C), and fresh water cooler (FWC). The fresh water circulated in FWC is cooled by seawater in a central cooler. Generally, intermediate fuel oil (IFO) or heavy fuel oil (HFO) is used for the main engine of marine power plants. In this system, heavy fuel oil is used as fuel for the main diesel engine. In the calculations, the lower heat value of the HFO is taken as 41.80 MJ/kg [28] . The T/C is cooled by lubricating oil which is also used for bearing lubrication.

The sensor values observed from 24 data points are listed in Table 3. Besides the calculated results of specific exergy (*e*) and total exergy (*E*) are shown in the same table. The seawater is the main cooling environment for marine power plants. Therefore, the temperature and salinity are important parameters for calculating the enthalpy, entropy and specific exergy. The salinity and temperature of the seawater is assumed to be 0.3% and 20°C, respectively. The specific heat of the lube oil under constant pressure ($c_{p,LO}$) which is required for enthalpy, entropy and exergy calculations is calculated from the following equation (Brendel, 1988) [29]:

$$c_{p,LO} = 4.19\left(\frac{0.402+0.00081T}{\sqrt{d_{15}}}\right)\left(\frac{kJ}{kg.K}\right) \tag{17}$$

where T is temperature of lube oil (°*C*), d_{15} is the density of lube oil at 15°C (g/cm^3).
The component based exergy efficiency analysis of the system is carried out based on Fuel and Product exergy principle. The plant consists of nine components which are, exhaust gas turbine (T), air compressor (C), air cooler (AC), main diesel engine (ME), fresh water cooler (FWC), condenser (CON), lube oil cooler (LOC), turbo generator (TG) and

Table 3. Exergy calculations of marine power plant with respect to state points in Fig.2

State point	Fluid type	m (kg/s)	T (°C)	P (bar)	h (kJ/kg)	s (kJ/kg.K)	e (kJ/kg)	E (MW)
0	Air	-	30	1	303.5	5.716	0	-
0	Seawater	-	20	1	99.8	0.35	0	-
0	Fresh water	-	45	1	125.1	0.434	0	-
0	Fuel oil	-	45	1	-	-	0	-
0	Lube oil	-	45	1	-	-	0	-
1	Air	87	45	1	318.7	5.765	0	0.00
2	Air	87	181.2	2.92	456.3	6.817	100.9	8.78
3	Air	87	41	2.92	314.7	5.444	98.89	8.60
4	Fuel oil	2.16	132	10.75	148.7	0.6449	42924	92.72
5	Exhaust gas	88	403	2.59	415.5	6.76	279.8	24.62
6	Exhaust gas	88	272.5	2.39	303.1	6.384	196.5	17.29
7	Exhaust gas	88	163.85	1.02	185.15	7.105	146.9	12.93
8	Fresh water	113.78	85	1,94	356.2	1.134	21.49	2.45
9	Fresh water	113.78	70.1	4.28	293.8	0.9559	13.03	1.48
10	Fresh water	290	32	3.67	134.4	0.4639	2.794	0.81
11	Fresh water	290	42.3	2.43	177.3	0.6028	3.658	1.06
12	Fresh water	76	32	3.67	134.4	0.4639	2.794	0.21
13	Fresh water	76	44	2.43	184.4	0.6253	3.953	0.30
14	Fresh water	2.32	25	0.05	104.8	0.367	2.579	0.01
15	Seawater	177.70	20.00	2.57	77.81	0.2952	0.5	0.09
16	Seawater	177.70	26.68	2.57	104.1	0.3892	0.433	0.08
17	Seawater	383	20	3.57	77.81	0.2952	0.5	0.19
18	Seawater	383	34.4	3.57	134.5	0.4951	0.972	0.37
19	Saturated steam	2.32	31.44	0.05	2294	7.522	21.28	0.05
20	Superheated steam	2.32	265.85	8.42	2989	7.079	847.4	1.97
21	Lube oil	200.55	67	6.98	67	1.454	3.78	0.76
22	Lube oil	200.55	45	5.06	27.2	0.734	0.65	0.13
23	Lube oil	10.25	70	4.38	72.4	1.534	4.39	0.05
24	Lube oil	10.25	74.5	4.38	80.5	1.646	5.39	0.06

waste heat boiler for heat recovery (WHR). The exergy equations of each component of combined marine power plant are shown in Table 4.

Table 4. Fuel and Product exergy of components

Components	E_F (MW)	E_P (MW)
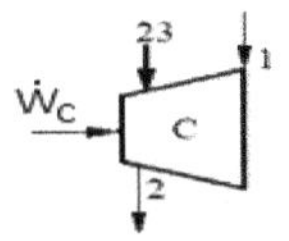	W_C	$(E_2\text{-}E_1)\text{-}E_{23}$
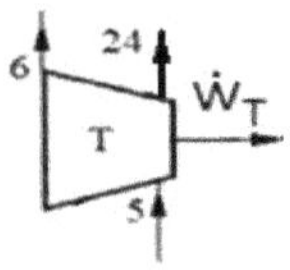	$(E_5\text{-}E_6)\text{-}E_{24}$	W_T
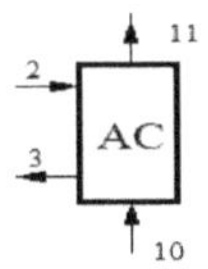	$(E_2\text{-}E_3)$	$(E_{11}\text{-}E_{10})$
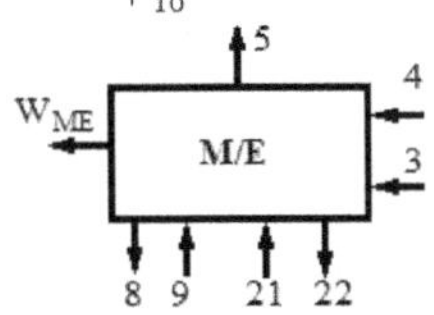	$(E_3+E_4+E_9+E_{21})$	$(E_5+E_8+E_{22}+W_{ME})$
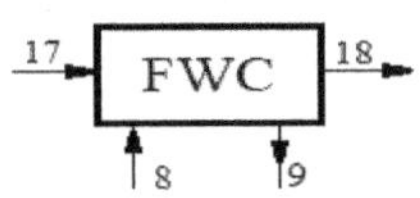	$(E_8\text{-}E_9)$	$(E_{18}\text{-}E_{17})$
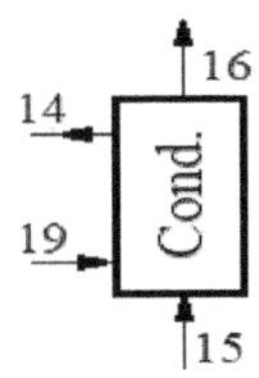	$(E_{19}\text{-}E_{14})$	$(E_{15}\text{-}E_{16})$
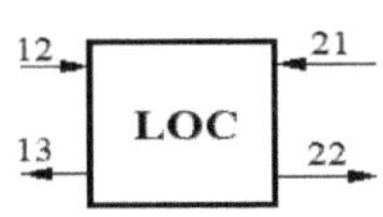	$(E_{21}\text{-}E_{22})$	$(E_{13}\text{-}E_{12})$
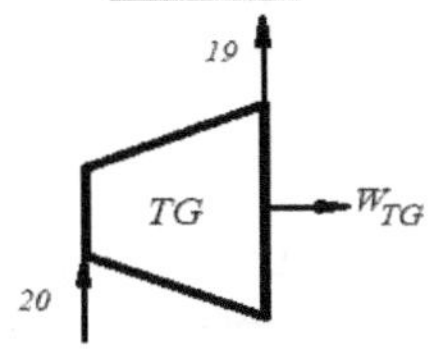	$(E_{20}\text{-}E_{19})$	(W_{TG})
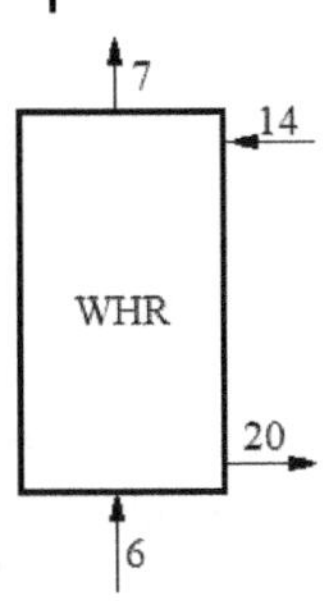	$(E_6\text{-}E_7)$	$(E_{20}\text{-}E_{14})$

RESULT and DISCUSSION

The component based exergy calculations are carried out using the equations shown in Table 4. The results of these calculations are listed in Table 5. The exergy destruction ratios (y_D and y_D^*) and overall system exergy efficiency (ε) can be seen from this table which is calculated as the ratio which is Ep divided by E_F.

Table 5. Component based exergy outputs of marine power plant

Components	E_F (kW)	E_P (kW)	E_D (kW)	y_D (%)	y_D^* (%)	ε (%)
C	11971	8730	3241	3.50	7.45	72.9
T	7270	9891	2621	2.83	6.02	136.1
AC	250	180	70	0.08	0.16	72.0
ME	103560	70050	33510	36.14	77.02	67.6
FWC	970	180	790	0.85	1.82	18.6
CON	40	10	30	0.03	0.07	25.0
LOC	630	90	540	0.58	1.24	14.3
TG	1920	1612	308	0.33	0.71	84.0
WHR	4360	1960	2400	2.59	5.52	45.0
Overall system	90288	44462	45826	46.93	100.00	49.20

In the results, it is observed that ME has the greatest fuel exergy value with 103560 kW. However, it has also the greatest exergy destruction value which is 33510 kW (Figure 3) due to heat loses through exhaust gases, jacket cooling water, and lube oil cooling. As a result, the exergy efficiency of the ME is 67.6% which is shown in Figure 4. The exergy destruction ratio (y_D) value also verify this result.

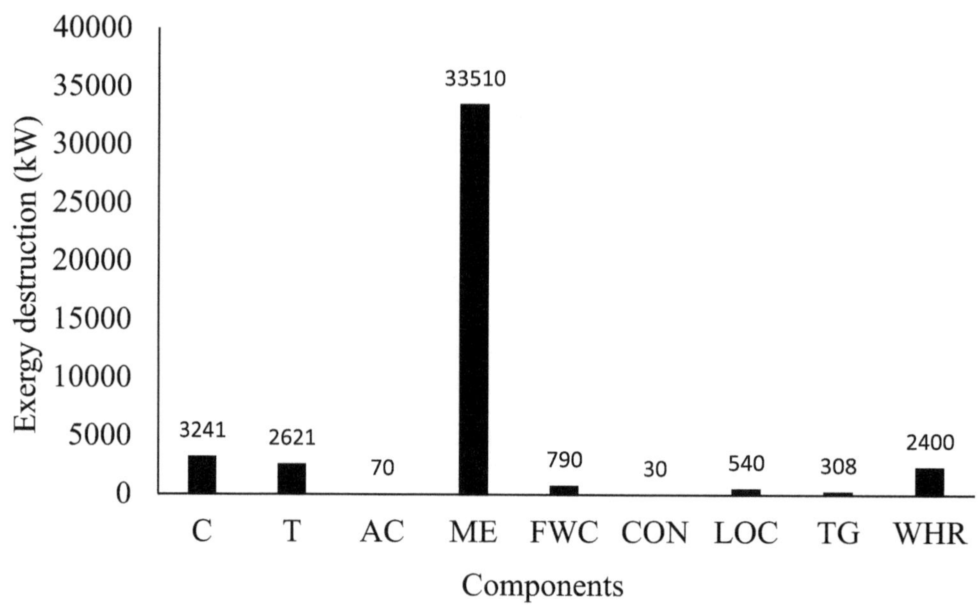

Figure 3. Exergy destructions of components in the combined marine power plant

In the components, TG has the highest exergy efficiency with 84.1%. The highest exergy values after TG are Turbine and Compressor of T/C which are 73.6% and 73.0%, respectively. The exhaust gas has higher heat energy at T/C. However, it is observed that TG has greater exergy efficiency value than T/C. The main reason of this situation is difference of utilization method of heat energy. T/C uses the velocity of exhaust gases whereas TG uses the heat energy of exhaust gases through WHR. That's why the exergy efficiency of TG is higher. The main reasons for high exergy efficiency values of TG and T/C comparing to other components are high input energy, type of input energy (waste heat), and no external work input to these components. In the components such as

FWC, CON, and LOC the energy transfer rate is very low due to low temperatures. That is why the exergy efficiency rates are comparatively low. In combined power systems, the significant indicator is overall system exergy efficiency. In this study, the overall system exergy efficiency is reached to 49.2% which is an acceptable result comparing to studies in the literature.

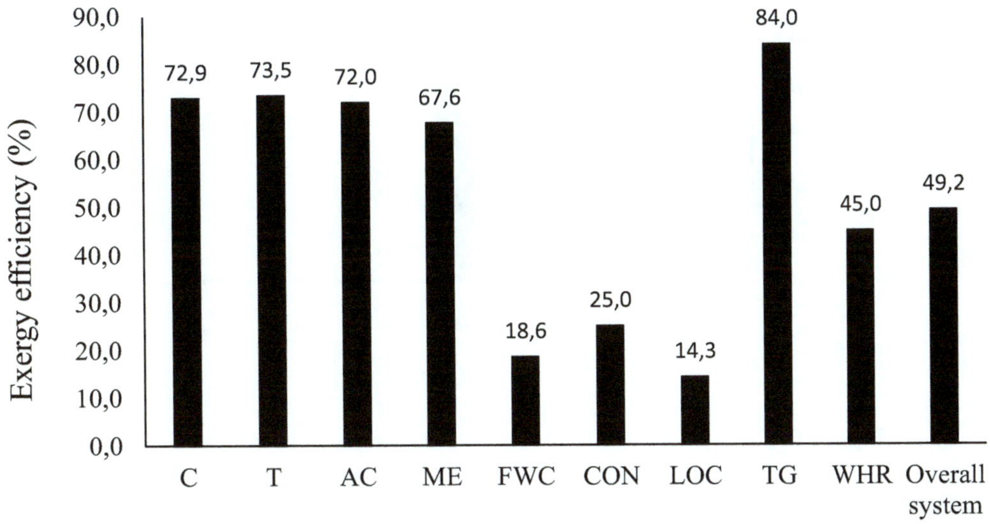

Figure 4. Exergy efficiency of components in the combined marine power plant.

In Figure 5, the exergy values of fuel and product of components and overall system is displayed. Here, ME has the highest fuel and product exergies due to fuel input to the ME and high power output which are 103560 kW and 70050 kW, respectively. ME is the main component determining the fuel and product exergy of the overall system which can be seen in the figure.

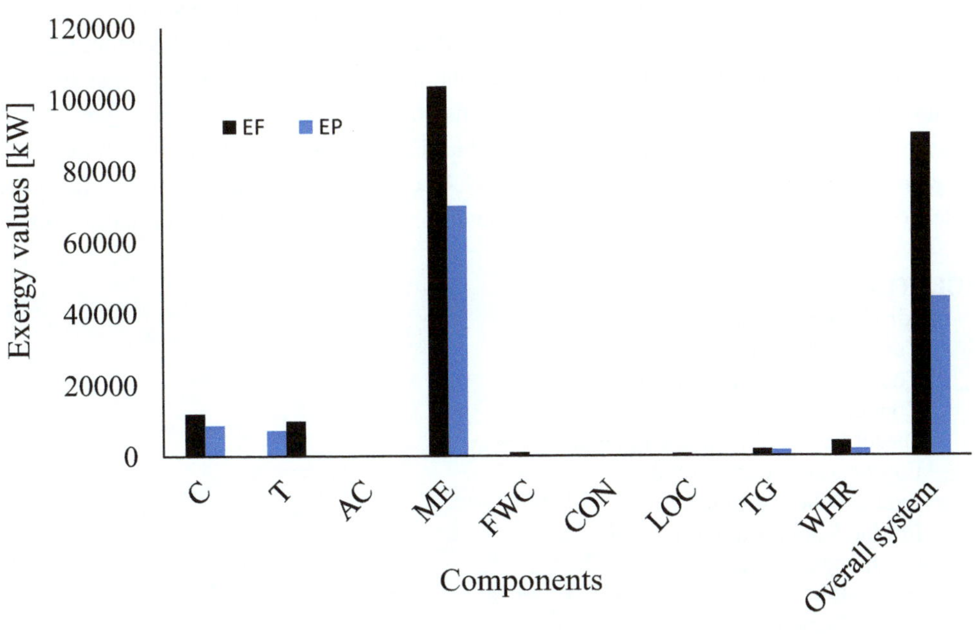

Figure 5. Exergy of fuel and product of components in the combined marine power plant.

In Figure 6, exergy destruction ratios are shown. The ME has highest exergy destruction values as fuel and product exergy values shown before. The main power input to the system is carried out from the engine with high exhaus temperatures and high energy transfers. In ME, the loss energy through various systems such as lubricating oil, jacket cooling water, mechanical friction loses is also higher comparing to other systems. That is why the exergy destruction ratio is higher in ME.

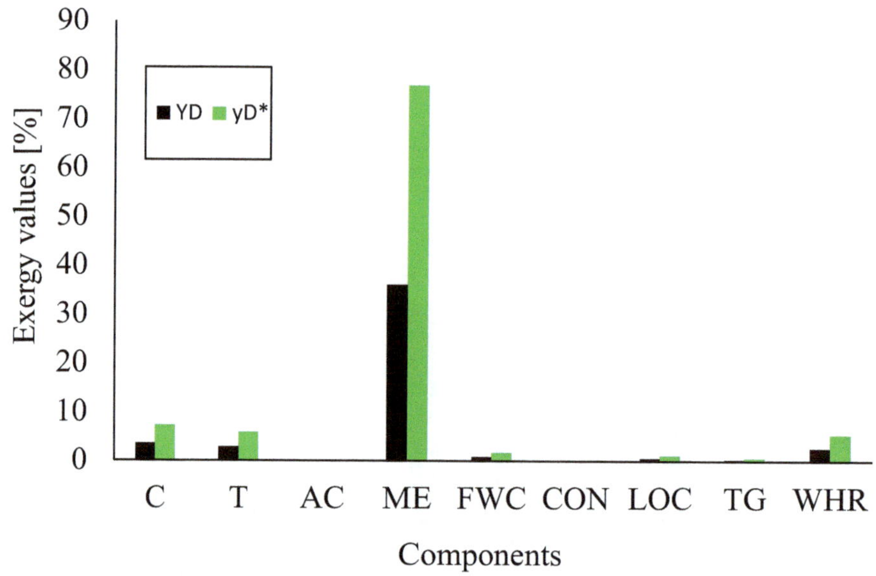

Figure 6. Exergy destruction ratios of components in the combined marine power plant.

CONCLUSION

In this study, thermodynamic performance analysis of a combined power system of a container ship is carried out by exergy method. The data's used in the calculations are obtained from a realistic full mission engine room simulator. It is assumed that the ship is fully loaded and at full ahead condition. In the analysis, the values of the exergy destruction and exergy values of each component and the overall system are calculated and the results are given as a table. As a result of the analysis, it is found that the largest exergy losses are through the exhaust gases and then in jacket cooling and lubricating oil cooling components, respectively. As a result, it is observed that the application of industrial energy recycling methods to ships increases the exergy efficiency up to 49%. This value is acceptable comparing to similar power plants exergy efficiency values [30]. Exergy efficiency can be further increased by the reduction of thermal losses and the integration of novel technologies. For example, the heat energy released to surrounding by jacket water can be recovered. One of the most studied technologies in recent years is Organic Rankine Cycle (ORC). It seems possible to increase the exergy efficiency by integrating an ORC system to waste heat recovery system of a ship. Furthermore, an absorption chiller can also be installed to decrease lost energy and increase exergy efficiency. These studies are planned as future studies. As a result, it is possible to say that exergy method is a useful in determining the energy losses in marine vessels.

NOMECLATURE

A/C	Air Cooler
CC	Combined Cycle
CHP	Combined Heat And Power
CON	Condenser
DEPC	Diesel Engine Powered Cogeneration System
EES	Engineering Equation Solver
EGT	Exhaust Gas Turbine
ERS	Engine Room Simulator
FWC	Fresh water cooler
GEN	Generator
HFO	Heavy Fuel Oil
HRSG	Heat Recovery Steam Generator
IC	Internal Combustion
ICE	Internal Combustion Engines
IFO	Intermediate Fuel Oil
IMO	International Maritime Organization
LOC	Lubricating Oil Cooler
M/E	Main Engine
MARPOL	International Convention for the Prevention of Pollution from Ships

ORC	Organic Rankine Cycle
ST	Steam Turbine
T/C	Turbocharger
T/G	Turbo-generator
TEG	Thermoelectric Generators

REFERENCES

[1] "IMO", Resolution MEPC.203(62) Amendments to the annex of the protocol of 1997 to amend the international convention, 2011.

[2] Feng, X, Cai, Y.N., Qian, L.L. (1998). A new performance criterion for cogeneration system. Energy Conversion and Management, 39 (15), 1607-1998.

[3] Franco, A., Russo, A. (2002). Combined cycle plant efficiency increase based on the optimization of the heat recovery steam generator operating parameters, International Journal of Thermal Sciences, 41, 843–859.

[4] Yılmaz T, (2004). Optimization of cogeneration systems under alternative performance criteria, Energy Conversion and Management, 45 (6), 939–945.

[5] Danov, S. N., Gupta, A. K. (2004). Modeling the performance characteristics of diesel engine based combined-cycle power plants-Part I: Mathematical Model. Journal of Engineering for Gas Turbine and Power, 126, 28-34.

[6] Danov, S. N., Gupta, A. K. (2004). Modeling the performance characteristics of diesel engine based combined-cycle power plants-Part II: Results and Applications. Journal of Engineering for Gas Turbine and Power, 126, 35-39.

[7] Tien, W.K., Yeh, R.H., Hong, J.M. (2007). Theoretical analysis of cogeneration system for ships. Energy Conversion and Management, 48, 1965-1974.

[8] Ebadi, M. J., Bandpy, M. G. (2005). Exergetic analysis of gas turbine plants. International Journal of Exergy, 2 (1), 31-39.

[9] Kamate, S. C., Gangavati, P. B. (2008). Exergetic, thermal, and fuel savings analyses of a 20.70 MW bagasse-based cogeneration plant. Cogeneration&Distributed Generation Journal, 23(3), 45-58.

[10] Abusoglu, A., Kanoglu, M. (2009). Exergetic and thermoeconomic analyses of diesel engine powered cogeneration: Part 1 – Formulations, Applied Thermal Engineering 29, 234–241.

[11] Abusoglu, A., Kanoglu, M. (2009). Exergetic and thermoeconomic analyses of diesel engine powered cogeneration: Part 2 – Applications, Applied Thermal Engineering 29, 242–249.

[12] MAN Diesel & Turbo. Waste heat recovery system (WHRS) for reduction of fuel consumption, emissions and EEDI. Copenhagen, Denmark; 2012

[13] Saidur, R., Rezaei, M., Muzammil, W. K., Hassan, M. H., Paria, S., Hasanuzzaman, M. (2012). Technologies to recover exhaust heat from internal combustion engines. Renewable and Sustainable Energy Reviews, 16(8), 5649-5659.

[14] Shu, G., Liang, Y., Wei, H., Tian, H., Zhao, J., Liu, L. (2013). A review of waste heat recovery on two-stroke IC engine aboard ships. Renewable and Sustainable Energy Reviews, 19, 385-401.

[15] Grljušić, M., Medica, V., Račić, N. (2014). Thermodynamic analysis of a ship power plant operating with waste heat recovery through combined heat and power production, Energies 7,7368-7394.

[16] Man B&W Diesel A/S. 2007. Thermo efficiency system (TES) for reduction of fuel consumption and CO2 emission. MAN Diesel brochure, Denmark.

[17] Baldi F, Gabrielii, C. (2015). A feasibility analysis of waste heat recovery systems for marine applications. Energy 80:654–65.

[18] Seyyedvalilu, M. H., Mohammadkhani, F., Khalilarya, S. (2015). A parametric study on exergy and exergoeconomic analysis of a diesel engine based combined heat and power system, International Journal of Engineering (IJE), TRANSACTIONS A: Basics 28 (4), 608-617.

[19] Benvenuto G., Trucco, A., Campora, U. (2016). Optimization of waste heat recovery from the exhaust gas of marine diesel engines. Proceedings of the Institution of Mechanical Engineers, Part M: Journal of Engineering for the Maritime Environment, 230(1), 83-94.

[20] Marty, P., Hétet, J. F., Chalet, D., Corrignan, P. (2016). Exergy analysis of complex ship energy systems. Entropy, 18(4), 127.

[21] Baldi, F., Ahlgren, F., Nguyen, T-V., Gabrielii, C., Andersson, K. (2015). Energy and exergy analysis of a cruise ship. In Proceedings of ECOS 2015: 28th International Conference on Efficiency, Cost, Optimization, Simulation and Environmental Impact of Energy Systems.

[22] Baldi, F., Johnson, H., Gabrielii, C., & Andersson, K. (2015). Energy and exergy analysis of ship energy systems–the case study of a chemical tanker. International Journal of Thermodynamics, 18(2), 82-93.

[23] Singh, D. V., Pedersen, E. (2016). A review of waste heat recovery technologies for maritime applications. Energy conversion and management, 111, 315-328.

[24] Kotas T.J. The exergy method of thermal plant analysis. Great Britain: Anchor Brendon Ltd, 1985.

[25] Ameri, M., Ahmadi, P. (2007). The study of ambient temperature effects on exergy losses of a heat recovery steam generator, in: Proceedings of the 15th international conference on Mechanical Engineering, no.2053, Tehran,Iran, 2007.
[26] Bejan, A., Tsatsaronis, G., Moran, M, Thermal design and Optimization, Wiley, New York, 1996.
[27] C. A. Frangopolous, Exergy, Energy System Analysis and Optimization, Eolss Publishers Co. Ltd., Oxford-United Kingdom, 2009.
[28] http://www.globalcombustion.com/oil-fuel-properties/
[29] Brendel H. (1988). Schmierstoffe, Wissensspeicher Tribotechnik, VEB Fachbuchverlag, Leipzig.
[30] Kanoglu, M., Dincer, I. (2009). Performance assessment of cogeneration plants. energy conversion and management, 50(1), 76-81.

METHOD OF INDIRECT DETERMINATION OF THERMAL LOAD PARAMETER

Yu. Baidak[1,*], I. Vereitina[1]

ABSTRACT

An indirect method for determining the effectiveness of the transformers operating in a distribution network with variable load schedules, according to the parameter of their thermal load as a ratio of heat accumulated by transformer oil to the maximum permissible heat is proposed provided the thermal stability of the insulation. a mathematical model is developed which allows establishing analytically the forced temperature value and time invariable of heating or cooling on the section of the time dependence of the oil temperature, as well as an algorithm for calculating the amount of heat accumulated by the oil and the coefficient of heat transfer from the oil to the transformer tank is compiled. approbation of the indirect method indicates that the divergence between the transformer oil heating time invariable which was obtained by solving the differential equation of thermal balance and calculated using the part of temperature dependence of its heating for two minutes, is 3.6%, and at determination of the forced temperature value - 0.8%.

Keywords: *Transformer, Thermal Subsystem, Energy Efficiency, Thermal Load, Timing Diagrams of Temperatures*

INTRODUCTION

Transformers involved in power distribution networks are designed and manufactured according to the minimum of the additive criterion of the combined costs of production and during operation, since these components are considered by manufacturers and consumers as the most influential in the composition of power efficiency.

Unfortunately, not all technical requirements that a transformer must meet, are satisfied, as well as those partial criteria that are taken into consideration are reasonable technically and economically. As a result, material

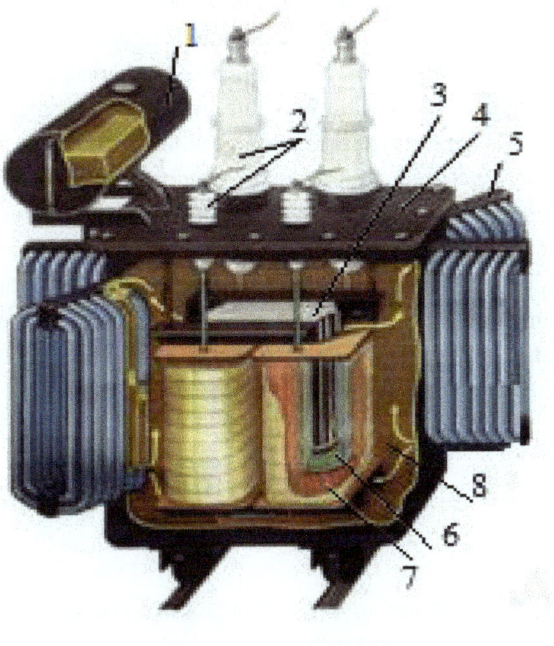

Figure 1. Single-phase transformer of distribution networks in the cross-section, in which: 1 - expander tank, 2 - insulators of windings ends, 3 - magnetic core, 4 - tank cover, 5 - natural cooling radiators, 6,7 - low and high voltage windings, 8 - transformer oil

and human costs inherent in the transformer is not used in full, including the parameter of its load.

This paper was recommended for publication in revised form by Regional Editor Jovana Radulovic
[1]Department of Thermodynamics and Renewable Energetics, Odessa national Academy of Food Tech., Odessa, UKRAINE
**E-mail address: yuribaydak@ukr.net*

With regard to the prospects for the additional design and development of existing transformers, the availability of opportunities for their economic justification, modification or development of new technical requirements, they are considered based on the actual analysis of the most important parameter of the present - cost estimation of performance. Typically, a feasibility study is performed on the existing equipment and on the basis of case-studies; although it is known that the greatest effect can be achieved at the last stage. This approach allows reconciling the functional resources of the transformer with the actual operating conditions. The modern industry has accumulated considerable experience in the development and production of transformers, which provide economic effect and economical transformation of energy, but provided they work continuously and at full capacity.

The parameters of optimization in them are: the diameter of the core of the magnetic circuit, the current density in the windings, temperature and dimensional parameters, but among them there is still no parameter that would take into account the level of load of the transformer during operation.

In the case of power outages and rolling blackouts or temporary overload of some of the transformers it becomes a significant factor.

Since the loss of electrical energy in the elements of the transformer constructions leads to their heating, and they are dependent on the load, the transformer itself can be regarded as a thermal system, and the heating temperature is an indirect sign of its load, which is adequate to determine the operational efficiency index - the thermal load.

THERMAL LOAD PARAMETER SIMULATION METHOD

Transformation of electric energy into a useful work is always accompanied by its partial loss with transformation into heat [1]. Heat transfer Q from the surface of the body is carried out simultaneously by all its types. Therefore, when calculating the coefficient of heat transfer, a generalized heat transfer coefficient (heat remove) k_T is used.

For the case of a stable mode, when all P power losses are allocated to the environment $Q = P$,, the heat remove is calculated from the Newton equation $Q = k_T F(\theta - \theta_0) = k_T F \tau_y$ or in the form of the Ohm's law $Q = \frac{\tau_y}{R_T}$ in which $R_T = \frac{1}{k_T F}$ is the thermal resistance, τ_y is the excess of the set temperature. Consequently, the main problem in installing heat transfer is to determine the heat transfer coefficient from the surfaces, which depends on many factors, and in liquids it is also determined by convective heat exchange. In the design practice the coefficients of heat transfer are determined collectively, with the use of empirical coefficients [2] [3].

Therefore, solving the problem of refining the value of the heat transfer coefficient, at any stage of its application, remains relevant [4]. It is known that the most substantial sources of heat in the transformer are: the steel magnet wire - with loss of idle speed P_0 and conductor windings materials - with losses of short circuit P_k. According to the design of the transformer, Figure 1, the magnetic core and the windings of the active part are in the cooling medium - transformer oil, which is to be the conductor of the heat flow to the tank casing, and from it into the environment. Consequently, its temperature should be initiated to establish a heat load.

The complete solution of the inhomogeneous differential equation of the thermal balance compiled for the cooling oil medium and such that is located inside the tank whose surface is in contact with the environment at a temperature θ_0 and absorbs the heat from the magnetic core and the transformer windings, is presented in [5] in the form

$$\theta_{oil} = \theta_{oil.pr}\left(1 - e^{\frac{1}{-T_{oil}}\cdot t}\right) + \theta_0 e^{\frac{1}{-T_{oil}}\cdot t} \tag{1}$$

where $\theta_{oil.pr}$ - constant or forced temperature of the oil, which it reaches during the long operation of the transformer without changing the load, that is, at $t = \infty$.

Figure 2 shows the time dependence of the oil heating temperature for (1) at $\theta_0 = 0$ (continuous line) when the transformer is switched on for a nominal load within 1 hour 30 minutes, as well as its normalized value according to Standards (interruptible line). Exceeding of the rationed value of $\theta_{St} = 60$°C adversely affects the dielectric resistance of winding insulation and is unacceptable as for its heat resistance class. The given dependence can also be obtained experimentally, according to the indications of the thermometer, which is set at the reference point of temperature measuring of the upper oil layer under the tank cover.

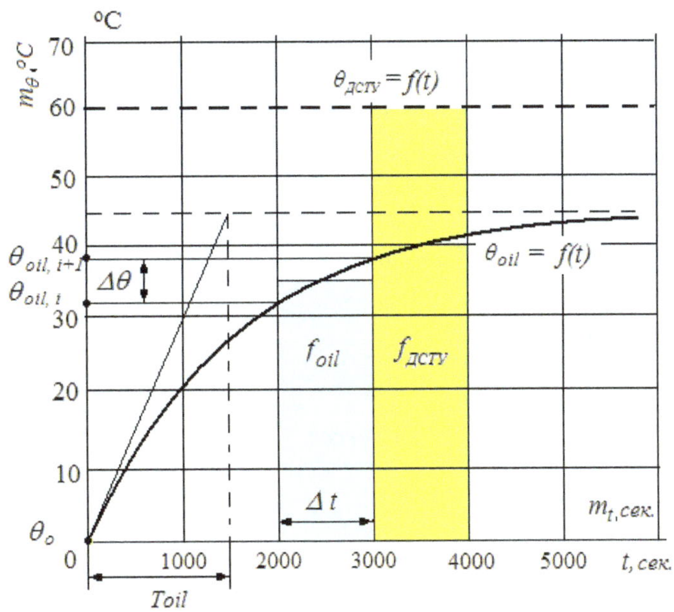

Figure 2. Time diagram of temperature for determining the equivalent of heat flow $f_{oil} m_t m_\theta$ and $f_{St} m_t m_\theta$

If, for the purpose of transformer thermal load determining, the ratio of the amount of heat absorbed by the oil by heat transfer and convection from the heat sources to its permissible value at the temperature established by the Standard will be applied, then, during the transformer operation time, even with a variable load, they are proportional to the areas of the surfaces formed in the lower part of the surface by the environment temperature θ_0, and in the upper part by the variable temperature values of the upper layer of oil θ_{oil} and, consequently, by the constant temperature according to the θ_{St}. In the general case, we have the equivalents of heat fluxes $f_{oil} m_t m_\theta$, $f_{St} m_t m_\theta$, in which m_t, m_θ are the scales on the abscissa and ordinate axes in the time charts, respectively (in h/mm) and (°C/mm); f_{oil} and f_{St}- surfaces areas (mm^2) formed by temperature and time in the coordinate system $\theta_{oil} = f(t)$. Thus, the actual amount of heat removed by the transformer oil and its permissible value will be

$$Q_{oil} = k_T F_{fr} \int_0^t (\theta_{oil} - \theta_0) e^{-\frac{1}{T_{oil}} \cdot t} dt = k_T F_{fr} \sum_{i=1}^{n} f_{oil,i} m_t m_\theta$$

(2)

$$Q_{St} = k_T F_{fr} \int_0^t (\theta_{\text{Д}St} - \theta_0) dt = k_T F_{fr} \sum_{i=1}^{n} f_{St} m_t m_\theta .$$

The construction of an algorithm for heat amount calculation according to the oil temperature monitoring data and the known time dependence of the change in its temperature during heating at the transformer load activation, requires a numerical solution of the equations (2) according to the experimentally obtained fraction of the temporal temperature diagram and the presence of known values of the working time t from the beginning of integration; the time constant of the transitional heating process $-T_{oil}$; the coefficient of heat transfer and the heat flow to the inner surface of the tank- k_T; the total surface area of tank cooling - F_{fr}

The invariable of integration can be calculated analytically by solving the equation of the thermal balance compiled for the oil in the cooling system or determined graphically by the experimentally obtained dependence $\theta_{oil} = f(t)$. Such approach in some way complicates the construction of a technical device - a tester of thermal load and, therefore, requires an analytical solution.

THE ALGORITHM OF THE THERMAL LOAD PARAMETER MODEL

The reason for thermal load parameter mathematical model creation is the presence of a section of the time dependence of the oil temperature in the cooling system, which can be obtained in any way. To determine the forced temperature of the transformer oil $\theta_{oil.forced}$ in (1) we use the indirect method. Initially, we establish the dependence of the oil temperature change on the time of the transition process progressing, that is, $\frac{\Delta\theta_{oil}}{\Delta t} = f(t)$. For example, after the differentiation of equation (1) we obtain

$$\frac{d\theta_{oil}}{dt} = -\frac{1}{T_{oil}}\left(\theta_{oil.forced}e^{-\frac{t}{T_{oil}}} - \theta_0 e^{-\frac{t}{T_{oil}}}\right) \tag{3}$$

in which, from the same expression (1),

$$\theta_{oil.forced}e^{-\frac{t}{T_{oil}}} = \theta_{oilforced} - \theta_{oil} - \theta_0 e^{-\frac{t}{T_{oil}}} \tag{4}$$

By substituting (4) into (3), we have

$$-T_{oil}\frac{d\theta_{oil}}{dt} = \theta_{oilforced} - \theta_{oil} + \theta_0 e^{-\frac{t}{T_{oil}}} - \theta_0 e^{-\frac{t}{T_{oil}}} = \theta_{oil.forced} - \theta_{oil},$$

from where the oil temperature will be determined by the linear equation

$$\theta_{oil} = \theta_{oil.forced} - T_{oil}\frac{d\theta_{oil}}{dt} \tag{5}$$

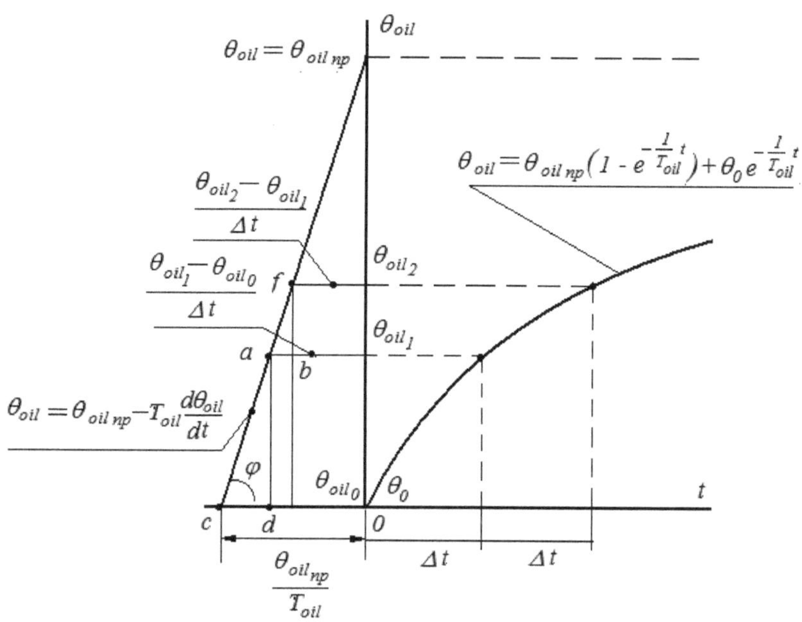

Figure 3. To the calculation of the values of the forced temperature of the oil heating and its time invariable by indirect method

Figure 3 shows the fraction of the time dependence of the oil heating temperature $\theta_{oil,i}(t)$, constructed on the basis of numerical calculations of the equation of its thermal balance and on which the linear dependence (5) was constructed on the left to determine the forced value of oil temperature and the time invariable at the load of the transformer. As follows from the drawing, the angle of inclination of the dependence (5) is invariable and can be defined from the triangle *abf* as follows:

- a section of temperature change rate after the second measurement of it

$$\overline{ab} = \frac{\theta_{oil,1}-\theta_{oil,0}}{\Delta t} - \frac{\theta_{oil,2}-\theta_{oil,1}}{\Delta t} = \frac{2\theta_{oil,1}-\theta_{oil,0}-\theta_{oil,2}}{\Delta t}; \quad (8)$$

- the angle of inclination of linear dependence (5)

$$\varphi = \tan^{-1}\frac{\theta_{oil,2}-\theta_{oil,1}}{\overline{ab}}; \quad (9)$$

- the period of temperature change after the first measurement

$$\overline{cd} = \frac{\theta_{oil,1}-\theta_{oil,0}}{\tan\varphi} = \frac{(\theta_{oil,1}-\theta_{oil,0})}{(\theta_{oil,2}-\theta_{oil,1})}\overline{ab}; \quad (10)$$

- the ratio of the forced temperature to the heating time variable at the beginning of the transformer load

$$\overline{c0} = \frac{\theta_{oil.forced}}{T_{oil}} = \frac{\theta_{oil,1}-\theta_{oil,0}}{\Delta t} + \overline{cd}; \quad (11)$$

- tangent of linear dependence (5) inclination angle from (11)

$$\tan\varphi = \frac{\theta_{oil.forced}-\theta_{oil,0}}{\frac{\theta_{oil.forced}}{T_{oil}}}, \quad (12)$$

where the forced oil temperature will be

$$\theta_{oil.forced} = \theta_{oil,0} + \frac{\theta_{oil.forced}}{T_{oil}}\tan\varphi, \quad (13)$$

and the heating time invariable from the expression (5)

$$T_{oil} = \frac{\theta_{oil.forced}-\theta_{oil}}{\frac{\Delta\theta_{oil}}{\Delta t}}. \quad (14)$$

As a parameter of thermal load (TL) it is possible to apply the ratio of the heat flow to the oil at the actual load of the transformer to its permissible value according to the Standard

$$TL = Q_{oil}/Q_{St}, \quad (15)$$

where

$$Q_{St} = k_T F_{fr}\int_0^t(\theta_{St}-\theta_0)dt = k_T F_{fr}(\theta_{St}-\theta_0)t.$$

APPROBATION OF THE HEAT LOAD DEFINITION METHOD

The calculation of the thermal load parameter must begin immediately after the transformer has been switched on to the mains and carried out at identical time intervals $\Delta t = (t_i - t_{i-1})$, Figure 2, a countdown of the temperature values at the beginning of $\theta_{oil,i=0} = \theta_0$ and at the end of each time interval $\theta_{oil,i+1}$.
According to current temperatures for each time interval $i = 1, \dots t/\Delta t$ are calculated:
- temperature average value (step approximation on time interval Δt, Figure 2)

$$\Delta\theta_{oil.mid} = 0{,}5(\theta_{oil.i} + \theta_{oil.i-1}); \tag{16}$$

- the equivalent of the heat flow amount, which oil gives to the transformer tank casing in one time interval

$$f_{\frac{oil}{St},i} m_t m_\theta = \frac{\Delta t}{3600}\Delta\theta_{oil.mid} + f_{\frac{oil}{St},i-1} m_t m_\theta, \tag{17}$$

where $f_{\frac{oil}{St},i=0} m_t m_\theta = 0$, and $i = 1,2,\dots,\frac{t}{\Delta t}$;

- using expressions (8 ... 14) the forced value of the oil temperature and the time invariable of the temperature change transition process can be numerically calculated:

$$T_{oil} = \left(\theta_{oil.forced} - \theta_{oil,i}\right)\frac{\Delta t_i}{(\theta_{oil,i} - \theta_{oil,i-1})}; \tag{18}$$

- coefficient of heat transfer from oil to the tank

$$k_T = \frac{c_i G_{oil}}{T_{oil} F_{fr}}, \tag{19}$$

where c_i is the specific heat of oil, its approximation, as a linear function, has the form of equation (20), G_{oil} - the weight of oil in the tank and radiators (is indicated on the plate of the investigated transformer)

$$c = 1{,}549 + 0{,}005938\theta_{oil,i}; \tag{20}$$

- amount of heat that is transferred by oil from the active part of the transformer to the tank casing during the time t and according to the equation (2)

$$Q_{oil} = -T_{oil} k_T F_{fr} \sum_{i=1}^{n} f_{oil,i} m_t m_\theta; \tag{21}$$

- allowable amount of heat that can be transferred from the active part of the transformer to the tank casing during the operation time t and according to the class of applied insulation

$$Q_{St} = t k_T F_{fr}(\theta_{St} - \theta_0). \tag{22}$$

The coefficient of thermal load is calculated as the ratio of the amount of heat removed from the useful volume of the tank into the environment to the allowable amount of heat over the test time by the equation (15). The higher the value of the obtained parameter, the more efficient the transformer is used.

For approbation of the indirect method of thermal load determination, a transformer of TM 1000 kVA 10 / 0.4 kV type is considered. Obtained for it in [5] numerical calculations are: forced temperature of the cooling oil $\theta_{oil.forced} = 40{,}3$ °C; time invariable $-T_{oil} = 870{,}8$ *сек*. As a result of calculations, the time chart of the transformer under consideration oil heating temperature is drawn in Figure 4. Additionally, it is known that the weight of the transformer oil according to the catalog is $G_{oil} = 700$ *кg*, and the design value of the area of the cooling surface of the tank with radiators (SB) per phase is $SB = 18\ m^2$.

In accordance with the algorithm of the transformer thermal load parameter model and based on the table values of time and temperature of the oil in Figure 4, for intervals $i = 2,3$, we have:

- the average value of temperature, with the degree approximation at the second interval of time $\Delta t = 50$ *sec*.

$$\Delta\theta_{oil.mid} = 0{,}5(4{,}372 + 2{,}249) = 3{,}288\ °C; \tag{23}$$

- temperature change rate after the second measurement

$$\overline{ab} = = \frac{2\theta_{oil,1}-\theta_{oil,0}-\theta_{oil,2}}{\Delta t} = \frac{2*2{,}249-0-4{,}372}{50} = 0{,}00252\frac{°C}{sec};\qquad(24)$$

- the angle of inclination of linear dependence

$$\varphi = \tan^{-1}\frac{\theta_{oil,2}-\theta_{oil,1}}{\overline{ab}} = \tan^{-1}\frac{4{,}372-2{,}249}{0{,}00252} = 89{,}93^{0};\qquad(25)$$

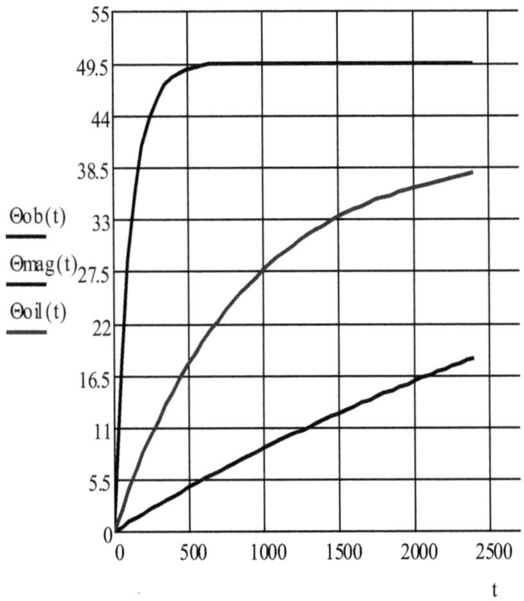

t =	Θoil (t) =
0	0
50	2.249
100	4.372
150	6.377
200	8.27
250	10.057
300	11.744
350	13.338
400	14.842
450	16.263
500	17.604
550	18.87
600	20.066
650	21.195
700	22.261
...	...

Figure 4. Time chart of heating temperatures for a period of 40 minutes after switching on the transformer on full power (windings - higher, magnetic core - lower, oil – average line)

- change in temperature after the first measurement

$$\overline{cd} = \frac{\theta_{oil,1}-\theta_{oil,0}}{\tan\varphi} = \frac{2{,}249-0}{818{,}5} = 0{,}00275;\qquad(26)$$

- the ratio of the forced temperature to the heating time invariable

$$\frac{\theta_{oil.forced}}{T_{oil}} = \frac{\theta_{oil,1}-\theta_{oil,0}}{\Delta t} + \overline{cd} = \frac{2{,}249-0}{50} + 0{,}00275 = 0{,}04773\ \frac{°C}{sec.};\qquad(27)$$

- forced oil temperature value

$$\theta_{oil.forced} = \theta_{oil,0} + \frac{\theta_{oil.forced}}{T_{oil}}\tan\varphi = 0 + 0{,}04773 * 818{,}5 = 39{,}96°C,;\qquad(28)$$

- oil heating time invariable

$$T_{oil} = \frac{\theta_{oil.forced}-\theta_{oil}}{\frac{\Delta\theta_{oil}}{\Delta t}} = \frac{39{,}96-2{,}249}{0{,}04498} = 838{,}4\ sec.;\qquad(29)$$

- specific heat of oil

$$c = 1,549 + 0,00593 * 2,249 = 1,562 \left(\frac{kW*sec}{kg°C}\right); \tag{30}$$

- coefficient of heat transfer from oil to the tank

$$k_T = \frac{c_i G_{oil}}{T_{oil} F_{fr}} = \frac{1562*700}{838,4*(18*3)} = 24,14 \left(\frac{W}{m^2 °C}\right); \tag{31}$$

- the equivalent of the amount of heat flow, which oil is transferred to the tank during time $t = 700\ sec.$

$$\sum_{i=1}^{14} \frac{f_{oil,i} m_t m_\theta}{3600} = \frac{50}{3600}(1,1245 + 3,288 + 5,3745 + 7,3235 + 9,1635 + 10,9005 + 12,541 + 14,115 + 15,5525 + 16,9335 + 18,237 + 19,468 + 20,6305 + 21,728) = 50 * 176,38 = 2,45\ (°C \cdot h), \tag{32}$$

where

$$\frac{f_{\frac{oil}{St},i=0} m_t m_\theta}{3600} = 0, \text{ and } i = 1,2, \ldots, \left(\frac{700}{50} = 14\right); \tag{33}$$

- amount of heat transferred by oil from the active part of the transformer to the tank surface for $t = 700\ sec.$ (operating time is determined by the limited table data given in Figure 4)

$$Q_{oil} = k_T F_{fr} \sum_{i=1}^{14} f_{oil,i} m_t m_\theta == 24,14 * (3 * 18) * 2,45 = 3193,7\ (W \cdot h); \tag{34}$$

- the allowable amount of heat that can be carried by the oil to the transformer tank during the operating time $t = 700\ sec.$ according to the class of applied winding insulation

$$Q_{St} = k_T F_{fr} \sum_{i=1}^{14} \frac{f_{St,i} m_t m_\theta}{3600} = 24,14 * (3 * 18) * 11,67 = 15212,5\ \ (W \cdot h); \tag{35}$$

- thermal load parameter (TL)

$$\mathrm{TL} = \frac{Q_{oil}}{Q_{St}} 100 = \frac{3193,7}{15212,5} 100 = 21\%\,. \tag{36}$$

Due to the known power losses in the phase of windings LV (low voltage) $P_{1LV} = 1736\ W$, HV (high voltage) $P_{1HV} = 2114\ W$ and in the core of the magnetic system $P_{20} = 467\ W$, and those which are placed in the basis of calculating of time dependences of the transformer temperature, given on Figure 4, and in general are

$$\Delta P = P_{1LV} + P_{1HV} + P_{20} = 4317\ W \tag{37}$$

we obtain the efficiency of the applied system of transformer cooling within one hour of its operation after switching on

$$\eta_{co} = \frac{Q_{oil}}{\Delta P \cdot t} 100 = \frac{3193,7}{4317 \cdot 1} 100 = 74\ \%. \tag{38}$$

CONCLUSIONS

1. An indirect method for determining the effectiveness of the transformers operating in a distribution network with variable load schedules, according to the parameter of their thermal load as a ratio of heat accumulated by transformer oil to the maximum permissible heat is proposed provided the thermal stability of the insulation.

2. A mathematical model is developed which allows establishing analytically the forced temperature

value and time invariable of heating or cooling on the section of the time dependence of the oil temperature, as well as an algorithm for calculating the amount of heat accumulated by the oil and the coefficient of heat transfer from the oil to the transformer tank is compiled.

3. Approbation of the indirect method indicates that the divergence between the transformer oil heating time invariable which was obtained by solving the differential equation of thermal balance and calculated using the part of temperature dependence of its heating for two minutes, is 3.6%, and at determination of the forced temperature value - 0.8%

REFERENCES

[1] Yildiran, R., Atas, S., Kahraman, M. (2016). Design and performance analysis of a district heating system utilizing waste heat of a thermal power plant. Journal of Thermal Engineering, Vol. 2, No. 5, 940-954.

[2] Lienhard, John H. IV; Lienhard, John H. V (2008). A Heat Transfer Textbook (3rd ed.). Cambridge, Massachusetts: Phlogiston Press. ISBN 978-0-9713835-3-1.

[3] Incropera, F. et al. (2012). Fundamentals of heat and mass transfer (7th ed.). Wiley. 2012. 710 p. ISBN 978-0-470-64615-1

[4] Maurya, R.S, Singh, S.(2017).Numerical investigation of isothermal flow around impingement plates in a shell and tube exchanger. Journal of Thermal Engineering, Vol. 3, No. 5. 1442-1452.

[5] Matukhno V., Baidak Yu., Tomlein P. (2016). Thermal subsystem of voltage distribution transformer. Refrigeration engineering and technology. Vol 52(6) URL: file:///C:/Users/User/Downloads/477-1152-1-PB (4).pdf

13

THERMODYNAMIC PERFORMANCE ANALYSIS OF GAS LIQUEFACTION CYCLES FOR CRYOGENIC APPLICATIONS

C. Yilmaz[1,*], T. H. Cetin[2], B. Ozturkmen[2], M. Kanoglu[2]

ABSTRACT

In this paper presents an analysis of the thermodynamic cycles the most commonly used for the liquefaction of gases in order to evaluate and compare their performance under given working conditions and system component efficiencies. The cycles considered are simple Linde-Hampson cycle, precooled Linde-Hampson cycle, Claude cycle, and Kapitza cycle. First and second law relations are investigated for each cycle and performance parameters are evaluated. Thermodynamically performances criteria are compared of cycles with respect to the each other. Cycles are model in the computer environment and analyzed with Engineering Equation Solver (EES) software program. Cycles of the liquefaction fractions, coefficient of performances and second law of efficiencies are calculated for the liquefaction of different gases. Second law efficiencies are calculated as 13.4%, 21.8%, 62.9%, and 77.2% for simple Linde-Hampson cycle, pre-cooled Linde-Hampson cycle, Claude cycle, and Kapitza cycle, respectively. Claude and Kapitza cycles give better performance but simple and precooled Linde-Hampson cycle has the advantages of the simplicity of their setup.

Keywords: *Cryogenic Cycles, Linde-Hampson, Claude, Kapitza, J-T Expansion, Thermodynamic Analysis*

INTRODUCTION

Liquefaction is a key branch of cryogenics and has wide application areas including producing commercial liquid gases and cryogenic engineering applications. The main scope of cryogenic engineering is the design, development and improvement of low temperature systems and components. In today's world, cryogenics and low-temperature refrigeration are taking on increasingly significant roles. From applications in the food industry, energy and medical technologies to transportation and the space shuttle, requirements exist for cryogenic liquids to be stored and transferred.

Cryogenics is the science that involves study of very low temperatures. Low temperatures usually obtained by using liquefied gases such as liquid nitrogen or liquid helium. Therefore, liquefaction of gases is an important area for cryogenic technologies. There are three main methods for producing refrigeration; liquid expansion, Joule-Thomson expansion and expansion engine. Liquid expansion cycles are conventional refrigeration cycles and suitable for obtaining temperatures around -70°C. Kanoglu [1] performed an exergy analysis of a cascade refrigeration system for producing LNG.

The second method is the Joule-Thomson expansion and utilizes the Joule-Thomson effect to produce low temperatures. Gas is compressed to a high pressure, cooled in the counter-current heat exchanger, and expanded through a J-T valve. In 1895, Carl von Linde and William Hampson developed a cycle for the liquefaction of air [2]. However, in 1966 Andrew Barron found that the Linde and Hampson cycle could also be used to liquefy hydrogen by incorporating a precooling process using liquid nitrogen. Similarly, the Claude cycle, invented in 1902 by Georges Claude, was developed to liquefy air but it can also be used to liquefy hydrogen. Using liquid nitrogen for precooling improves the cycle efficiency compared to a pre-cooled Linde-Hampson cycle [3].

Thomas et al. [4] presented, with the help of a commercial process simulator Aspen HYSYS, the exergy analysis of a helium liquefaction system based on modified Claude cycle. Atrey [5] presented a cycle simulation for the Collins helium liquefaction cycle with six heat exchangers and two reciprocating expanders. It highlights the concept of an optimum mass flow rate through expanders for the liquefier. Wang et al. [6] presented a thermodynamic review on cryogenic refrigeration cycles for the liquefaction process of natural gas. The main

This paper was recommended for publication in revised form by Regional Editor Tolga Taner
[1]Department of Mechanical Engineering, Afyon Kocatepe University, Afyonkarahisar, Turkey
[2]Department of Mechanical Engineering, University of Gaziantep, Gaziantep, Turkey
**E-mail address: ceyhunyilmaz@aku.edu.tr*

concept of this study is to search the thermodynamic analysis of different cycles and provide a theoretical basis for selecting a cycle in accordance with different design conditions. Mehrpooya and Ansarinasab [7] considered a single mixed refrigerant Linde and Air Products liquefaction system. A thermoeconomic model based on energy and exergy analyses and an economic model according to the total revenue requirement are developed. Minimizing the unit cost of the refrigeration effect as a product of liquefaction plant is performed using the optimum working conditions. Bisht [8] performed a thermodynamic analysis of a nitrogen liquefaction system based on Kapitza cycle which is a modified Claude cycle.

Fundamentals and thermodynamic analysis of liquefaction cycles are given in Barron [9] and Timmerhaus and Flynn [10]. Maytal [11] optimized simple Linde-Hampson cycle and determined optimum flow rates of cryocoolers sharing same size recuperator in order to maximize liquid production. Nandi and Sarangi [12] discussed performance of several cycles for hydrogen liquefaction. Hilal [13] optimized Claude cycle for helium liquefiers. Kanoglu et al. [14] performed exergy analysis and developed an expression for the minimum work requirement with the help of a Carnot refrigerator.

Rashidi et al. [15] analyzed the ejector refrigeration cycle using the artificial neural network. In this study, they described the results of the ejector refrigeration cycle using R600 as a working fluid. In this study, they are used EES software for calculating the refrigerant fluid different thermodynamic properties. In the other paper [16], they proposed the thermodynamic performances of a thermal system which combines an organic Rankine cycle and an ejector refrigeration cycle. The performance of different refrigeration working fluids is investigated using classical and finite-size thermodynamics for a case for which the power to refrigeration ratio.

Most conventional cryogenic systems operate with pure fluids, the major exception being natural gas liquefiers that use mixed refrigerant processes. Previous studies on the liquefaction systems have primarily concentrated on performance assessment of a single conventional storage system for a special work. There has also been some work on performance analyses and comparisons of the different liquefaction cycles. Hundreds of patents exist on different aspects of system processes for the liquefaction of natural gas and the composition of mixtures for Linde–Hampson and other cycles. However, it is difficult to piece together the existing information to choose an appropriate process and an optimum composition for a given application.

This study presents an analysis of the thermodynamic cycles commonly used for the liquefaction of gases in order to evaluate and compare their performance under given working conditions and system component efficiencies. The cycles considered are simple Linde-Hampson cycle, precooled Linde-Hampson cycle, Claude cycle, and Kapitza cycle. The main objectives of the study are as follows: (i) First and second laws of thermodynamic relations are developed for each cycle. (ii) Liquefaction cycles performance parameters are evaluated. (iii) Liquefaction fractions, coefficient of performances COP and second law of efficiencies η_{II} are calculated for the liquefaction of different gases. (iv) Effect of inlet gas temperature on the liquefaction fraction, net work input, and second law efficiencies are investigated.

Also, the novelty of this study is due to the configuration of liquefaction system consisting of the integrated part and heat exchangers unit as well as thermodynamic analysis based working conditions of the overall system. Thermodynamic analysis of the liquefaction system cycles are performed and simulated in EES software program and the analysis are performed using the program. Demonstrates thermodynamic analysis of the cycles that results of the cycles analyses for the gas liquefaction, including estimates of liquefied mass rates are presented. The effects of the gas inlet temperature on the liquefaction fraction, unit work requirement, and the efficiencies are investigated and evaluated with the figures. As a result of this study, it is possible to use the gas liquefaction cycles for liquid mass production and storages with the best working conditions and at optimum operating conditions. Parametric studies are performed at varying inlet gas temperatures while optimum operating conditions that minimize the unit gas liquefaction work are obtained in optimization studies.

THERMODYNAMIC PERFORMANCE ANALYSIS OF CYCLES

An important parameter to be analyzed in any gas liquefaction system is thermodynamic efficiency, since it directly affects liquefied mass fraction of the gas. It is advantageous to know effects of different parameters on the efficiency, so that optimum parameters that result in the best working conditions can be selected. There are four basic components in any gas liquefaction cycle. These are compressors, expanders, heat exchangers, and Joule-Thomson (J-T) valves. The parameters that will be considered for cycle performance analysis are (i) feed gas compressor discharge pressure, (ii) temperature difference between hot and cold flows,

(iii) lowest cooling temperature, (iv) number of the heat exchangers, and (v) number of expanders. Following assumptions are made in order to simplify the analysis. (i) Pressure drop through the cycle is negligible. (ii) Compressor's isothermal efficiency is 70%. (iii) Isentropic efficiency of the expander is 80%. (iv) Heat losses to the surroundings are negligible.

There are three important parameters for the performance of the liquefaction cycles: work required per unit mass of gas compressed $\frac{-\dot{W}}{\dot{m}}$, work required per unit mass of gas liquefied $\frac{-\dot{W}}{\dot{m}_f}$, and fraction of the gas that is liquefied $y=\frac{\dot{m}_f}{\dot{m}}$. In any liquefaction cycle, we want to minimize the work requirements and maximize the liquefied gas fraction.

Simple Linde Hampson cycle

Simple Linde-Hampson cycle has the simplest setup among all liquefaction cycles, but besides its simplicity simple Linde-Hampson cycle can liquefy most of permanent gases except hydrogen and helium. Linde-Hampson cycle consist of a compressor, a heat exchanger, J-T valve and liquid reservoir. At state 1, uncondensed portion of gas and make-up gas is mixed and compressed isothermally to high pressures at the state 2. The high pressure gas is cooled in heat exchanger by uncondensed portion of gas and at state 3, and the gas is throttled in J-T valve. The liquid portion is collected as desired product. The gas portion is send back to the cycle to cool high pressure gas, and the cycle is repeated. The schematic representation of the simple Linde-Hampson cycle and temperature entropy (*T-s*) diagram of the cycle are shown in Figs. 1 and 2.

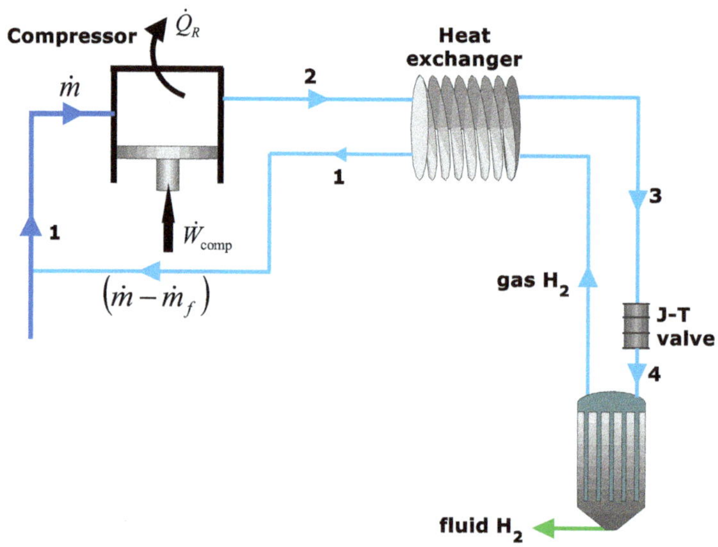

Figure 1. Schematic diagram for a simple Linde-Hampson cycle.

The refrigeration effect per unit mass of liquefaction can be defined as the heat removed from inlet of the compressor state 1 and turn into liquid state. Heat transfer can be expressed as

$$q_l = h_1 - h_6 = h_1 - h_f \quad (1)$$

where h_f is the enthalpy of saturated liquid. From energy balance of the cycle, heat transfer per unit mass of gas can be expressed as in same manner

$$q_l = h_1 - h_2 \quad (2)$$

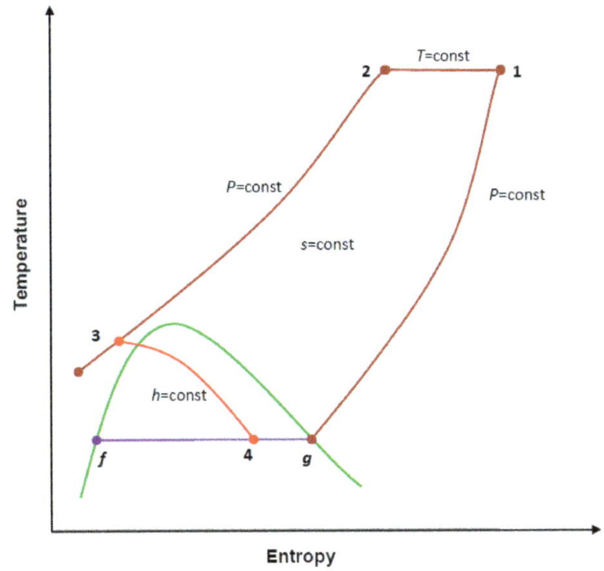

Figure 2. Temperature-entropy (T-s) diagram for a simple Linde-Hampson cycle.

The ratio of Eq. 2 to Eq. 1 is the fraction of liquefied gas in the cycle

$$y = \frac{h_1 - h_2}{h_1 - h_f} \tag{3}$$

From Eq. 3, the amount of gas liquefied in the cycle depends on the pressure and temperatures in the ambient condition. Compressor outlet pressure P_2 is an important parameter for the performance of the cycle.

The energy balance for heat exchanger can be written as

$$h_3 = h_2 - x(h_1 - h_g) \tag{4}$$

$$h_1 = h_g + \varepsilon(h_1' - h_g) \tag{5}$$

where h_g is the enthalpy of the saturated vapor and x is quality of state 4. x also can be written as

$$x = 1 - y \tag{6}$$

The energy balance for compressor can be written as

$$w_{in} = h_2 - h_1 - T_1(s_2 - s_1) \tag{7}$$

$$h_2 = h_1 - \eta(h_1 - h_2') \tag{8}$$

Joule-Thomson throttling process is isenthalpic so we can write energy balance as

$$h_3 \cong h_4 \tag{9}$$

The coefficient of performance of this cycle given by

$$\text{COP}_{\text{act}} = \frac{q_l}{w_{\text{in}}} = \frac{h_1 - h_2}{h_2 - h_1 - T_1(s_2 - s_1)} \tag{10}$$

Reversible work provides useful data for comparison between minimum work requirement for liquefaction and actual work requirement. By using second law of thermodynamics, we can develop relations between actual work and reversible work for liquefaction processes.

$$w_{\text{act}} = w_{\text{rev}} + T_0 s_{\text{gen}} \tag{11}$$

The reversible work for the simple Linde-Hampson cycle can be expressed by the exergy difference between state 1 and 6 as

$$w_{\text{rev}} = ex_6 - ex_1 = h_6 - h_1 - T_1(s_6 - s_1) \tag{12}$$

The second law of efficiency can be defined as

$$\eta_{II} = \frac{w_{\text{rev}}}{w_{\text{act}}} = \frac{h_6 - h_1 - T_1(s_6 - s_1)}{\left(\frac{1}{y}\right)[h_2 - h_1 - T_1(s_2 - s_1)]} \tag{13}$$

We assume that heat exchanger has an effectiveness of 100%. The J-T valve is isenthalpic, and it's assumed that there is no heat leakage and pressure drop in the cycle. Taking inlet state of compressor to be 25°C and 1 atm and outlet pressure to be 200 atm, we obtain various performance parameters for air, nitrogen, oxygen, argon, methane, and fluorine. They are given in Table 1.

Table 1. Performance parameters of various fluids for a simple Linde-Hampson cycle.

Gas	y	w_{in} (kJ/kg)	w_{rev} (kJ/kg)	COP_{act}	COP_{rev}	η_{II}
Air	0.0828	5485	733.2	0.07741	0.5791	0.1337
Nitrogen	0.07551	6207	762.1	0.06949	0.566	0.1228
Oxygen	0.1071	3758	629.4	0.1076	0.6426	0.1675
Fluorine	0.0763	4460	564.5	0.07702	0.6086	0.1266
Argon	0.1152	2801	492.6	0.1003	0.57	0.1759
Methane	0.1987	3892	1081	0.2338	0.8421	0.2777

Precooled Linde-Hampson cycle

The simple Linde-Hampson cycle is simple, it can liquefy many gases, but it is inefficient. The cycle requires high work input per unit liquid yield. There are several methods for improving performance of the cycle. Precooled Linde-Hampson cycle improve the performance of the simple Linde-Hampson cycle by decreasing temperature of the gas entering the main heat exchanger. The temperature lower than ambient is provided by auxiliary refrigeration cycle using a fluid such as carbon dioxide, ammonia, or a freon compound used to cool the main gas stream. The schematic representation and *T-s* diagram of the Precooled Linde-Hampson cycle is shown in Figs. 3 and 4, respectively.

Thermodynamic analysis of precooled Linde-Hampson cycle can be done in similar manner as simple Linde-Hampson cycle with some modifications. In the first heat exchanger, for 100% effectiveness, T_3 and T_6 are the same and cannot be lower than the boiling point of the auxiliary refrigerant. Refrigeration effect per unit gas can be written as

$$q_l = h_1 - h_2 + r(h_a - h_b) \tag{14}$$

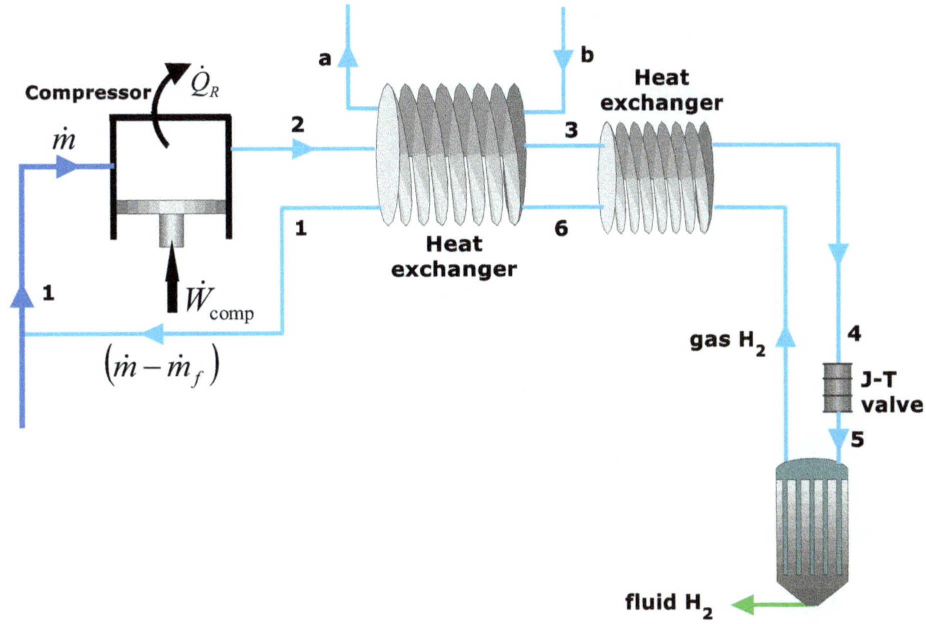

Figure 3. Schematic representation of a precooled Linde-Hampson cycle.

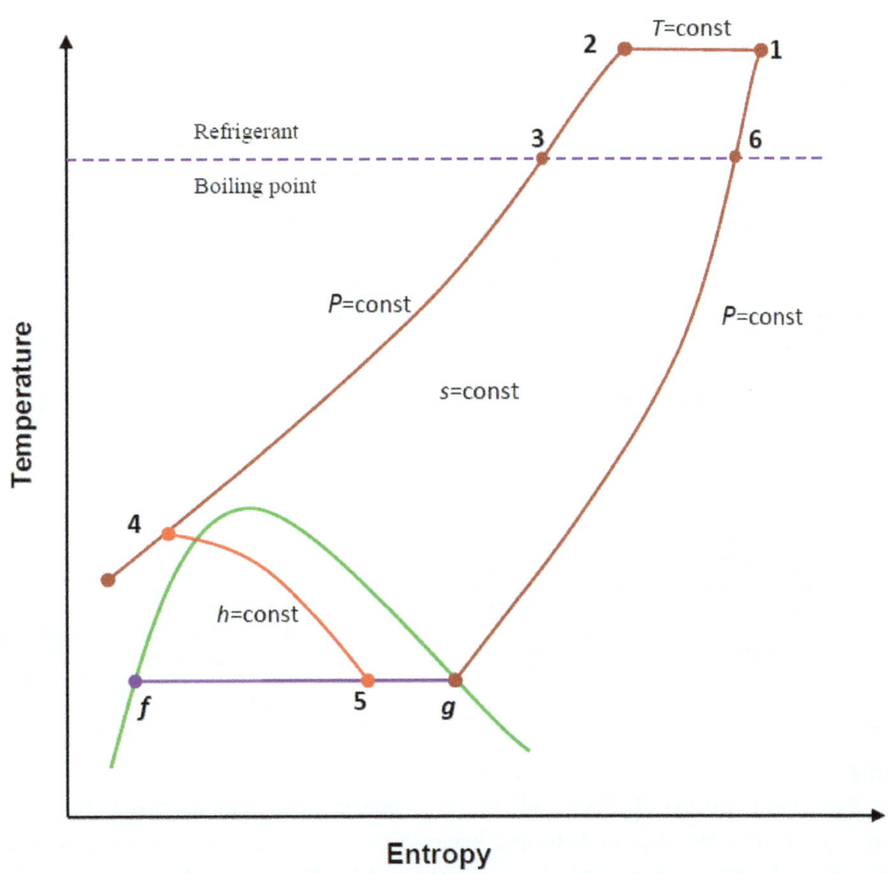

Figure 4. Temperature-entropy (T-s) diagram for a precooled Linde-Hampson cycle.

Liquid yield y for the precooled Linde-Hampson cycle can be expressed as

$$y = \frac{h_1 - h_2}{h_1 - h_f} + r\left(\frac{h_a - h_b}{h_1 - h_f}\right) \tag{15}$$

where r is refrigerant flow rate ratio $\frac{\dot{m}_r}{\dot{m}}$ and h_a and h_b are the inlet and outlet of the auxiliary refrigeration cycle's compressor. For suitable value of refrigerant flow-rate ratio, equation (15) can be modified as

$$y_{\max} = \frac{h_8 - h_3}{h_8 - h_f} \tag{16}$$

Cycle net liquefaction work input can be written as

$$w_{\mathrm{im}} = w_{\mathrm{comp}} + \mathrm{w}_{\mathrm{auxilary}} = h_2 - h_1 - T_1(s_2 - s_1) - r(h_a - h_b) \tag{17}$$

where w_{auxilary} is the auxiliary refrigeration cycle's work input.

Performance of the precooled Linde-Hampson cycle using R-134a as an auxiliary refrigerant is investigated for air, nitrogen, oxygen, fluorine, methane, and argon. The temperatures T_3 and T_8 are equal to the R-134a's boiling point (-26.1°C). Auxiliary refrigeration cycle takes freon-134a at 1 atm and compress it to 1.38 atm. Thermodynamic analysis results of the cycle are given in Table 2. Effects of various auxiliary refrigerants on the performance of the cycle are given in Table 3.

Table 2. Performance parameters of various fluids for a Precooled Linde-Hampson cycle using R-134a as an auxiliary refrigerant.

Gas	y	w_{in} (kJ/kg)	w_{rev} (kJ/kg)	$\mathrm{COP}_{\mathrm{act}}$	$\mathrm{COP}_{\mathrm{rev}}$	η_{II}
Air	0.143	3171	733.2	0,1339	0.5791	0.2312
Nitrogen	0.130	3552	762.1	0,1214	0.566	0.2144
Oxygen	0.1836	2192	629.4	0,1845	0.6426	0.2871
Fluorine	0.1363	2506	564.5	0,1371	0.6086	0.2252
Argon	0.1939	1664	492.6	0,1687	0.57	0.2959
Methane	0.3358	2304	1081	0,3949	0.8421	0.4689

Table 3. Performance parameters of Precooled Linde-Hampson cycle with different auxiliary refrigerants.

Refrigerant	y	w_{in} (kJ/kg)	$\mathrm{COP}_{\mathrm{act}}$	η_{II}
Ammonia	0.1546	4138	0.1026	0.1771
CO_2	0.2697	1934	0.2195	0.3789
R12	0.1485	3171	0.1339	0.2312
R22	0.1698	2853	0.1488	0.257
R134a	0.1430	3171	0.1339	0.2312

Claude cycle

The expansion process through J-T valve is economically suitable due to its simplicity, but from thermodynamic point of view, it is undesirable because of the irreversibilities. Thermodynamically, expansion valve can be replaced with a turbo expander, as discussed by Kanoglu [17] for LNG cycles. In Claude cycle, expansion engine is used for removing energy from gas stream and producing some work.

The schematic representation of the Claude cycle and the *T-s* diagram of Claude cycle are given in Figs. 5 and 6, respectively. In Claude cycle, gas stream is compressed to around 40 atm, and then passed through first heat exchanger at state 2. After first heat exchanger at state 3, approximately 70-80 percent of the gas diverted from mainstream and expanded through turbo expander. This portion of the gas returns the stream at the entrance of the second heat exchanger in low pressure line. Undiverted gas stream passes through second and third heat exchangers (states 4 and 5) and expand through J-T valve.

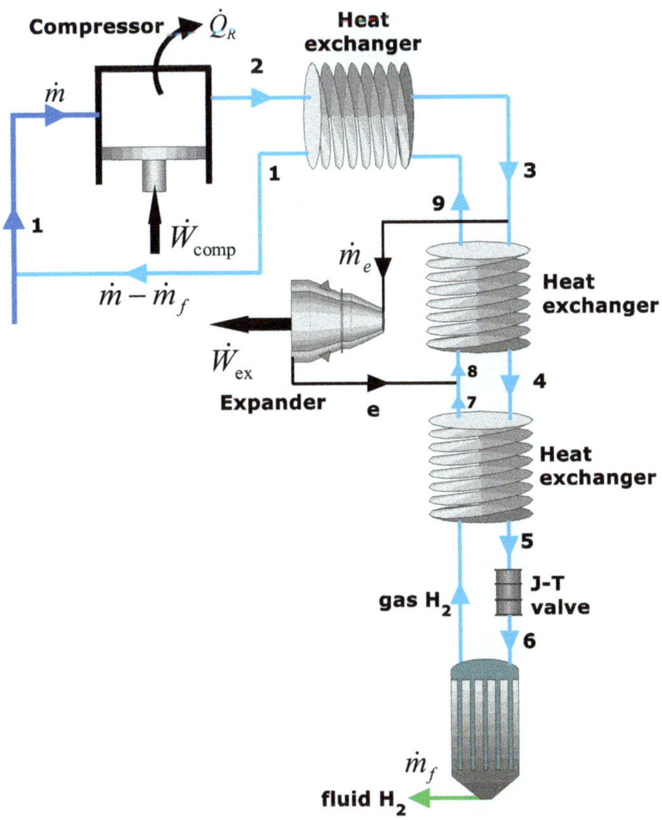

Figure 5. Schematic representation of a Claude Cycle.

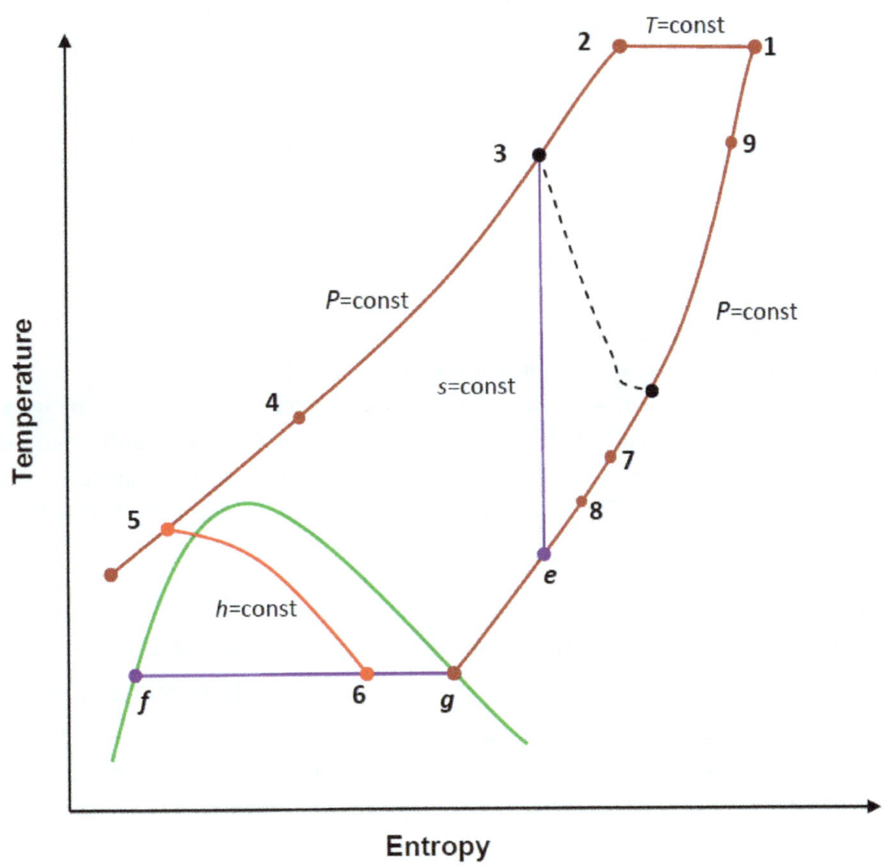

Figure 6. Temperature-entropy (*T-s*) diagram for a Claude Cycle.

Refrigeration effect per unit gas for the Claude cycle can be expressed as

$$q_l = h_1 - h_2 + z(h_3 - h_e) \tag{18}$$

where z is the ratio of gas diverted through the expander to the total flow rate. It can be expressed as

$$z = \frac{\dot{m}_e}{\dot{m}} \tag{19}$$

The expander can be treated as isentropic and work produced by expander can be written as

$$s_3 = s_e \tag{20}$$

$$h_e = h_3 - \eta_{ex}(h_3 - h_e') \tag{21}$$

$$w_e = h_3 - h_e \tag{22}$$

Liquid yield can be expressed as

$$y = \frac{h_1 - h_2}{h_1 - h_f} + z\left(\frac{h_3 - h_e}{h_1 - h_f}\right) \tag{23}$$

Eq. 23 states that, for a constant inlet temperature of the expander T_3, increasing the expander flow would increase the cooling effect and consequently decrease the inlet temperature of the J-T valve, so more liquid yield is obtained. But when liquefaction increases, the amount of gas return to the cycle is decreased and this reduces the precooling effect. So there is an optimum flow rate diverted through the expander for the cycle.

The work produced by the expander is used in the cycle. Therefore, the net work input is

$$w_{\text{in}} = w_{\text{comp}} - w_e = [(h_1 - h_2) - T_1(s_1 - s_2)] - z(h_3 - h_e) \tag{24}$$

The Kapitza cycle is a modified version of Claude cycle in which the coldest (third) heat exchanger is removed from the system. First heat exchanger is actually a regenerator which combines purification and pre-cooling. Another notable difference from Claude cycle is that the reciprocation expander is replaced with rotary expander which is more efficient. Kapitza cycle operates at relatively low pressures (7-10 atm). For Kapitza cycle high pressure states 4 and 5, and low pressure states 7 and 8 are identical.

Performance of Claude and Kapitza cycles are investigated for air, nitrogen, oxygen, argon, fluorine and methane. The compressor inlet state is taken 25°C and 1 atm for both cycles but the compressor exit pressure is taken 40 atm for Claude cycle and 7 atm for Kapitza cycle. Expanders are assumed to be isentropic. Diverted fraction through the expander $z = 0.7$. The results are given at Table 4 and Table 5 for Claude cycle and Kapitza cycle, respectively.

RESULTS AND DISCUSSION

Gas liquefaction process is a heat rejection process from the gas between inlet state of the compressor and outlet state of the J-T valve. Due to this reason the refrigeration effect per unit mass of the liquefied gas q_l are equal for all cycles. Hence, equation 1, 14 and 18 yield same results. The data obtained from Tables 1, 2, 4 and 5 represents performance parameters of the simple Linde-Hampson, precooled Linde-Hampson, Claude and Kapitza cycle for various fluids. The different gases have different performance parameters due to differences in their thermophysical properties.

Table 4. Performance parameters of various fluids for Claude Cycle.

Gas	y	w_{in} (kJ/kg)	w_{rev} (kJ/kg)	COP_{act}	COP_{rev}	η_{II}
Air	0.2488	1165	733.2	0.3644	0.5791	0.6292
Nitrogen	0.2493	1203	762.1	0.3585	0.566	0.6333
Oxygen	0.2343	1122	629.4	0.3607	0.6426	0.5613
Fluorine	0.2330	950.6	564.5	0.3614	0.6086	0.5938
Argon	0.2694	733.6	492.6	0.3629	0.57	0.6366
Methane	0.2594	1955	1081	0.4655	0.8421	0.5527

Table 5. Performance parameters of various fluids for Kapitza Cycle.

Gas	y	w_{in} (kJ kg^{-1})	w_{rev} (kJ kg^{-1})	COP_{act}	COP_{rev}	η_{II}
Air	0.1628	949.2	733.2	0.4473	0.5791	0.7724
Nitrogen	0.1585	1055	762.1	0.4090	0.566	0.7226
Oxygen	0.1546	907.2	629.4	0.3607	0.4459	0.6938
Fluorine	0.1536	769.5	564.5	0.4465	0.6086	0.7336
Argon	0.1623	694.5	492.6	0.4043	0.57	0.7093
Methane	0.1429	1959	1081	0.4647	0.8421	0.5518

The high liquid yield is related with the temperature at the inlet of the Joule-Thompson valve. Precooled Linde-Hampson cycle use auxiliary refrigeration cycle to lower the temperature at the inlet of the valve but have high work input because of auxiliary refrigeration cycle. Claude and Kapitza systems use expansion engine to make pre-cooling effect. As Tables 3 and 4 show, expansion engine help reduce work requirement in the cycle.

Table 1 and Table 2 represent performance parameters of the simple Linde-Hampson cycles and precooled Linde-Hampson cycle. It's clear from the results that simple Linde-Hampson cycle has the lowest efficiency among all liquefaction cycles. By adding auxiliary refrigeration cycle to the simple Linde-Hampson cycle second law efficiency of the cycle increase by about 70% and work consumption decrease by about 40%.

Table 3 represents effects of different refrigerants used in auxiliary refrigeration cycle on the performance of Precooled Linde-Hampson cycle. Performance of the auxiliary refrigeration cycle and boiling temperature of the auxiliary refrigerant directly affects the overall performance of the cycle. CO_2 has the lowest boiling temperature (-78.4°C) and yield best results among selected auxiliary refrigerants.

Table 4 and 5 represent performance of the expansion engine cycles: Claude and Kapitza cycles. These cycles use expansion engine besides Joule-Thompson valve. Expansion engine provides some refrigeration effect. These cycles have lower work consumption and higher liquid yield. Kapitza cycle operates at lower pressures and provides about 20% better performance than Claude cycle.

Effect of gas inlet temperature on the liquid yield is shown in Fig. 7 for different cycles. For lower inlet temperatures, greater fractions of gas can be liquefied. This effect is significant in simple Linde-Hampson, precooled Linde-Hampson, and Claude cycle but rather insignificant for Kapitza cycle.

Figs. 8 and 9 show effect of gas inlet temperature on the net work consumptions and second law efficiencies for the liquefaction cycles. As the inlet temperature increases, net work input increases and the second-law efficiency decreases. Net work input for simple Linde-Hampson cycle is the highest and second-law efficiency is the lowest because of low liquid yield. Kapitza and Claude cycles have a better performance than Linde-Hampson cycles. Kapitza and Claude cycles have close work input values.

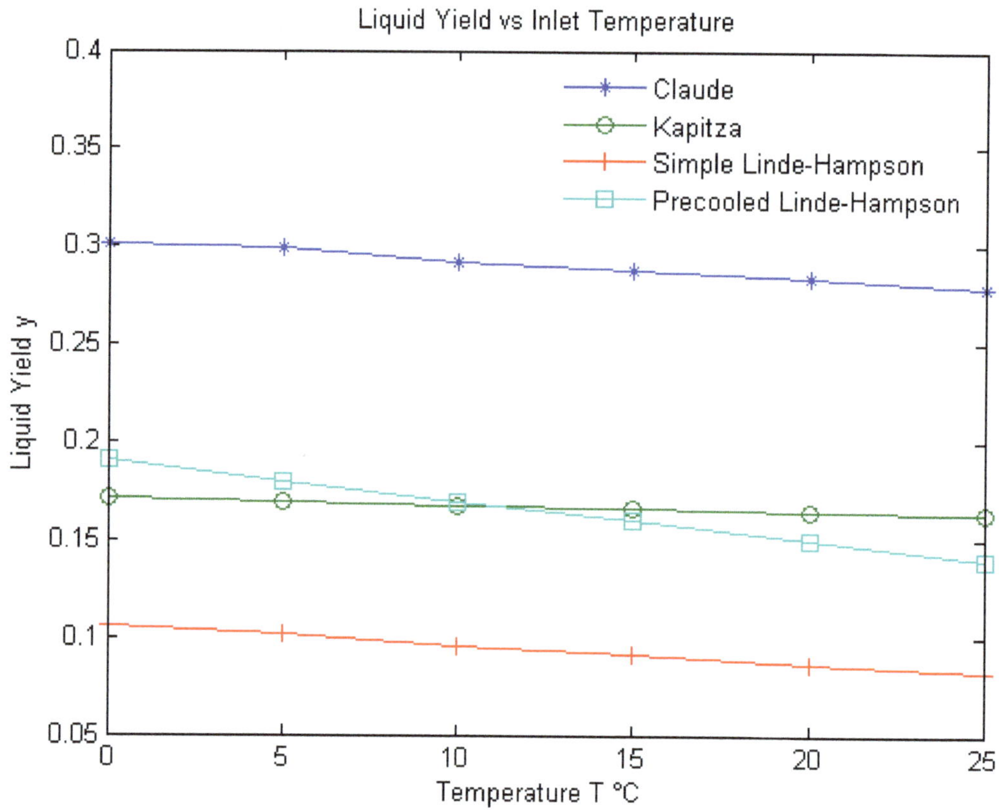

Figure 7. Liquid fraction with respect to the gas inlet temperature.

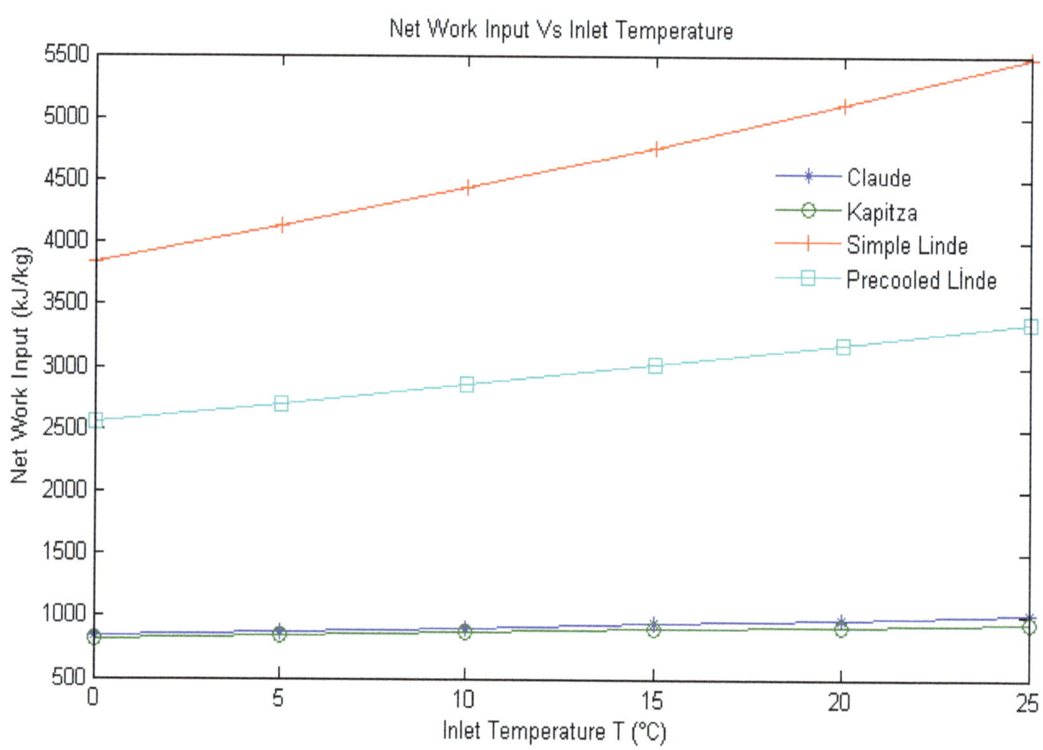

Figure 8. Cycles liquefaction work input with respect to the gas inlet temperature.

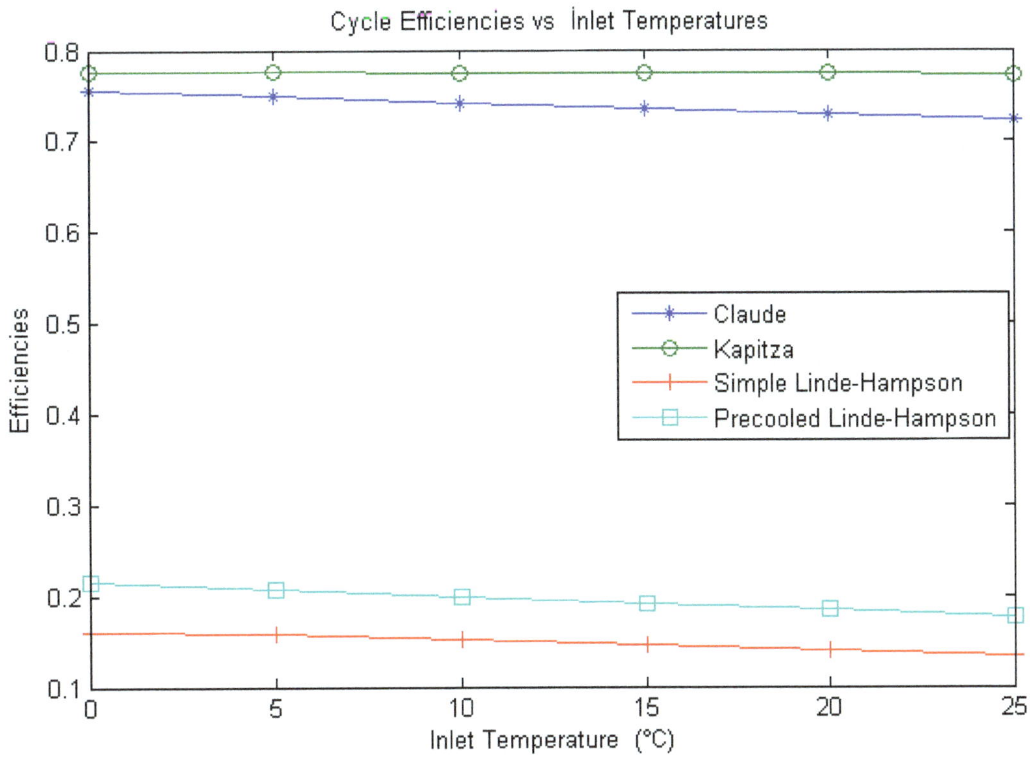

Figure 9. Second law efficiencies of the cycles with respect to the gas inlet temperature.

Table 6. Comparison of the liquefaction systems using air as the working fluid for each cycle.

Gas	Inlet pressure P_1 (atm)	y	w_{in} (kJ/kg)	COP_{act}	η_{II}
Simple Linde-Hampson	200	0.08228	5485	0.07741	0.1337
Precooled Linde-Hampson	200	0.1423	2118	0.1265	0.2184
Claude	40	0.2488	1165	0.3607	0.6292
Kapitza	7	0.1628	949.2	0.4473	0.7724

CONCLUDING REMARKS

The cycles with J-T systems have high work inputs and corresponding low efficiencies but have a simple setup which is desirable. Expansion engine systems have higher efficiencies and lower work inputs but have complex setup. Table 6 gives performance comparison of the Simple Linde-Hampson, Precooled Linde-Hampson, Claude and Kapitza cycles for air. The work inputs for these liquefaction cycles are 5485, 2118, 1165, and 949 kJ kg^{-1}, respectively. Claude cycle has the best liquid yield and Kapitza cycle has the best second law efficiency under given operating conditions.

The following conclusion may be drawn from the thermodynamic performance analysis of the cycles for different gases:

- Simple Linde-Hampson cycle and precooled Linde-Hampson cycle have simple setup but they have low efficiencies.
- Expansion engine cycles are much more efficient than J-T expansion cycles since the power output generated by expander is used to help compression of the gas.

- Refrigerant used to precool liquefied gas in the auxiliary refrigeration cycle of precooled Linde-Hampson cycle directly affects the performance of the cycle.
- Mass fraction diverted to the expander in Claude and Kapitza cycles is directly proportional to the refrigeration effect produced by the expander. By increasing the mass fraction diverted to the expander more cooling and more liquid yield can be obtained. But the amount of returning gas decrease and which decreases precooling effect in the heat exchangers. So optimum mass fraction diverted to the expander should be selected. This value usually varies between 0.6 and 0.8.
- Methods to resolve the challenges of efficient liquefaction include proposing completely new configurations and efficient gas liquefaction cycles coupled with improved efficiencies of the main cycle components such as compressors, expanders, and heat exchangers.

NOMENCLATURE

COP	Coefficient of performance
ex	Exergy (kJ/kg)
h	Specific enthalpy (kJ/kg)
$\dot{m}$	Mass flow rate (kg s^{-1})
P	Pressure (atm)
q	Specific heat transfer per unit mass (kJ/kg)
r	Fraction of auxiliary refrigerant
s	Specific entropy (kJ/kg)
T	Temperature (°C)
w	Specific work (kJ/kg)
x	Quality
y	Fraction of gas liquefied
z	Fraction of gas divert into expander

Greek symbols

ε	Effectiveness of the heat exchanger
η	Efficiency

Subscripts

act	Actual
comp	Compressor
e	Expander exit
ex	Expander
f	Fluid
g	Gas
gen	Generation
rev	Reversible

REFERENCES

[1] Kanoglu, M. (2002). Exergy analysis of multistage cascade refrigeration cycle used for natural gas liquefaction. International Journal of Energy Research, 26(8), 763-774.

[2] Krasae-in, S., Stang, J. H., & Neksa, P. (2010). Development of large-scale hydrogen liquefaction processes from 1898 to 2009. International Journal of Hydrogen Energy, 35(10), 4524-4533.

[3] Flynn, T. (2004). Cryogenic Engineering, revised and expanded. CRC Press.

[4] Thomas, R. J., Ghosh, P., & Chowdhury, K. (2011). Exergy analysis of helium liquefaction systems based on modified Claude cycle with two-expanders. Cryogenics, 51(6), 287-294.

[5] Atrey, M. D. (1998). Thermodynamic analysis of Collins helium liquefaction cycle. Cryogenics, 38(12), 1199-1206.

[6] Wang, M., Khalilpour, R., & Abbas, A. (2014). Thermodynamic and economic optimization of LNG mixed refrigerant processes. Energy Conversion and Management, 88, 947-961.

[7] Mehrpooya, M., & Ansarinasab, H. (2015). Exergoeconomic evaluation of single mixed refrigerant natural gas liquefaction processes. Energy Conversion and Management, 99, 400-413.

[8] Bisht, V. S. Thermodynamic Analysis of Kapitza Cycle based on Nitrogen Liquefaction. IOSR Journal of Engineering (IOSRJEN). Vol. 04, Issue 05 (May. 2014), ||V6|| PP 38-44.

[9] Barron, R. F. (1999). Cryogenic Heat Transfer (Vol. 1). Philadelphia, PA: Taylor & Francis.

[10] Timmerhaus, K. D., & Flynn, T. M. (1989). Properties of solids. In Cryogenic Process Engineering (pp. 39-102). Springer US.

[11] Maytal, B. Z. (2006). Maximizing production rates of the Linde–Hampson machine. Cryogenics, 46(1), 49-54.

[12] Nandi, T. K., & Sarangi, S. (1993). Performance and optimization of hydrogen liquefaction cycles. International Journal of Hydrogen Energy, 18(2), 131-139.

[13] Hilal, M. A. (1979). Optimization of helium refrigerators and liquefiers for large superconducting systems. Cryogenics, 19(7), 415-420.

[14] Kanoglu, M., Dincer, I., & Rosen, M. A. (2008). Performance analysis of gas liquefaction cycles. International Journal of Energy Research, 32(1), 35-43.

[15] Rashidi M.M., Aghagoli A., Raoofi R. (2017). Thermodynamic analysis of the ejector refrigeration cycle using the artificial neural network. Energy 129, 201- 215.

[16] Habibzadeh A., Rashidi M.M., Galanis N. (2013). Analysis of a combined power and ejector-refrigeration cycle using low temperature heat. Energy Conversion and Management 65, 381- 391.

[17] Kanoğlu, M. (2001). Cryogenic turbine efficiencies. Exergy, An International Journal, 1(3), 202-208.

14

MULTI-OBJECTIVE OPTIMIZATION OF A R744/R134A CASCADE REFRIGERATION SYSTEM: EXERGETIC, ECONOMIC, ENVIRONMENTAL, AND SENSITIVE ANALYSIS (3ES)

M.M. Keshtkar[1*], P. Talebizadeh[1]

ABSTRACT

This work presents the optimization of a two stage-cascade refrigeration system (TS-CRS), based on exergetic, economic, environmental, and sensitive analysis (3ES). R134a and R744 are considered as the refrigerants of high and low temperature circuits, respectively. Two single-optimization strategies including exergetic and economic optimizations and a multi-objective optimization are applied on the problem. In the first step, a comprehensive performance evaluation of different effective parameters, based on the genetic algorithm, used to indicate the optimum operative conditions in single objective strategies. In the next step, a multi-objective optimization is performed with considering a decision-making strategy based on the Pareto frontier using TOPSIS method. The higher exergetic efficiency and lower cost found in the exergetic and economic single-optimization, respectively. The multi-objective optimization results demonstrate that, the total system cost and the exergetic efficiency increase 28.6% and 99.5%, respectively, compared to the base design, and 46.6% higher energy can be saved in the compressors.

Keywords: *Cascade Refrigeration, Thermo-economic Optimization, Exergetic, Economic, Environmental, Sensitive Analysis*

INTRODUCTION

The cascade refrigeration system (CRS) is a freezing system that uses two kinds of refrigerants having different boiling points, which circulate through their own independent refrigeration cycle and are thermally coupled with each other through a cascade condenser [1]. This system is employed to obtain temperatures of -40 to -80°C or ultra-low temperatures [2]. At such ultra-low temperatures, a common single-refrigerant two-stage compression system limits the low-temperature characteristics of the refrigerant to a considerably poor level, making the system significantly inefficient [3]. The efficiency is improved by combining two kinds of refrigerants having different temperature characteristics. Different refrigerants with suitable characteristics for specific applications can be employed in CRSs [4]. Devotta et al. [5] carried out an experimental work based on the replacement of propane (R290) with R22. They concluded that R290 has more COPs and less energy consumption with respect to R22. Chen [6] replaced R22 with R410A in a residential air conditioner and showed an improvement of about 4% on cooling capacity and 13.9% on performance coefficient. Lee et al. [7] performed a thermodynamic investigation for a R744–R717 CRS. The optimal condensing temperature was obtained with the aim of maximizing the COP and minimizing the exergy destruction for the cascade condenser. Bhattacharyya et al. [8] studied a cascade cycle and determined the optimum intermediate temperature for R744/R290. Mafi et al. [9] performed an exergy analysis for multistage CRS and found that the minimum work depends only on the properties of incoming and outgoing process streams cooled or heated with refrigeration system and the ambient temperature.

Another researchers showed that, thermodynamic optimization is widely used for the optimization of energy systems [10]. Modeling and optimization of R717/R134a ice thermal energy storage air conditioning systems using two multi-objective optimization algorithms (NSGA-II and MOPSO) studied by Rahdar et al. [11]. The optimal design parameters which lead to the optimal objective functions (exergy efficiency and total cost rate) achieved. By using TOPSIS decision-making method, the optimum point from Pareto frontier of each optimization algorithm selected for both refrigerants. Najjar et al. [12] carried out a thermoeconomic analysis by combining the energy and exergy analysis with economic analysis to evaluate the total operating cost on a novel

This paper was recommended for publication in revised form by Regional Editor Alibakhsh Kasaeian
[1]Department of Mechanical Engineering, Kerman Branch, Islamic Azad University, Kerman, Iran.
**E-mail address: keshtkar@iauk.ac.ir*

inlet air cooling system with gas turbine engines using cascaded waste-heat recovery. In this work, a thermoeconomic analysis was done using specific exergy costing method by combining the energy and the exergy analysis with the economic analysis to evaluate the total operating cost of the system over a wide range of operating variables. Rezayan and Behbahaninia [13] studied a thermo-economic optimization of a R744/R717 cascade refrigeration cycle. The optimum values were calculated based on a trade-off between the capital and input exergy costs. Raja et al. [14] used a multiobjective heat transfer search (MOHTS) algorithm and investigated thermo-economic and thermodynamic optimization of a plate–fin heat exchanger. Effectiveness and accuracy of the proposed algorithm were evaluated by analyzing application examples of a PFHX. The obtained results using the proposed algorithm for thermo-economic considerations were compared with the NSGA-II and TLBO. Dubey et al. [15] performed a thermodynamic optimization of a transcritical R744 /R1270cycle for heating and cooling applications. Parekh et al. [16] studied a thermo-economic and cost optimizations using R404A-R508B and determined the cost of the products. Keshtkar [17] modeled and optimized a R744/R717 cascade refrigeration system using multi-objective optimization. Toghyani et al. [18] investigated the efficiency and the power loss due to pressure drop into the heat exchangers for a Stirling system using non-ideal adiabatic analysis and the second-version Non-dominated Sorting Genetic Algorithm. The optimized answers were chosen from the results using three decision-making methods. There are some other works in the literature that studied the optimization of thermal systems [19-22].

At this work, the optimum operative condition of a R744-R134a TS-CRS has been evaluated based on thermo-economic performance using genetic algorithm and TOPSIS decision-making procedure with Pareto frontier. For this purpose, an exergy investigation has been performed based on thermo-economic analysis. Then, two different optimization strategies according to exergy, economic and a multi-objective optimization as equilibrium between cost and exergetic optimizations have been carried out. Various parameters of the system including evaporating and condensing temperatures as well as the superheating and sub-cooling degrees of both stages have been optimized. Exergetic, economic, environmental, and sensitive analyses (3ES) have not been carried out so far for R134a/R744 cascade refrigeration system.

MATERIAL AND METHOD

Figure 1 and Figure 2 display a schematic diagram of a CRS using R744/R134a as the refrigerants and the corresponding P–h diagram, respectively. The R134a and R744 are the working fluid of higher and lower stage cycles, respectively. A detailed description of the circuits has been expressed in [19]; however, the difference is using the R134a as a higher stage working fluid in the current study. The basic design parameters used as a case study for the CRS have been listed in Table 1.

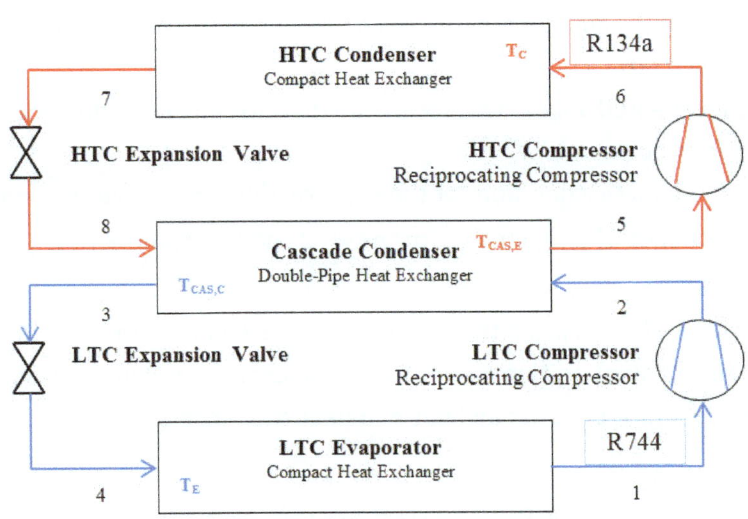

Figure 1. Schematic diagram of R744/R134a system.

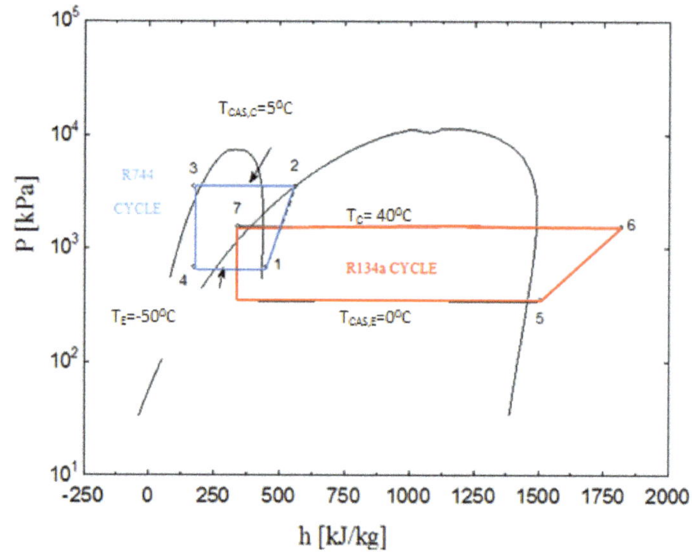

Figure 2. Pressure-enthalpy diagram of R744/R134a CRS.

Table 1. Input parameters used for simulation of CRS at a basic design

Refrigeration capacity $(\dot{Q}_L)$	60kW
Condensing temperature (T_C)	40°C
Evaporating temperature (T_E)	−50°C
Temperature difference in cascade condenser $(\Delta T_{Cas.cond})$	5°C
Ambient temperature (T_0)	26°C
Cold room temperature (T_{CL})	−27°C

The following assumptions have been taken into account:

- Considering sub-cooled liquid state for the condenser and cascade condenser outlet conditions at LT circuit.
- Considering superheated vapor state for the evaporator and cascade condenser outlet conditions at HT circuit.
- Considering the adiabatic condition for the compression process within compressor
- Considering the values of $\eta_m \eta_e$ and η_s for the HTC and LTC compressors equal to 0.93 and 0.73, respectively.
- Negligible heat transfer between the heat exchangers and the ambient.
- Considering isenthalpic condition for the throttling process within expansion valve.
- Neglecting the pressure drop in pipelines and heat exchangers.

Energy and Exergy Analysis

For energy and exergy study of CRS, the needed equations have been specified in Table 2. In addition, the performance coefficient is given as:

$$\mathrm{COP}=\frac{\dot{Q}_E}{\dot{W}_{HTC}+\dot{W}_{LTC}}=\frac{(h_5\text{-}h_8)(h_1\text{-}h_4)}{(h_6\text{-}h_5)(h_2\text{-}h_3)+(h_5\text{-}h_8)(h_2\text{-}h_1)} \tag{1}$$

The total inlet exergy to the system is given as [23]:

$$\dot{E}x_{in}=\dot{W}_{LTC,comp}+\dot{W}_{HTC,comp}+\dot{W}_{fan,evap}+\dot{W}_{fan,cond} \tag{2}$$

The outlet exergy (exergy of products) can be determined as:

$$\dot{E}x_{out}=\left(\frac{T_0}{T_{CL}}-1\right)\dot{Q}_L \tag{3}$$

The overall system exergy destruction and exergetic efficiency are given as [24]:

$$\dot{E}x_{D,total}=\dot{E}x_{in}-\dot{E}x_{out} \tag{4}$$

$$\eta_{II}=\frac{\dot{E}x_{out}}{\dot{E}x_{in}}=1-\left(\frac{\dot{E}x_{D,total}}{\dot{E}x_{in}}\right) \tag{5}$$

Table 2. Energy and exergy study of CRS components

Component	Energy Balance	Exergy Balance
Evaporator	$\dot{Q}_L=\dot{m}_L(h_1-h_4)$	$\dot{E}x_{D,evap}=\left(1-\left(\frac{T_0}{T_{cl}}\right)\right)\dot{Q}_L+(\dot{E}x_4-\dot{E}x_1)+\dot{W}_{fan,evap}$
LTC compressor	$\dot{W}_{LTC,Comp}=\frac{\dot{m}_L(h_{2s}-h_1)}{\eta_s\eta_m\eta_e}=\frac{\dot{m}_L(h_2-h_1)}{\eta_m\eta_e}$	$\dot{E}x_{D,LTC,comp}=(\dot{E}x_1-\dot{E}x_2)+\dot{W}_{LTC,comp}$
HTC compressor condenser	$\dot{W}_{HTC,Comp}=\frac{\dot{m}_H(h_{6s}-h_5)}{\eta_s\eta_m\eta_e}=\frac{\dot{m}_H(h_6-h_5)}{\eta_m\eta_e}$	$\dot{E}x_{D,LTC,comp}=(\dot{E}x_1-\dot{E}x_2)+\dot{W}_{LTC,comp}$
Cascade condenser	$\dot{Q}_{Cas.cond}=\dot{m}_L(h_2-h_3)=\dot{m}_H(h_5-h_8)$, $\frac{\dot{m}_H}{\dot{m}_L}=\frac{h_2-h_3}{h_5-h_8}$	$\dot{E}x_{D,Cascond}=((\dot{E}x_2+\dot{E}x_8)-(\dot{E}x_3+\dot{E}x_5))$
Condenser LTC expansion valve HTC expansion valve	$\dot{Q}_H=\dot{m}_H(h_6-h_7)$ $h_3=h_4$ $h_7=h_8$	$\dot{E}x_{D,cond}=\left(\frac{T_0}{T_0}-1\right)\dot{Q}_H+(\dot{E}x_6-\dot{E}x_7)+\dot{W}_{fan,cond}$ $\dot{E}x_{D,LTC,exp}=(\dot{E}x_3-\dot{E}x_4)$ $\dot{E}x_{D,HTC,exp}=(\dot{E}x_7-\dot{E}x_8)$

Thermo-economic Analysis

Thermo-economic is the combination of exergy analysis with economic constraints. For an energy system, the thermo-economic analysis allows one to calculate the cost rate of all streams based on thermodynamic and economic analysis. In thermo-economic analysis, all of the stream costs are determined through the exergy costing principles [25]. The exergy costing equation includes the cost balances for every component. For mth component, a cost balance indicates that the total cost values related to all streams is identical to the total cost of all streams entering and the suitable charge because of annual capital cost $\left(\dot{z}_m^{CI}\right)$, maintenance cost and annual operating $\left(\dot{z}_m^{OM}\right)$, and annual penalty cost $\left(\dot{z}_m^{ENV}\right)$ in terms of US$ per year. Note that in this analysis, as mentioned, the input exergy cost of each component is equal to the output rate of previous component. Therefore, the unknown quantity in the equation is the product's exergy cost. The system's total cost is given as [25]:

$$\dot{C}_{total}=\sum c_{out}\dot{E}x_{out}+c_w\dot{W}=\sum c_{in}\dot{E}x_{in}+c_Q\dot{E}x_Q+\sum_m \dot{Z}_m \tag{6}$$

In Eq. (6), the place of heat and work can change as production or input according to the system conditions. In the exergy costing equation, $\dot{C}_{total}$ is the sum yearly cost of the system; c_i and c_o are the unit costs of the input and product exergy, respectively. $\dot{E}x_{in}$ and $\dot{E}x_{out}$ are the annual exergy rates from internal sources and output products, respectively. In TS-CRS, the input exergy is only the electrical energy, while the product is cooling capacity. The capital cost of each component $\left(z_m^{CI}\right)$ is given as [26]:

$$Z_{HTC,comp}=9624.2\,\dot{W}_{HTC,comp}^{0.46} \tag{7}$$

$$Z_{LTC,comp}=10167.5\,\dot{W}_{LTC,comp}^{0.46} \tag{8}$$

$$Z_{cond}=1397A_{cond}^{0.89}+629.05\,\dot{W}_{fan,cond}^{0.76} \tag{9}$$

$$Z_{evap}=1397A_{evap}^{0.89}+629.05\,\dot{W}_{fan,evap}^{0.76} \tag{10}$$

$$Z_{cas,cond}=2382.9A_{cas,cond}^{0.68} \tag{11}$$

In engineering economics, by introducing the capital recovery factor (CRF), the time interval unit considered for the capital cost calculation is commonly considered as a year and determined as [25]:

$$CRF=\frac{i(1+i)^n}{(1+i)^n-1} \tag{12}$$

At the present work, the rate of interest (i) and the system life (n) are assumed 14% and 15 years, respectively. According to the investment cost z_m $(\$)$, the common function for the cost rate, $\dot{z}_m$ $(\$/s)$, related to capital cost and the maintenance cost for the mth is given as:

$$\dot{Z}_m=\frac{Z_m\times CRF\times f}{N\times 3600} \tag{13}$$

At this work, the values of the maintenance factor (ϕ), the system yearly hours of operation (N), and CRF are considered 1.06, 8000 h and 18.2%, respectively. After obtaining the cost rate for all components, the total annual cost rate is given as:

$$\dot{C}_{tot}=\left(\dot{Z}_{HTC,comp}+\dot{Z}_{LTC,comp}+\dot{Z}_{cond}+\dot{Z}_{evap}+\dot{Z}_{cas,cond}\right)+$$
$$\left(\dot{W}_{HTC,comp}+\dot{W}_{LTC,comp}+\dot{W}_{Fan,LTC}+\dot{W}_{Fan,HTC}\right)\times\left[\frac{1}{3}\frac{c_{elec,peak}}{3600}+\frac{2}{3}\frac{c_{elec,mid\text{-}peak}}{3600}\right]+$$
$$\frac{\left(m_{Co_2}/1000\right)}{N}\times\frac{c_{Co_2}}{3600} \tag{14}$$

where value of $c_{elec,peak}$, is 0.07 kWh^{-1}, and $c_{elec,mid-peak}$ is 0.04 kWh^{-1}, which are the unit costs of electricity during the working hours for the peak and the mid-peak time, respectively. Note that the 1/3 and 2/3 coefficients show the electricity use in the peak and the mid-peak. $m_{Co_2}(kg)$ is the amount of CO_2 emission and is determined from the annual electricity consumption $[kWh]$, using the emission conversion factor equal to $0.968\,kg\,kWh^{-1}$. At the present work, the unit environmental cost of emission of carbon dioxide (c_{Co_2}) is assumed $90(US\,\$/ton\,Co_2)$ [26].

Heat Exchanger Design

Since the annual cost of CRS depends on the pressure drop and thermal area of heat exchangers, the design of the employed heat exchangers including evaporator, air-cooled, and cascade condensers is important [27]. Evaporator and air-cooled condenser are compact air-cooled heat exchangers. The cascade condenser is a double-pipe heat exchanger. For designing of heat exchangers, the number of tube rows, tube diameter, and tube thickness are important. At the present work, for heat transfer and pressure drop calculations, the library functions in EES (Engineering Equation Solver) were used [28]. The heat transfer area for every heat exchanger is given as:

$$A_o=\dot{Q}/U_o\left(LMTD\right) \tag{15}$$

The overall coefficient of heat transfer (U_o) regarding the external heat transfer area and the logarithmic temperature are given as:

$$\frac{1}{U_oA_o}=\frac{1}{h_iA_i}+\frac{1}{h_oA_o}+\frac{\ln\left(d_o/d_i\right)}{2\pi Lk},\qquad LMTD=\frac{\left(T_{hot,i}-T_{cold,o}\right)-\left(T_{hot,o}-T_{cold,i}\right)}{\ln\frac{\left(T_{hot,i}-T_{cold,o}\right)}{\left(T_{hot,o}-T_{cold,i}\right)}} \tag{16}$$

Figure 3 shows a compact tube-fin, cross-flow heat exchanger used in the present work for the evaporator, and air-cooled condenser. The geometric parameters have been shown in Fig. 3. The fins have been made from copper with a thickness of th_{fin} and a fin pitch of p_{fin}. The heat exchanger includes 10 rows of tubes in two columns connected in series. The vertical and horizontal spacing between adjacent tubes is s_v and s_h, respectively. The heat exchanger length in the direction of the airflow is *L*, and the width and height of the front face are *W* and *H*, respectively. The tubes have been made from copper with outer diameter of D_{out} and wall thickness, th. The roughness of the tube inner surface is $e=10\,\mu m$. Clean dry air passes through the heat exchanger perpendicular to the tubes with a volumetric flow rate of $\dot{V}_c=0.06\,m^3/s$. Table 3 shows the geometric parameters of the condenser and evaporator. The pressure drop of the air regarding the compact tube-fin heat exchanger is given as:

$$\Delta p=\frac{G^2}{2\rho_{in}}\left[f\frac{4L_{flow}}{D_h}\frac{\rho_{in}}{\bar{\rho}}+\left(1+\sigma^2\right)\left(\frac{\rho_{in}}{\rho_{out}}-1\right)\right], \qquad \frac{1}{\bar{\rho}}=\frac{1}{2}\left(\frac{1}{\rho_{in}}+\frac{1}{\rho_{out}}\right) \tag{17}$$

G is mass velocity, and ρ_i , ρ_o , and $\bar{\rho}$ are, the inlet, outlet, and mean air density, respectively. f is the coefficient of friction presented in [29]. More information on the subject of air-side and tube-side pressure drop and heat transfer calculations for all heat exchangers are found at [30-32]. It should be noted that, selection of cascade condenser model has been carried out based on pre-defined models in EES library [28]. At the double-pipe heat exchanger, the fluid in inner pipe is considered R744 and in annual space is R134a.

Table 3. Geometry specifications of evaporator and condenser at this work.

Component	Outside diameter (m)	Tube thickness (mm)	Fin thickness (mm)	Horizontal pitch (mm)	Vertical pitch (mm)	Number of tube rows
Evaporator	0.0102	0.9	0.33	22	25.4	10
Condenser	0.0159	0.9	0.33	22	25.4	10

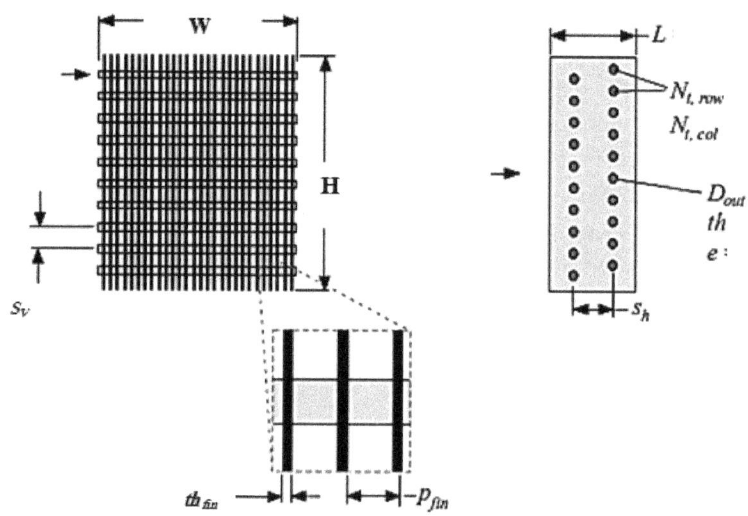

Figure 3. Schematic of evaporator and condenser.

SYSTEM OPTIMIZATION WITH NSGA-II

For the thermo-economic optimization, six independent parameters have been shown in Table 4. These parameters are chosen as the decision making variables within the defined range. The maximum and minimum values for each design parameter have been applied as the constraints. The loaded optimization limitations of the key design parameters have been indicated according to the proposed rates from the literature [25].

Two different objective functions have been employed as the optimization strategies; the total cost rate $(\dot{C}_{tot})$ and the exergy efficiency (η_{II}) of CRS. The exergy efficiency represents the thermodynamic irreversibility due to the heat transfer extracted by the variation of temperature in various components. The following multi-linear regression equations are fitted on the EES achieved data:

$$\eta_{II}=a_0+a_1\Delta T_{sub,H}+a_2\Delta T_{sub,H}^2+a_3\Delta T_{sub,L}+a_4\Delta T_{sub,L}^2+a_5\Delta T_{sup,H}+a_6\Delta T_{sup,H}^2$$
$$+a_7\Delta T_{sup,L}+a_8\Delta T_{sup,L}^2+a_9 T_C+a_{10} T_C^2+a_{11} T_E+a_{12} T_E^2 \tag{18}$$

$$\dot{C}_{total}=b_0+b_1\Delta T_{sub,H}+b_2\Delta T^2_{sub,H}+b_3\Delta T_{sub,L}+b_4\Delta T^2_{sub,L}+b_5\Delta T_{sup,H}+b_6\Delta T^2_{sup,H} +b_7\Delta T_{sup,L}+b_8\Delta T^2_{sup,L}+b_9\,T_C+b_{10}\,T_C^2+b_{11}\,T_E+b_{12}T_E^2 \quad (19)$$

The multi-linear regression coefficients for Eqs. (18) and (19) have been given in Table 5. The root mean square errors and coefficients of correlations have been calculated equal to 1.076E-03 and 99.48% for Eq. (18) and 4.309E-05 and 97.15% for Eq. (19), respectively [33].

Table 4. Decision variables for TS-CRS and their corresponding range of variation.

Decision variables	**Range of variation**
Condensing temperature of R134a	$38°C<T_C<48°C$
Evaporating temperature of R744	$-60°C<T_E<-40°C$
Degree of superheating at R744circuit	$0°C<\Delta T_{sup,LTC}<20\ °C$
Degree of subcooling at R744circuit	$0°C<\Delta T_{sub,LTC}<10\ °C$
Degree of superheating at R134a circuit	$0°C<\Delta T_{sup,HTC}<20\ °C$
Degree of subcooling at R134a circuit	$0°C<\Delta T_{sub,HTC}<10°C$

Table 5. Coefficients of Eqs. (18 and 19) using Multi-linear regression.

η_{II}	a_0	a_1	a_2	a_3	a_4	a_5	a_6	a_7	a_8	a_9	a_{10}	a_{11}	a_{12}
Values	-1.6468 E+00	1.5169 E+00	3.4059 E-03	9.1870 E-01	2.6264 E-01	-1.0013 E-01	-8.2059 E-03	-1.2031 E+00	-9.5615 E-03	1.3121 E-01	3.3807 E-03	-7.8029 E-04	1.6868 E-03
$\dot{C}_{tot}$	b_0	b_1	b_2	b_3	b_4	b_5	b_6	b_7	b_8	b_9	b_{10}	b_{11}	b_{12}
Values	5.8974 E-03	-1.1183 E-04	-5.2061 E-05	6.0660 E-04	-1.1655 E-04	2.8431 E-05	1.7375 E-05	-9.5376 E-05	5.0526 E-05	-3.5672 E-05	-4.2426 E-07	-1.2371 E-04	2.2200 E-06

It is noticeable that for optimization of the system, the exergy efficiency (Eq. 18) should be maximized, while the total cost rate (Eq. 19) should be minimized. To optimization of the CRS, a NSGA-II genetic algorithm method has been used. In engineering optimization, genetic algorithms can establish reasonable solutions with random initialization [34]. Genetic algorithm is a non-deterministic optimization technique with general-purpose search method based on the principles of evolution including the crossover, selection, and mutation operators with the aim of presenting the optimal solution in a population to achieve an identified criterion of termination. The setting parameters in GA optimization method have been given in Table 6.

Table 6. The setting parameters in the optimization process

Setting parameters	**value**
Population size	500
Maximum number of generation	400
Minimum function tolerance	10^{-5}
Probability of crossover	90%
Probability of mutation	1%
Number of crossover point	2
Selection process	Tournament
Tournament size	2

Note that an in-house computer code software has been developed in MATLAB to achieve the Pareto frontier. The Pareto frontier solution indicates the optimal solutions, and one of these solutions is selected as the best solution in this work using a decision-making technique. TOPSIS is a decision-making technique for finding the closest result to the ideal solution defined in the following argument.

In this method since the dimension of various objectives might be different, first, the scales and dimension of objectives space must be non-dimensionalized. So, a non-dimensionalized objective, F_{ij}^{n}, is introduced as:

$$F_{ij}^{n} = \frac{F_{ij}}{\sqrt[2]{\sum_{i=1}^{m} (F_{ij})^2}} \tag{20}$$

Decision making in the TOPSIS method is performed by calculating each solution's distance on the Pareto frontier from the best point indicated by d_i^+ and from the non-ideal point indicated by d_i^- as following [35]:

$$d_i^+ = \sqrt{\sum_{j=1}^{n} (F_{ij} - F_j^{Ideal})^2} \tag{21}$$

$$d_i^- = \sqrt{\sum_{j=1}^{n} (F_{ij} - F_j^{Non-Ideal})^2} \tag{22}$$

where n shows the objectives; i stands for every solution on the frontier of Pareto ($i=1,2,....,m$). F_j^{Ideal} and $F_j^{Non-Ideal}$ are the ideal and non-ideal values for j_{th} objective calculated from a single-objective optimization, respectively. In the non-ideal point, every objective possesses its worst value. Therefore, a Cl_i variable is also explained as:

$$Cl_i = \frac{d_i^-}{d_i^+ + d_i^-} \tag{23}$$

Finally, a solution with minimum Cl_i is chosen as a suitable solution.

RESULTS AND DISCUSSION

To find the effect of various parameters on the COP of the system, the exergy destruction value of all components $(\dot{E}x_D)$, the exergetic efficiency (η_{II}), and the overall system's cost are first analyzed. A sensitive analysis is carried out before the optimization process. The evaporation, condensation, sub-cooling, and superheating temperature in both cycle have been selected as the most effective parameters in CRS cycle. Furthermore, their variations are much more important than other parameters.

Figure 4 shows the effect of evaporating temperature in R134a circuit $(T_{CAS,E})$ on the COP of the system at the given operating conditions. Figs.4-a and 4-b display the variation of COP at different values of sub-cooling and superheating degrees, respectively. It can be seen that the COP_{MAX} increased as the degree of sub-cooling increased. Furthermore, increasing the sub-cooling degree from $\Delta T_{sub} = 0$ to $\Delta T_{sub} = 10\,^{\circ}C$ enhanced in the COP_{MAX} from 1.342 to 1.445, i.e. a 10.3% increase with respect to its value without sub-cooling. On the contrary (Fig. 4-b), increasing the superheating degree from $\Delta T_{\sup} = 0$ to $\Delta T_{\sup} = 20\,^{\circ}C$ decreases the COP_{MAX} from1.346 to 1.275 (7.1%).

Figures 5-a and 5-b show the COP and mass ratio variations as a function of condensing and evaporation temperature and variation in temperature in cascade condenser, respectively. As the temperature in condenser or cascade condenser increases, the COP of the cascade system decreases; however, the elevation in the evaporating temperature results in a rise in the COP of the system. As the condensing temperature in HTC circuit increases, the pressure ratio of the R134a compressor increases, leading to an increase in the electric power in the HTC compressor, and therefore, the COP decreases. As observed in Fig. 5-b, as the evaporating temperature increases, the mass ratio of CRS decreases; however, the rise in condensing temperature causes a higher mass ratio.

The exergy destruction of all components for two cases without sub-cooling and both sub-cooling /superheating for $60kw$ cooling load, have been specified in Table 7. As observed in this figure, the condenser has the maximum rate of exergy destruction compared with other components in the cycle.

As listed, condenser has the most exergy destruction in the system, and the evaporator is at the next level. According to the first law of thermodynamic, a significant improvement exists in the condenser. Note that that reducing the sub-cooling causes a higher but negligible effect on exergy destruction values of the components.

Figures 6-a and 6-b show the variation of R744 evaporating temperature, T_E , on the components' cost for various sub-cooling and superheating values in both cycles. The higher cost of air-cooled condenser was achieved for lower values of T_E since the logarithmic temperature variation between the condensation and ambient temperatures decreased. However, the cost of the LT compressor decreased because of the decrease in the LTC pressure ratio. The cost of air-cooled condenser decreased because of a decline in the removed heat from the condenser, and consequently, the cost of cascade condenser increased. Furthermore, the cost of HTC compressor varied very little because of a minor variation in the mass flow rate of R134a circuit. Comparing Figures 6-a and 6-b, the effect of ΔT_{sub} and $\Delta T_{\sup}$ on both cycles was clarified. Furthermore, by increasing ΔT_{sub} from 0 K to 8 K and $\Delta T_{\sup}$ from 10 K to 20 K, the evaporator cost decreased; however, the condenser and cascade condenser costs increased. Moreover, because of the negligible influence of superheating and sub-cooling degrees' variations on the compressors, the costs of LT and HT compressors remained constant.

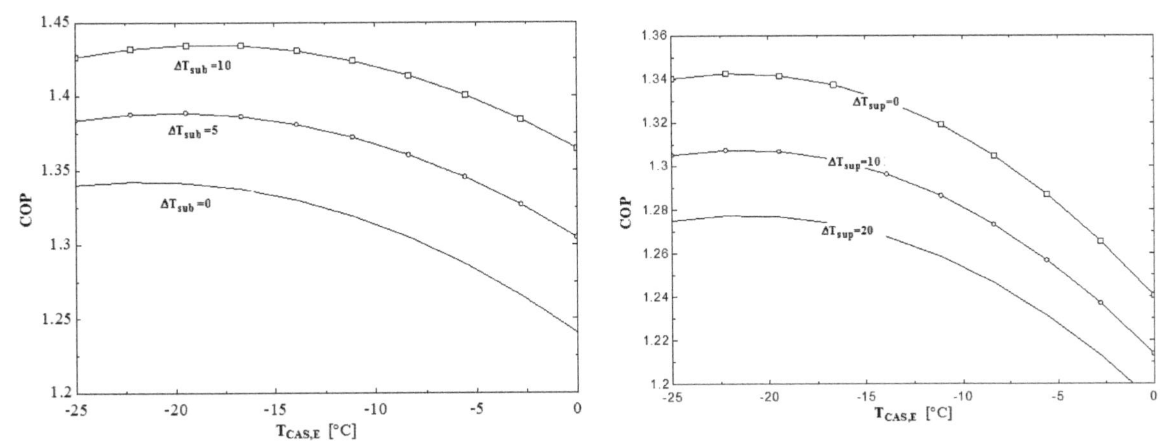

(a) $(T_C = 40°C, T_E = -50°C, \Delta T_{Cascond} = 5°C, \Delta T_{\sup} = 0°C)$

(b) $(T_C = 40°C, T_E = -50°C, \Delta T_{Cascond} = 5°C, \Delta T_{sub} = 0°C)$

Figure 4. Variation of system coefficient performance for different evaporating temperature in HT circuit with considering of (a) subcooling and (b) superheating on both cycles.

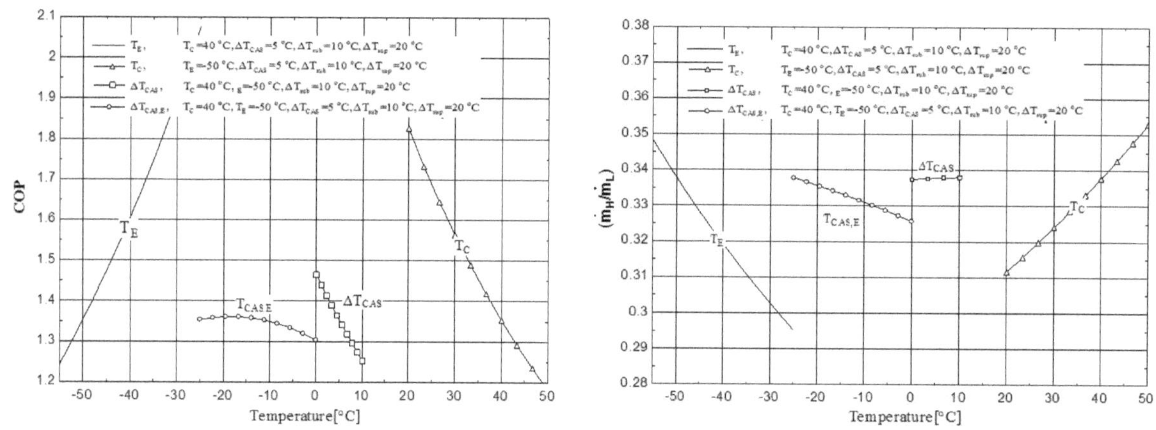

(a) (b)

Figure 5. Effect of evaporating (T_E), condensing (T_C), evaporating temperature in R134a cycle $(T_{CAS,E})$ and temperature difference in cascade condenser (ΔT_{CAS}) on (a) system performance and (b) mass ratio at various conditions.

Table7. The exergy destruction rates (kW) for cooling load of $60kw$.

Component	HTC Comp.	LTC Comp.	Condenser	Evaporator	Cascade Cond.	HTC Ex. Valve	LTC Ex. Valve
$(\Delta T_{sub}=8^{\circ}C, \Delta T_{sup}=10^{\circ}C)$	2.1	2.4	10.2	5.6	0.7	0.2	0.1
$(\Delta T_{sub}=0^{\circ}C, \Delta T_{sup}=10^{\circ}C)$	2.3	2.6	10.7	5.9	0.85	0.25	0.15

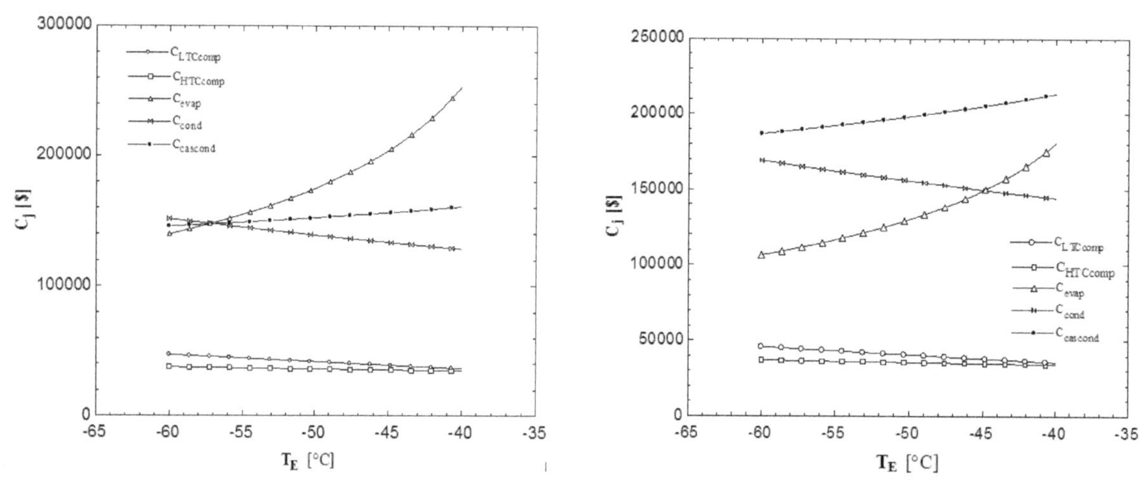

(a) $\Delta T_{sub}=0, \Delta T_{sup}=10^{\circ}C$ (b) $\Delta T_{sub}=8^{\circ}C, \Delta T_{sup}=20^{\circ}C$

Figure 6. Effect of R744 evaporating temperature on the costs of system components $(T_C=40°C,\ \Delta T_{Cascond}=5°C, T_{CAS,E}=-5°C)$

Figures 7-a and 7-b show the variation of $\Delta T_{sub,L}$ and $\Delta T_{sub,H}$, i.e. the sub-cooling degrees in low and high temperature circuits on the annual total cost and exergetic efficiency of the CRS. By increasing $\Delta T_{sub,L}$ or $\Delta T_{sub,H}$, the exergetic efficiency and annual overall cost rose. Furthermore, an optimum point for sub-cooling degree can be found by carrying out a trade-off between the annual total cost and the system's exergetic efficiency.

The variations of $\Delta T_{\sup,L}$ and $\Delta T_{\sup,H}$, i.e. the superheating degrees in low and high temperature circuits on the annual total cost and exergetic efficiency of the CRS have been shown in Figs.7-c and 7-d. Increasing $\Delta T_{\sup,L}$ or $\Delta T_{\sup,H}$ decreases the annual total cost and exergetic efficiency. However, comparing Figures 7-c and 7-d, optimum points for superheating degree can be obtained by performing a trade-off between the total cost and exergetic efficiency of the system.

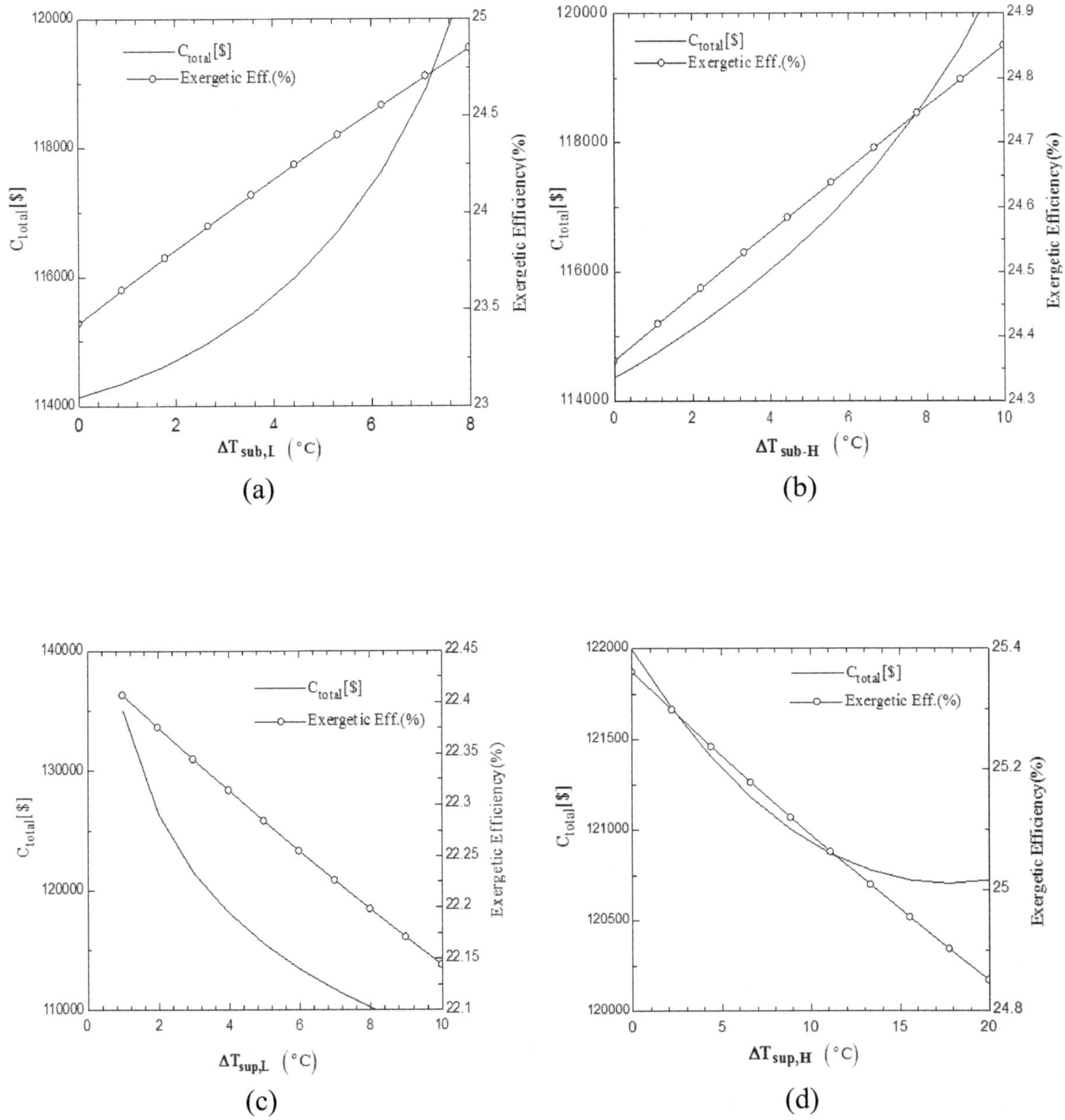

Figure 7. Variations of annual cost and exergetic efficiency of the system versus the (a- b) subcooling degree and (c-d) superheating degree in low and high temperature circuits ($T_C = 40°C, T_E = -50°C$, $\Delta T_{Cascond} = 5°C, T_{CAS,E} = -5°C$)

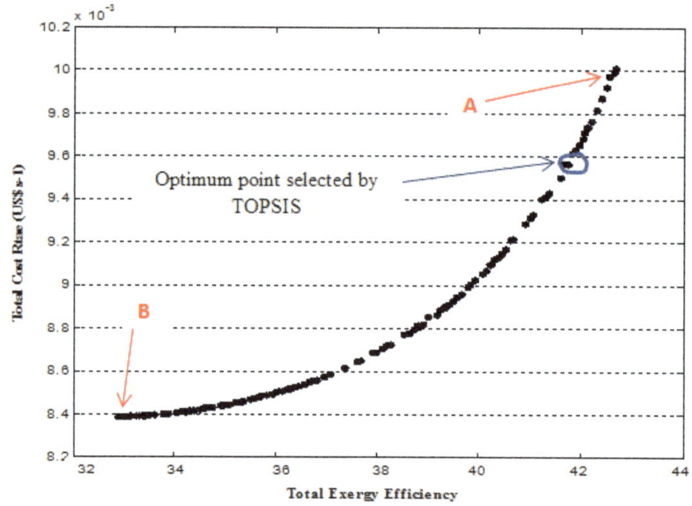

Figure 8. Pareto optimal frontier from multi-objective optimization of R134A/R744 CRS.

CONCLUSION

In the present work, R134a/R744 cascade refrigeration system, based on exergetic, economic, environmental, and sensitive analysis (3ES) was modeled and analyzed. A multi-objective optimization was also done regarding the Pareto frontier employing NSGA-II optimization and a decision-making strategy. The findings showed that by enhancing $\Delta T_{sub,L}$ or $\Delta T_{sub,H}$, the exergetic efficiency and the total annual cost of the CRS increased. Furthermore, according to single objective optimizations, the exergetic efficiency increased by 94.8%, resulting from 38.7% decrease in the total cost, using the exergetic optimization, in comparison to the base design. The total system cost decreased by 10.3%, using the cost optimization, and the exergetic efficiency increased by 20.6%. The higher cost of air-cooled condenser was achieved for lower values of T_E since the logarithmic temperature variation between the condensation and ambient temperatures decreased. However, the cost of the LT compressor decreased because of the decrease in the LTC pressure ratio. The cost of air-cooled condenser decreased because of a decline in the removed heat from the condenser, and consequently, the cost of cascade condenser increased. The findings of multi-objective optimization demonstrated that the exergetic efficiency and the overall annual system expense increased 99.5% and 28.6%, respectively, and 46.6% higher energy was saved in both compressors.

REFERENCES

[1] Getu, H. M., Bansal, P. K. (2006). Simulation model of a low-temperature supermarket refrigeration system. HVAC&R Research, 12(4), 1117-1139.

[2] Getu, H. M., Bansal, P. K. (2008). Thermodynamic analysis of an R744–R717 cascade refrigeration system. International journal of refrigeration, 31(1), 45-54.

[3] Messineo, A. (2012). R744-R717 cascade refrigeration system: performance evaluation compared with a HFC two-stage system. Energy Procedia, 14, 56-65.

[4] Pearson, A. (2001). New developments in industrial refrigeration. Ashrae Journal, 43(3), 54.

[5] Devotta, S., Padalkar, A. S., Sane, N. K. (2005). Performance assessment of HC-290 as a drop-in substitute to HCFC-22 in a window air conditioner. International Journal of Refrigeration, 28(4), 594-604.

[6] Chen, W. (2008). A comparative study on the performance and environmental characteristics of R410A and R22 residential air conditioners. Applied thermal engineering, 28(1), 1-7.

[7] Lee, T. S., Liu, C. H., Chen, T. W. (2006). Thermodynamic analysis of optimal condensing temperature of cascade-condenser in CO2/NH3 cascade refrigeration systems. International Journal of Refrigeration, 29(7), 1100-1108.

[8] Bhattacharyya, S., Bose, S., Sarkar, J. (2007). Exergy maximization of cascade refrigeration cycles and its numerical verification for a transcritical CO2–C3H8 system. International Journal of Refrigeration, 30(4), 624-632.

[9] Mafi, M., Naeynian, S. M., Amidpour, M. (2009). Exergy analysis of multistage cascade low temperature refrigeration systems used in olefin plants. International journal of refrigeration, 32(2), 279-294.

[10] Keshtkar, M. M., Talebizadeh, P. (2017). Multi-objective optimization of cooling water package based on 3E analysis: A case study. Energy, 134, 840-849.
[11] Rahdar, M. H., Heidari, M., Ataei, A., Choi, J. K. (2016). Modeling and optimization of R-717 and R-134a ice thermal energy storage air conditioning systems using NSGA-II and MOPSO algorithms. Applied Thermal Engineering, 96, 217-227.
[12] Najjar, Y. S., Abubaker, A. M. (2017). Thermoeconomic analysis and optimization of a novel inlet air cooling system with gas turbine engines using cascaded waste-heat recovery. Energy, 128, 421-434.
[13] Rezayan, O., Behbahaninia, A. (2011). Thermoeconomic optimization and exergy analysis of CO2/NH3 cascade refrigeration systems. Energy, 36(2), 888-895.
[14] Raja, B. D., Jhala, R. L., Patel, V. (2018). Multiobjective thermo- economic and thermodynamics optimization of a plate–fin heat exchanger. Heat Transfer—Asian Research, 47(2), 253-270.
[15] Dubey, A. M., Kumar, S., Agrawal, G. D. (2014). Thermodynamic analysis of a transcritical CO2/propylene (R744–R1270) cascade system for cooling and heating applications. Energy conversion and management, 86, 774-783.
[16] Parekh, A. D., Tailor, P. R., Sutaria, N. (2012). Thermoeconomic Optimization of Cascade Refrigeration System using Refrigerant Pair R404A-R508B. In Applied Mechanics and Materials (Vol. 110, pp. 677-684). Trans Tech Publications.
[17] Keshtkar, M. M. (2016). Effect of subcooling and superheating on performance of a cascade refrigeration system with considering thermo-economic analysis and multi-objective optimization. Journal of Advanced Computer Science and Technology, 5(2), 42-47.
[18] Toghyani, S., Kasaeian, A., Ahmadi, M. H. (2014). Multi-objective optimization of Stirling engine using non-ideal adiabatic method. Energy Conversion and Management, 80, 54-62.
[19] Sadatsakkak, S. A., Ahmadi, M. H., Ahmadi, M. A. (2015). Optimization performance and thermodynamic analysis of an irreversible nano scale Brayton cycle operating with Maxwell–Boltzmann gas. Energy Conversion and Management, 101, 592-605.
[20] Kaushik, S. C., Kumar, R., Arora, R. (2016). Thermo-economic optimization and parametric study of an irreversible regenerative Brayton cycle. Journal of Thermal Engineering (ICES2015).
[21] Heidarnejad, P. (2017). Exergy Based Optimization of a Biomass and Solar Fuelled CCHP Hybrid Seawater Desalination Plant. Journal of Thermal Engineering, 3(1), 1034-1043.
[22] Keshtkar, M. M. (2017). Performance analysis of a counter flow wet cooling tower and selection of optimum operative condition by MCDM-TOPSIS method. Applied Thermal Engineering, 114, 776-784.
[23] Wouagfack, P. A. N., Tchinda, R. (2011). Irreversible three-heat-source refrigerator with heat transfer law of QαΔ (T− 1) and its performance optimization based on ECOP criterion. Energy Systems, 2(3-4), 359-376.
[24] Kotas, T. J. (1986). Exergy method of thermal and chemical plant analysis. Trans IChemE, 64, 212-229.
[25] Bejan, A., Tsatsaronis, G., Moran, M., Moran, M. J. (1996). Thermal design and optimization. John Wiley & Sons.
[26] Aminyavari, M., Najafi, B., Shirazi, A., Rinaldi, F. (2014). Exergetic, economic and environmental (3E) analyses, and multi-objective optimization of a CO2/NH3 cascade refrigeration system. Applied Thermal Engineering, 65(1-2), 42-50.
[27] Rao, R. V., Saroj, A. (2018). Multi-objective design optimization of heat exchangers using elitist-Jaya algorithm. Energy Systems, 9(2), 305-341.
[28] Klein, S. A., Alvarado, F. L. (1992). EES: Engineering equation solver for the Microsoft Windows operating system. F-Chart software.
[29] Nellis, G., Klein, S. (2009). Mass transfer. Heat Transfer, Cambridge University Press, New York.
[30] Kays, W.M., London, A.L., "Compact Heat Exchangers", Krieger Publishing Company, (1984).
[31] Shah, R.K., Sekulic, D.P., "Fundamentals of Heat Exchanger Design", Wiley, (2003)
[32] Thirumaleshwar, M., "Software Solutions to Problems on Heat Transfer - Boiling and Condensation", (2013)
[33] Harrell Jr, F. E. (2015). Regression modeling strategies: with applications to linear models, logistic and ordinal regression, and survival analysis. Springer.
[34] Konak, A., Coit, D. W., Smith, A. E. (2006). Multi-objective optimization using genetic algorithms: A tutorial. Reliability Engineering & System Safety, 91(9), 992-1007.
[35] Holland, J. H. (1992). Adaptation in natural and artificial systems: an introductory analysis with applications to biology, control, and artificial intelligence. MIT press.

EXPERIMENTAL INVESTIGATION AND PERFORMANCE OPTIMIZATION OF A CROSS FLOW HEAT EXCHANGER BY ENTROPY GENERATION MINIMIZATION APPROACH

T. Mohapatra[1], S. K. Rout[1,*]

ABSTRACT

The exergy loss and performance optimization of a cross flow heat exchanger (CFHE) with air and hot water as working fluid have been experimentally investigated. Experiments are performed on various mass flow rates of hot water and air over a ranges of 0.015 $kg.s^{-1}$– 0.04 $kg.s^{-1}$ and 0.117 $kg.s^{-1}$ - 0.763 $kg.s^{-1}$ respectively. To validate the results of the present study, they are compared with available data in literature through which a reasonably a good agreement is obtained between them. This study demonstrates the successful application of Taguchi approach for optimal design of the (CFHE). The effects of design parameters and two different operating conditions such as the fin pitch, the inside tube diameter and the mass flow rate of water and air on exergy loss are investigated. In the Taguchi experimental design method, exergy loss is considered as performance parameter and it is revealed that parameter combinations such as$\dot{m}_h$=0.0163 $kg.s^{-1}$, $\dot{m}_a$=0.1175 $kg.s^{-1}$, d_i=0.0199 m and F_p=0.00259 m resulted optimum performance. Contribution ratios of each parameter on exergy loss are assessed. Mass flow rate of air is determined to be the most effective parameter on exergy loss with a contribution ratio of 51.26 % followed by mass flow rate of hot water, fin pitch, and diameter contribution ratios of 35.91 %, 7.02 % , 5.79 % respectively.

Keywords: *Experimental Investigation, Cross Flow Heat Exchanger, Performance Optimization, Exergy Analysis, Taguchi Method*

INTRODUCTION

In various industries, power plants, chemical process plants, petroleum refineries and air conditioning sectors etc. the use of cross flow heat exchangers is vast. The efficiency of the heat exchangers affects the performance of the systems significantly. Hence optimization of heat exchanger parameters and low cost have been important research topics since last two decades or more. Various approaches adopted by researchers are shape of heat exchanger, types of extended surfaces, mal distribution, Nano fluid, disturbance elements etc. to enhance the heat transfer using various methods like entropy generation principle[1], genetic algorithm[2], Taguchi method[3], exergy loss method [4] and so on. Sahin et al. [5] used the Taguchi method to conduct an experimental investigation for identifying the effects of the longitudinal fin pairs and lateral separations of consecutively on the thermos hydraulic performance. Yakut et al. [6] studied the performance of the hexagonal fins, span-wise and stream-wise distances between fins and flow velocity on heat transfer and pressure drop characteristics using Taguchi method. Kotcioglu et al. [7] experimentally analyzed the heat transfer and pressure drop in a cross-flow heat exchanger with hexagonal, square and circular pin-fin arrays. Gunes et al. [8] determined of the optimum values of the design parameters in a tube with equilateral triangular cross sectioned coiled wire inserts. Bahiraeiet al. [9] evaluated the entropy generation of a biological nanofluid containing silver nanoparticles in a liquid block heat sink for cooling of an electronic processor. They found from the analysis that irreversibility in the whole liquid block decreases with increasing either concentration or Reynolds number, which is a positive result based on second law of thermodynamics. Using CFD approach Bahiraeiet al. [10] studied the irreversibilities caused by heat transfer and friction for a power-law Nano fluid in a minichannel having chaotic perturbations were examined through calculation of entropy generation rates. Heshmatian et al [11] reported the irreversibilities caused by heat transfer and friction for the water-TiO_2 nanofluid flow in a circular microchannel by evaluating entropy generation rates. Heat transfer characteristics and second law of thermodynamics are evaluated by Bahiraei et al. [12] for the water–Al_2O_3 nanofluid flow in a triangular minichannel under constant wall heat flux. The convective heat transfers and analysis of the second law of

This paper was recommended for publication in revised form by Regional Editor Sandip Kale
[1]Department of Mechanical Engineering, C. V. Raman College of Engineering, Bhubaneswar, Odisha, India.
**Email: sachindra106@gmail.com*

thermodynamics for the non-Newtonian nanofluid flow containing TiO_2 nanoparticles within a narrow annulus was investigated by Bahiraei et al. [13].

Based upon the studies mentioned above, all of the process parameters affecting the exergy loss have not been investigated in detail. However, it has been reported that more investigations are required for cross flow of heat exchangers to optimize the design parameter based upon exergy loss. In the present investigation simultaneously both designing parameter and operating conditions are considered as process parameters for analysis using Taguchi approach where exergy loss is considered as the response parameter for designing and predicting of the CFHE.

OBJECTIVE

The objectives of the present work are

a) Validation and verification of the results of experimental approach by comparing the results of tube side convective heat transfer coefficients with the data available in the literature.

b) Performance optimization of the CFHE using entropy minimization approach.

EXPERIMENTS

Experimental setup and it's description

Experimental investigation and performance optimization of (CFHE) were carried out in the experimental set up shown in Fig. 1. The test section of the CFHE (16"x 22") is simply a fin and a tubetype cross flow heat exchanger. It consists 279 aluminum fins with a fin pitch of 2 mm and fin thickness of 0.3 mm.Tubes used in CFHE test sections were manufactured from "Type L" copper with inside tube diameter of 0.008525 m and number of tube rows was 2. Recent study was carried out for two different fluids i.e. hot water and air. Hot water was flowing in a closed loop through the tubes of CFHE test section transferring heat into air, which was forced through the fins of the CFHE test section in an open loop by a fan working as the cooling fluid.

Hot water loop consists of a steel tank, a centrifugal pump, a flow control valve, a rotameter, two digital temperature indicator and two pressure gauges. Hot water was prepared in a steel tank of 45 liter capacity equipped with an immersion heater of 1.5 kW and circulated along the hot water loop by a centrifugal pump of 0.5 HP capacity. For flow regulation and measurement of the hot water one flow control valve and the rotameter were connected after the pump. The operating range of the rotameter is 20 LPH to 200 LPH at a temperature is up to 90°C. Two digital temperature indicators (-50 °C to 200 °C) and two pressure gages are connected along the hot water loop at the inlet and outlet of the CFHE test section to measure the pressures and temperatures.

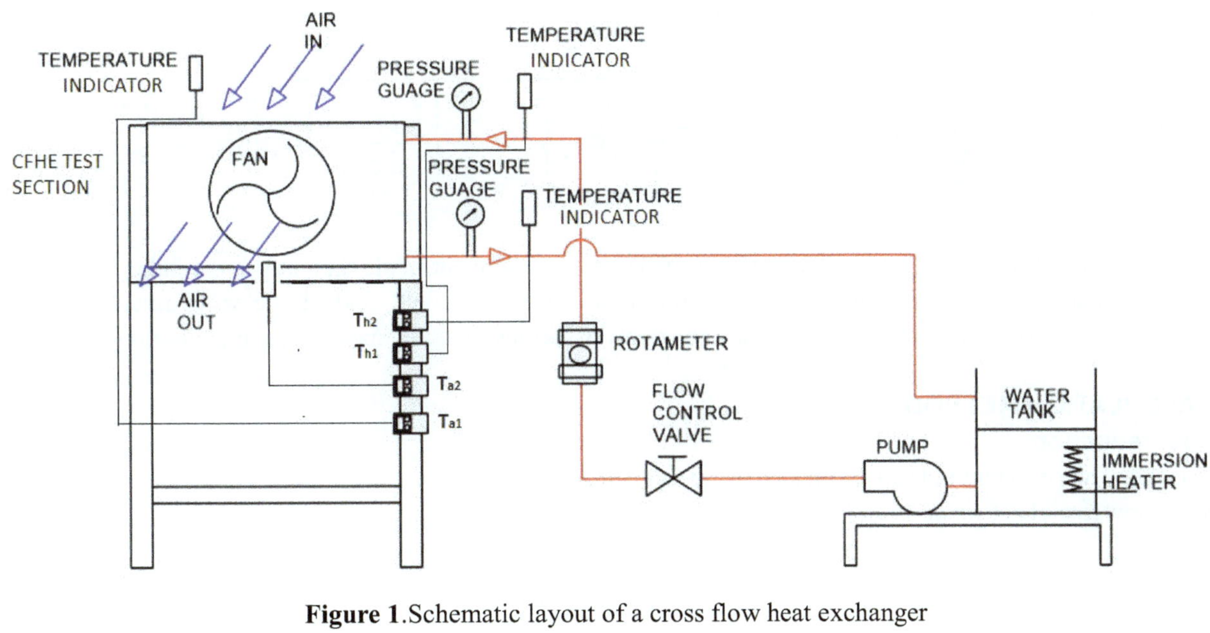

Figure 1.Schematic layout of a cross flow heat exchanger

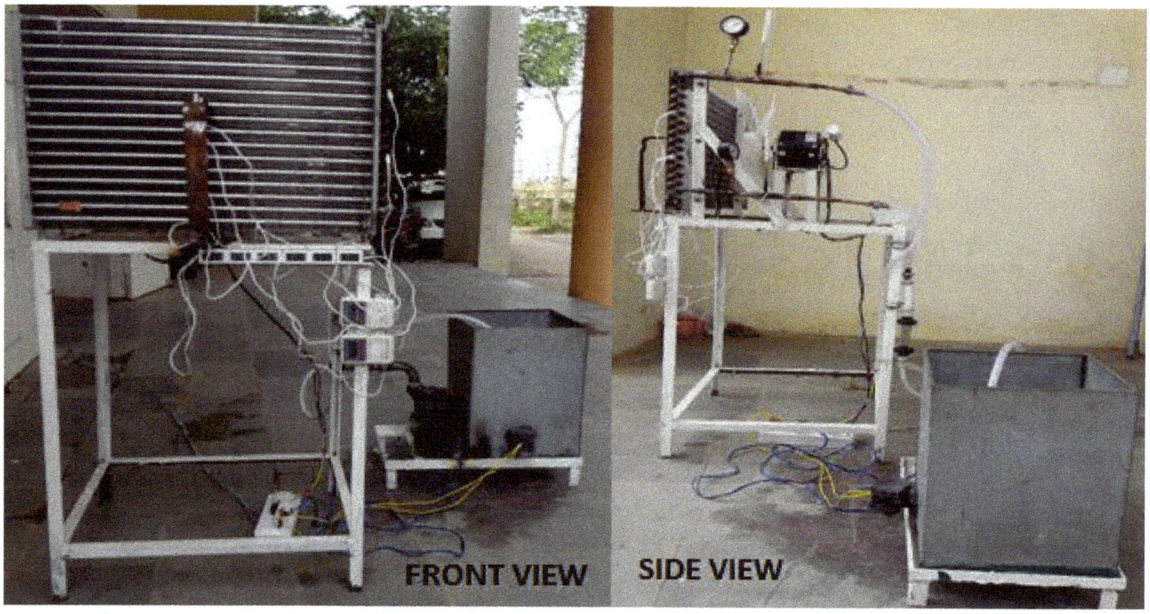

Figure 2. Experimental set up the (CFHE)

Air was forced through the CFHE test section by a fan of 14" diameter to take away the heat from the copper tubes through which hot water was flowing as shown in Fig. 2. A speed regulator was provided for the fan motor to supply air at three different speeds. Two digital temperature indicator (-50 oC to 200 oC) were connected along the open air loop to measure the temperatures of air at the inlet and outlet of the CFHE test section. An anemometer was used at the upstream i.e. in between supply fan and cross flow heat exchanger for the measurement of air velocity and at downstream air velocity decreases, when it passes over the fins and copper tubes of the CFHE.

UNCERTAINTY OF MEASUREMENT

In experimental studies, the results are typically associated with measurement errors. Uncertainty analysis is generally used to determine the experimental error and to quantify the degree of goodness of the measured values. The uncertainties in the experimental results or dependent variable (σr) can be expressed as a function of errors associated with different measuring instruments or independent variables (i.e. σx1, σx2, … ,σxn) as follows:

$$\sigma_r = \sqrt{\left[\left(\frac{\partial f}{\partial x_1}\right)\times\sigma_{x1}\right]^2 + \left[\left(\frac{\partial f}{\partial x_2}\right)\times\sigma_{x2}\right]^2 + \ldots + \left[\left(\frac{\partial f}{\partial x_n}\right)\times\sigma_{xn}\right]^2} \quad (1)$$

A type-B uncertainty associated) with temperature measurement in digital temperature indicator is ± 1.2 oC, volumetric flow rate measurement in rotameter is ± 0.324 L.min-1 and air velocity measurement in anemometer is ± 0.0176 m. s^{-1}. Therefore, the uncertainty associated with the measured exergy loss is (± 12.2%).

CALCULATION METHOD

Data reduction

Mass flow rate of both fluids are calculated using following formulas:
Mass flow rate of hot water:

$$\dot{m}_h = \rho_h.\dot{V}_h \quad (2)$$

Mass flow rate of air:

$$\dot{m}_a = \rho_a . \dot{V}_a \tag{3}$$

Heat transfer rate is calculated by
Hot water heat rejection rate:

$$\dot{Q}_h = \dot{m}_h . c_{p,h} . (T_{h,1} - T_{h,2}) \tag{4}$$

Air heat gain rate:

$$\dot{Q}_a = \dot{m}_a . c_{p,a} . (T_{a,2} - T_{a,1}) \tag{5}$$

Reynolds number are obtained from:
Hot water Reynolds number:

$$\mathrm{Re}_h = \frac{\rho_h . V_h . d_i}{\mu_h} \tag{6}$$

Air Reynolds number:

$$\mathrm{Re}_a = \frac{\rho_a . V_a . D}{\mu_a} \tag{7}$$

As it is observed hot water flow inside the tube is turbulent, the heat transfer coefficient, hh is calculated from Nusselt number correlation was provided by Petukhov-Popov.

$$Nu_h = \frac{(f/8)\,\mathrm{Re}_h\,\mathrm{Pr}_h}{K_1 + K_2 (f/8)^{1/2} (\mathrm{Pr}_h{}^{2/3} - 1)} \quad 10^4 < \mathrm{Re}_h < 5 \times 10^6 \; 0.5 < \mathrm{Pr} < 2000 \tag{8}$$

Now heat transfer co-efficient inside the condenser tube:

$$h_h = \frac{Nu_h . K_h}{d_i} \tag{9}$$

Similarly the heat transfer coefficient ha can be calculated from Nusselt number correlation for flow over cylindrical tube.

$$Nu_a = 0.027\,\mathrm{Re}_a{}^{0.805}\,\mathrm{Pr}_a{}^{0.334} \tag{10}$$

Now heat transfer co-efficient outside the condenser tube:

$$h_a = \frac{Nu_a . K_a}{d_o} \tag{11}$$

Overall heat transfer coefficient outside of the helical coil when heat transfer takes from hot water to air:

$$\frac{1}{U_o} = \frac{d_o}{h_h . d_i} + \frac{A_o . \ln\left(\frac{d_o}{d_i}\right)}{2\pi k_{cu} L} + \frac{1}{h_a} \tag{12}$$

ENERGY BALANCE

For the present study, we can write a steady state energy balance as follows.
From conservation of energy, it can be represented that
Heat rejected from hot water per unit time = Heat gain by air per unit time + Heat loss/time

$$\dot{Q}_w = \dot{Q}_a + \dot{Q}_{loss} \tag{13}$$

Where,

$$\dot{Q}_w = \dot{m}_w . c_{p,w} . (T_{w,1} - T_{w,2}) \tag{14}$$

$$\dot{Q}_a = \dot{m}_a . c_{p,a} . (T_{a,2} - T_{a,1}) \tag{15}$$

$\dot{Q}_{loss}$ = Heat loss/time (Which is may be radiation or any other unobserved heat loss/time)

It is observed during study that negligible heat loss/time takes place from CFHE to surrounding corresponding to increment in mass flow rates of hot water; therefore, the heat exchange process between hot water and air can be assumed to be adiabatic.

EXERGY BALANCE

Maximum useful work that is obtained reversibly in a specified environment is known as exergy. In the current study, assuming heat transfer process between hot water and air in CFHE is steady and adiabatic; overall exergy loss can be expressed as the summation of individual exergy loss related to the hot water and the air.
Considering $s = s(p,T)$, the change in specific entropy can be written as:

$$ds = \left(\frac{\partial s}{\partial T}\right)_p dT + \left(\frac{\partial s}{\partial p}\right)_T dp \tag{16}$$

Using thermodynamic relations, the change in specific entropy can also be written as:

$$ds = \frac{c_p}{T} dT - \left(\frac{\partial v}{\partial T}\right)_p dp \tag{17}$$

For a differential element the entropy variation of fluid can be given as:

$$dS = \dot{m}\left[\frac{c_p}{T} dT - \left(\frac{\partial v}{\partial T}\right)_p dp\right] \tag{18}$$

The total entropy generation in a heat exchanger is considered to be equal to the total entropy variation of both fluids. For a cross flow heat exchanger, integrating the above equation, i.e., Eq. (15), from the inlet to the outlet of air and hot water, the entropy generation rate can be written as follows:

$$S_g = \left[\int_1^2 \dot{m} c_p \frac{dT}{T}\right]_h + \left[\int_1^2 \dot{m} c_p \frac{dT}{T}\right]_a + \left[-\dot{m}\int_1^2 \left(\frac{\partial v}{\partial T}\right)_p dp\right]_h + \left[-\dot{m}\int_1^2 \left(\frac{\partial v}{\partial T}\right)_p dp\right]_a \tag{19}$$

Rearranging Eq. (16) yields:

$$S_g = S_{g,\Delta T} + S_{g,\Delta p} \tag{20}$$

Where, $S_{g,\Delta T}$ is the entropy generation contributed by heat transfer and $S_{g,\Delta p}$ is the entropy generation contributed by fluid friction.

As we know, the irreversibility of fluid flow due to viscous friction is negligible as compared to that of the irreversibility of heat transfer due to finite temperature difference. Therefore the irreversibility of fluid flow is neglected in this analysis and exergy [$T_o\left(S_g\right)$] balance performed.

$$E = E_h + E_a \tag{21}$$

Exergy losses related to the hot water flow inside the copper tubes can be written as:

$$E_h = T_o\left\{\dot{m}_h\left(\Delta S\right)_h\right\} \tag{22}$$

Where, $\left(\Delta S\right)_h = S_{h,2} - S_{h,1} = C_{p,h}\ln\left(\frac{T_{h,2}}{T_{h,1}}\right)$ (23)

Similarly, exergy loss related to air flow through the fins can be written as:

$$E_a = T_o\left\{\dot{m}_a\left(\Delta S\right)_a\right\} \tag{24}$$

Where, $\left(\Delta S\right)_a = S_{a,2} - S_{a,1} = C_{p,a}\ln\left(\frac{T_{a,2}}{T_{a,1}}\right)$ (25)

By substituting Eq. (20) in to Eq. (19) and Eq. (22) into Eq. (21), overall exergy loss can be expressed as $E = T_o\left\{C_{p,h}\ln\left(\frac{T_{h,2}}{T_{h,1}}\right) + C_{p,a}\ln\left(\frac{T_{a,2}}{T_{a,1}}\right)\right\}$ (26)

Dimensionless exergy loss can be determined:

$$e = \frac{E}{T_o C_{\min}} \tag{27}$$

Where, $C_{\min} = \text{Min}\left\{\dot{m}_h\, c_{p,h} \text{and}\, \dot{m}_a\, c_{p,a}\right\}$ (28)

TAGUCHI METHOD

Thermodynamic performance of CFHE can be enhanced considerably by using a reasonable optimization approach [13, 15]. Numbers of factors affect the thermal performance of the CFHE and out of these factors; the principal factors are determined by Taguchi method for optimum design. As reported by Gunes et al. [13]. Taguchi method is a standardized approach for determining the best combination of input parameters to get optimum result. This is accomplished through the design of experiments (DOE). It provides a method for quantitatively identifying just the right parameters that go together for optimal performance or service. Taguchi method is associated with three design stages i.e system design, parameter design and tolerance design. In this paper, the optimum values of system parameters or thermal performance parameters i.e. E is determined from parameter design stage. Signal to noise ratio (SNR or S/N ratio) measures the variations in performance parameters. S is the signal stands for mean and N is the noise stands for standard deviation in Taguchi approach. Three types of quality characteristics "larger is the better"; "nominal is the best" and "smaller is the better" are used to analyze S/N ratio. For the current study following steps of Taguchi technique are used.
Step 1 Identification of the objectives: Minimum exergy loss for the flow in the CFHE is identified as the objective of this work.

Table 1. Design factors and their levels

Control Factors	$\dot{m}_h$, kg.s^{-1}	$\dot{m}_n$, kg. s^{-1}	d_i, m	F_p, m
Level 1	0.0163	0.1175	0.0085	0.0016
Level 2	0.0273	0.2051	0.0138	0.002
Level 3	0.0382	0.0138	0.0199	0.0025

Step 2 Selection of characteristics: "Smaller is the better" quality characteristic is preferred for exergy loss, E. As available energy is higher at minimum exergy loss, hence "Smaller is the better" quality characteristics is preferred.Step 3 Selection of various factors and their levels: In this study, mass flow rate of hot water , mass flow rate of air , tube diameter (d_i) and fin pitch (F_p) are selected as the control factors for influencing exergy loss in the CFHE. The levels of control factors are represented in Table 1.

Table 2. L9 orthogonal array for control factors

No.	$\dot{m}_h$, kg.s^{-1}	$\dot{m}_h$, kg.s^{-1}	$d_{i,\,m}$	$F_{p,\,m}$
1	0.01638	0.1175	0.008525	1.64
2	0.01638	0.2051	0.0138	2
3	0.01638	0.7801	0.0199	2.59
4	0.0273	0.1175	0.0138	2.59
5	0.0273	0.2051	0.0199	1.64
6	0.0273	0.7801	0.008525	2
7	0.03822	0.1175	0.0199	2
8	0.03822	0.2051	0.008525	2.59
9	0.03822	0.7801	0.0138	1.64

Step 4 Selection of an orthogonal array: Taguchi method is well known for design of experiments by employing a standard table known as orthogonal array for study of various design parameters relative to single quality characteristics. A suitable orthogonal array is selected depending on the number of factors with their levels, degrees of freedom, objective of the study, available resources. Minitab 16.0 is used for the analysis. For four control factors with three levels, L9 orthogonal array is generated and shown in Table 2.
Step 5 Conduction of experiments and result analysis: Orthogonal array design provides the idea about the number of experiments to be conducted in Taguchi method analysis. In this study total 9 numbers of experiments were conducted and results are presented in Table 3 for analysis. The effect of control factors on exergy loss of the CFHE are measured by SNR or S/N (signal to noise) ratio.

For exergy loss, E following formula is used during the analysis of SNR or S/N ratio for "smaller is the better" quality characteristics.

$$SNR_S = -10\log_{10}\left(E^2\right) \tag{29}$$

The optimum levels of various design parameters are obtained according to the greatest value of S/N ratio and contribution ratios of these factors on exergy loss of the CFHE are calculated as shown in Table 4.

RESULTS AND DISCUSSIONS

Validation of experimental approach

The results of the experimental approach i.e. convective heat transfer coefficient of hot water, $h_h$were compared with literature for cross flow configuration and the results are shown in Fig. 3. For verification and validation of the results of this study, the calculated convective heat transfer of hot water, h_h, were compared with the results of correlation values presented by Gnielinski [14]. There is a good agreement observed between the results of the experimental approach with the literature and thus validated.

Table 3. Mean and S/N ratio of exergy loss, E

RUN	$\dot{m}_h$, kg.s^{-1}	$\dot{m}_n$, kg. s^{-1}	d_i,m	$F_{p,\,m}$	Exergy loss, *E*	Responses	
						Mean	S/N Ratio
1	0.01638	0.1175	0.008525	1.64	3.451105949	3.451105949	-10.75916585
2	0.01638	0.2051	0.0138	2	4.328802392	4.328802392	-12.72735522
3	0.01638	0.763	0.0199	2.59	5.867249014	5.867249014	-15.36869041
4	0.0273	0.1175	0.0138	2.59	4.29422205	4.29422205	-12.65768995
5	0.0273	0.2051	0.0199	1.64	5.900487087	5.900487087	-15.41775728
6	0.0273	0.763	0.008525	2	9.006400795	9.006400795	-19.09102539
7	0.03822	0.1175	0.0199	2	4.928097633	4.928097633	-13.85358607
8	0.03822	0.2051	0.008525	2.59	6.832268443	6.832268443	-16.69129843
9	0.03822	0.763	0.0138	1.64	11.73446402	11.73446402	-21.38926515

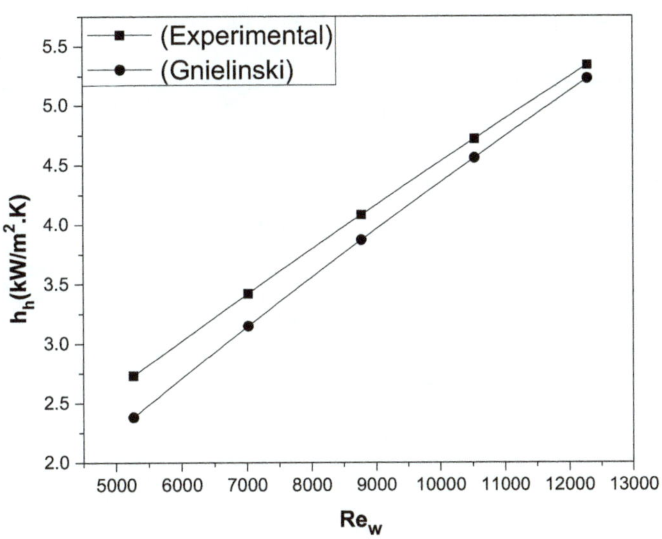

Figure 3. Validation of experimental approach

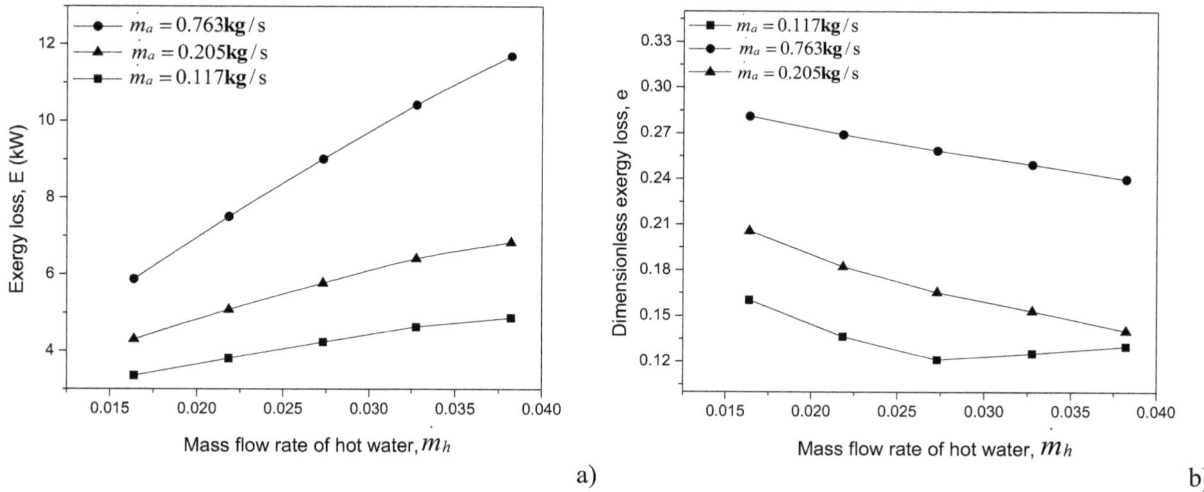

Figure 4. Effect of mass flow rate hot water and air on (a) exergy loss, (b) dimensionless exergy loss

Fig. 4 illustrates the effect of mass flow rate of hot water on exergy loss of the CFHE at different mass flow rates of air. The investigation is conducted at fix dimensions of heat exchanger for all cases. It is observed from Fig. 4 (a) that exergy loss increases with increment in mass flow rates of hot water and air, because more amount of hot water or air flow rate can cause more amount of Reynolds number through the cross flow heat exchanger. It is obvious that higher Reynolds number causes higher heat transfer rate. On the other hand, more heat transfer rate implies more finite temperature difference which is the reason of exergy loss.

In Fig. 4 (b), it is observed that dimensionless exergy loss decreases with the increase of hot water flow rate and increases with the increase of air flow rate. As mentioned in Eq. (23), Dimensionless exergy loss is dependent upon exergy loss and product of ambient temperature, minimum heat capacity. Therefore increment or decrement in dimensionless exergy loss is dependent upon the value of numerator and denominator of Eq. (23). It is observed during analysis that increment of denominator is dominant over increment of numerator for enhancement of hot water flow rate; whereas for enhancement of air flow rate it is reversed. Hence, enhancement of air flow rate increases the dimensionless exergy loss and the increment of hot water flow rate decreases the dimension less exergy loss.

The effect of fin pitch and tube diameter on exergy loss

Fig. 5 demonstrates the effect of fin pitch and tube diameter on exergy loss of the cross flow heat exchanger. The investigation is made at fix of other dimension of heat exchanger and air flow rate for all cases. It is observed from this figure that exergy loss and dimensionless exergy loss increases with increase in inside tube diameter, but exergy loss and dimensionless exergy loss negligibly change with increase in fin pitch.

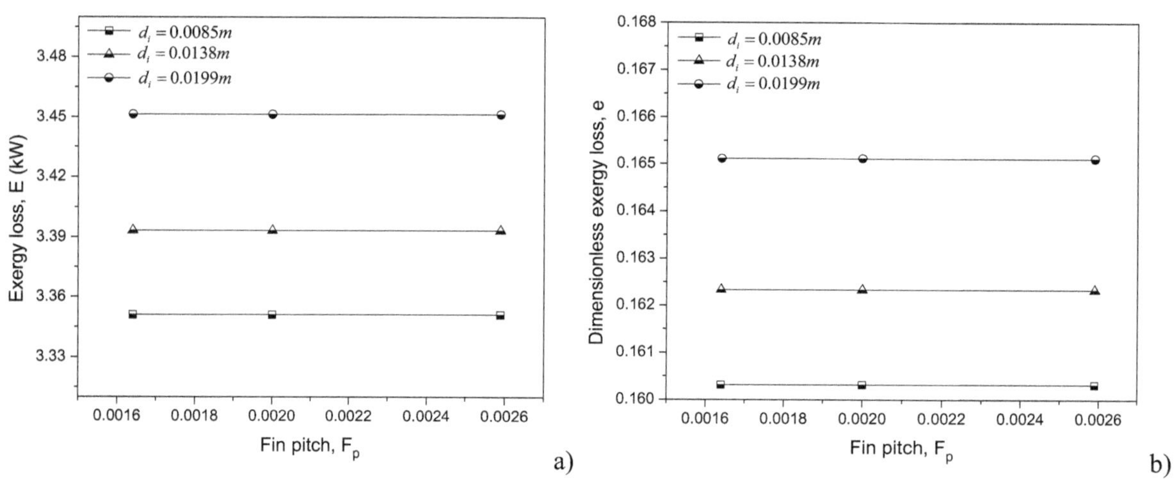

Figure 5. Effect of fin pitch and tube diameter on (a) exergy loss, (b) dimensionless exergy loss

Optimum heat transfer enhancement condition

The calculated SNR or S/N ratios for 9 numbers of experiments are shown in Table 3. The contribution of each factor on exergy loss of the CFHE is presented in Table 4, where R is the difference between the maximum and minimum S/N ratio for each factor, contribution ratio is the ratio of R value of any factor to the total R values of all factors and rank represents the orderly assignment of factor with respect to minimum exergy loss. The optimum condition is calculated for a combination of levels of control factors having the largest S/N ratio.

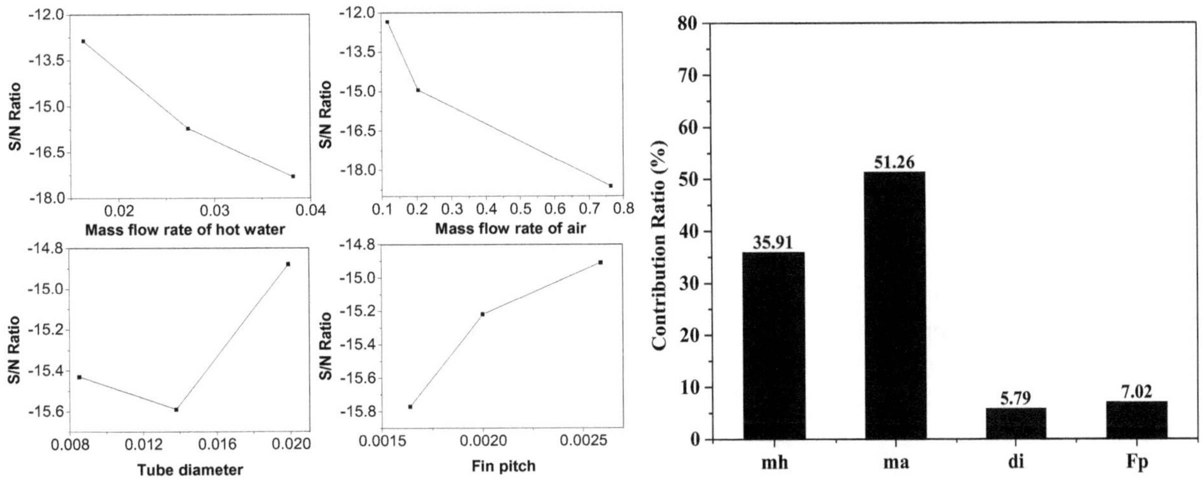

Figure 6. Effects and contribution ratio of each design parameter on exergy loss, E

The effect of each design parameters on exergy loss, E, is shown in Fig. 6. The optimum values of these design parameters for minimum exergy loss, E, is obtained for, $\dot{m}_h$=0.0163 kg.s^{-1}, $\dot{m}_a$=0.1175kg.s^{-1}, d_i=0.0199 m and F_p=0.00259 m according to "smaller is the better" quality characteristics. Using these optimum values another experiment is conducted and exergy loss is calculated about 3.35110719 kW. The contribution ratio of each parameter on exergy loss is shown in Fig. 6 and Table 4. It is clear from figure that the mass flow rate of hot water is the most effective parameter on exergy loss with a contribution ratio of 51.26 % followed by mass flow rate of air , fin pitch, Fp and tube diameter, di with the contribution ratio of 35.91 %, 7.02 % and 5.79 % respectively. It is observed from the results that tube diameter and fin pitches have the least effect on exergy loss.

Table 4. Contribution of design parameters of the CFHE on exrgy loss, E

	Level	$\dot{m}_h$	$\dot{m}_a$	d_i	F_p
S/N Ratio	1	-12.87	-12.34	-15.43	-15.77
-	2	-15.72	-14.95	-15.59	-15.22
-	3	-17.31	-18.62	-14.88	-14.91
R	-	4.4	6.28	0.71	0.86
Rank	-	2	1	4	3
Contribution Ratio %		0.3591	0.5126	0.0579	0.0702

CONCLUSIONS

In this study, experimental investigation and performance optimization of the CFHE was performed. Taguchi method was used for determination of optimum design parameters for minimum exergy loss in CFHE. The results of this work can be summarized as follows.

1. For minimum exergy loss in CFHE, the optimum value of different design parameters were obtained as mass flow rate of hot water is 0.0163 kg/s, mass flow rate of air is 0,117 kg/s, tube diameter is 0.0199 m and fin pitch is 0.00259 m.
2. Contribution ratios of each design parameter on exergy loss were determined. Mass flow rate of hot water is found out the most effective parameter on exergy loss with a contribution ratio of 51.26 % followed by mass flow rate of air , fin pitch, Fp and tube diameter, di with the contribution ratio of 35.91 %, 7.02 % and 5.79 % respectively.

NOMENCLATURE

A	Area [m^2]
c_p	Specific heat [kJ/kgK]
C	Heat capacity [kW/ K]
d	Diameter of the tube [m]
E	Exergy loss [kW]
F_p	Fins pitch [m]
h	Convective heat transfer coefficient [kW/m^2K]
K	Thermal conductivity [kW/mK]
L	Length of the copper tube [m]
$\dot{m}$	Mass flow rate [kg/s]
Nu	Nusselt number
p	Pressure [kPa]
Pr	Prandtl number
$\dot{Q}$	Rate of heat transfer [kW]
Re	Reynolds number
S	Entropy [kJ/K]
s	Specific entropy [kJ/kgK]
SNR	Signal to noise ratio
T	Temperature, K
U	Overall heat transfer coefficient [Kw/m^2K]
V	Velocity [m/s]
$\dot{V}$	Volumetric flow rate [m^3/sec]
ρ	Density [kg/m^3]
υ	Kinematic viscosity [m^2/s]
μ	Dynamic Viscosity [Ns/m^2]
1	Inlet
2	Outlet
a	air
Cu	Copper
h	Hot water
I	Inner
Min	Minimum
o	Outer
S	Smaller
w	Water

REFERENCES

[1] Ko, T.H. (2006). Thermodynamic analysis of optimal curvature ratio for fully developed laminar forced convection in a helical coiled tube with uniform heat flux. International Journal of Thermal Sciences, 45(7), 729-737.

[2] Imran, M., Pambudi, N.A., Farooq, M. (2017).Thermal and hydraulic optimization of plate heat exchanger using multi objective genetic algorithm. Case Studies in Thermal Engineering, 10, 570-578.

[3] Kotcioglu, I., Cansiz, A., Nasiri Khalaji, M. (2013). Experimental investigation for optimization of design parameters in a rectangular duct with plate-fins heat exchanger by Taguchi method. Applied Thermal Engineering, 50, 604-613.

[4] Ipek, O., Kılıc, B., Gurel, B. (2017). Experimental investigation of exergy loss analysis in newly designed compact heat exchangers. Energy, 124, 330-335.

[5] Sahin, B. (2005). Optimum design parameters of a heat exchanger. Applied Energy, 82(1), 90-106.

[6] Yakut, K. (2006). Experimental investigation of thermal resistance of a heat sink with hexagonal fins. Applied Thermal Engineering, 26(17), 2262-2271.

[7] Kotcioglu, I., Caliskan, S., Baskaya S. (2011). Experimental study on the heat transfer and pressure drop of a cross-flow heat exchanger with different pin–fin arrays. Heat and mass transfer, 47(9), 1133.

[8] Bahiraei, M., Heshmatian, S. (2017). Application of a novel biological nanofluid in a liquid block heat sink for cooling of an electronic processor: Thermal performance and irreversibility considerations. Energy Conversion and Management, 149, 155–167.

[9] Bahiraei, M., Gharagozloo, K., Alighardashi, M, Mazaheri, N. (2017). CFD simulation of irreversibilities for laminar flow of a power-law nanofluid within a minichannel with chaotic perturbations: An innovative energy-efficient approach. Energy Conversion and Management, 144, 374-387.

[10] Heshmatian, S., Bahiraei, M. (2017). Numerical investigation of entropy generation to predict irreversibilities in nanofluid flow within a microchannel: Effects of Brownian diffusion, shear rate and viscosity gradient. Chemical Engineering Science, 172, 52-65.

[11] Bahiraei, M., Mohammadi, S. (2016). Prediction of entropy generation for nanofluid flow through a triangular minichannel using neural network. Advanced Powder Technology, 27, 673-683.

[12] Bahiraei, M., Alighardashi, M. (2016). Investigating non-Newtonian nanofluid flow in a narrow annulus based on second law of thermodynamics. Journal of Molecular Liquids, 219, 117-127.

[13] Gunes, S., Manay, E., Senyigit, E., Ozceyhan, V. (2011). A Taguchi approach for optimization of design parameters in a tube with coiled wire inserts. Applied Thermal Engineering, 31, 2568-2577.

[14] Gnielinski, V. (1976). New equation for heat and mass transfer in turbulent pipe and channel flow, International Chemical Engineering, 16, 359-368.

[15]Rout S.K., Choudhury B. K, Sahoo R. K., Sarangi S. K. (2014). Multi-objective parametric optimization of Inertance type pulse tube refrigerator using response surface methodology and non-dominated sorting genetic algorithm.Cryogenics 62, 71-83.

PERFORMANCE ASSESSMENT OF PHOSPHORIC ACID FUEL CELL - THERMOELECTRIC GENERATOR HYBRID SYSTEM WITH ECONOMIC ASPECT

Fatma Gül Çelik[1], Emin Açıkkalp[1*], Hasan Yamik[1]

ABSTRACT

Purpose of this paper is to evaluate phosphoric acid fuel cell (PAFC) - thermoelectric generator hybrid system with economic and thermoeconomic point of view. Firstly, basic equations of PAFC, thermoelectric generator and hybrid system are described. Secondly, basic performance parameters like power output, energy efficiency, exergy efficiency and exergy destruction rates are investigated. Finally, cost equations are set up to determine economic results of the considered system, in addition to that, these system are considered by using EXCEM analysis. According to results, the maximum total cost of the hybrid system is obtained j= 10900 am^{-2}, exergy loss ratio to capital cost (*ec*) of the hybrid system increases dramatically after the point where is j= 11000 am^{-2}. Maximum power density, maximum energy efficiency and of the hybrid system are 8735.340 wm^{-2}, 81.35% and 86.6% respectively.

Keywords: *Phosphoric Acid Fuel Cell, Thermoelectric Generator, Hybrid System, Economic Evaluation, EXCEM Analysis.*

INTRODUCTION

Fuel cell technology is the conspicuous technology because of their nearly harmless to environment, relatively higher efficiency and their flexible fuel option. Nowadays, one of the most important problem of engineers and scientists are design to systems that convert to energy as most efficiently that enable us environmental friendly usage of them. Efficiency of a heat engine are restricted by the Carnot efficiency, however, this limitations is not valid for fuel cells. Because, fuel cell generates electricity in fuel cell by electrochemical reaction. Irreversible fuel cells including Solid oxide fuel cells (SOFC), molten carbonate fuel cell (MOFC), proton exchange membrane (PEM) fuel cell, phosphoric acid fuel cell (PAFC) and direct carbon fuel cell (DCFC) are investigated [1-5]. Zhang et al. considered PAFC in [5], they investigated cell voltage for different parameters including temperature, resistances, power and efficiency of the fuel cell according to current density, load resistance and mole fraction. They defined the optimum operating conditions.

Another opportunity offered by fuel cells are that heat rejected by the fuel cell can be used in another process and fuel cell - heat engine/refrigeration hybrid systems may be designed. In last decade, irreversible fuel cell and hybrid systems have been gained attention. In ref. [6], a hybrid PAFC- absorption refrigerator system was investigated. Power densities, efficiencies are studied under different conditions like temperature, pressure and phosphoric acid mole fraction. General performance characteristics and optimum criteria of the hybrid system are shown. PAFC-heat driven refrigerator cycle was taken into account in ref. [7]. System was researched for power density and efficiency and variable parameters were defined according to current density and refrigeration temperature, finally optimum conditions and performance characteristics were obtained. Performance of different fuel cell hybrid systems can be found literature; in refs. [8-10] fuel cell- Stirling, in refs. [11-15] fuel cell – heat engine, in refs. [16-25] fuel cell- Brayton, in refs. [26,27] fuel cell-Braysson hybrid systems were investigated for different conditions and performance criteria. In addition, novel hybrid systems including fuel cell- thermionic generator and fuel cell – Super critical CO_2 – Organic Rankine cycle can be seen in refs. [28, 29].

Thermoelectric generators may be an alternative for creating hybrid system with fuel cells. They have several advantageous like; they don't have moving part, noiseless and they are not required maintenance

---

This paper was recommended for publication in revised form by Regional Editor Hakan Demir
[1]Department of Mechanical Engineering, Engineering Faculty, Bilecik S.E. University, Bilecik, Turkey
E-mails: eacikkalp@gmail.com, emin.acikkalp@bilecik.edu.tr

frequently. A thermoelectric generator is used heat to generate electricity based on the Seebeck effect and similarly they can be utilized as cooler on the principle of the Peltier effect. Some authors have been researched about optimization of thermoelectric devices [28-45].In the open literature, one can find papers about fuel cell-thermoelectric generators hybrid systems. Chen et al, analyzed PAFC-thermoelectric hybrid system for the maximum power output conditions [46]. Zhao et. al. investigated direct carbon fuel cell-thermoelectric generator hybrid system [47]. Chen et al. analyzed a PEM- thermoelectric hybrid system in terms of power output and efficiency [48]. Another alternative to evaluate and design thermal cycles is finite time thermodynamics and constructional optimization. Some example of these may be found in [49-53] and these methods can be applied for the fuel cell hybrid systems widely.

In the open literature, authors are no aware of any study about PAFC- thermoelectric hybrid system in terms of economical and thermoeconomic aspects. Such techniques seek to determine of the appropriate allocation of economic resources so as to optimize the design and/or operation of a system; and/or the economic feasibility and profitability of a system [54]. In this paper, this shortage is tried to be completed. Firstly, thermodynamic relations are expressed for the fuel cell and thermoelectric generator. Secondly, PAFC-thermoelectric generator hybrid system is analyzed with regard to power output, energy efficiency, exergy efficiency and exergy destruction rate. After these, cost of the hybrid system and thermoeconomic analysis are performed. EXCEM method [54] is chosen for the thermoeconomic evaluation. Results are obtained numerically, some important outcomes are discussed and suggestions are made.

THERMODYNAMIC ANALYSIS

PAFC is made of phosphoric acid as electrolyte and two electrodes which are anode and cathode. Air is provided to cathode and hydrogen is provided to anode and electrochemical reaction occurred in the fuel cell is $H_2+1/2\ O_2 \longrightarrow H_2O$ + heat + electricity. This produced heat resulted from the electrochemical reaction that can be used in another process. Operating temperature of PACF is about 150-200 °C which can be used in cogeneration applications for buildings and low temperature processes. Schematic of the hybrid system is shown in fig.1. In this paper, PAFC - thermoelectric hybrid system is analyzed in terms of economical aspect. Firstly, equations for fuel cell are presented. Maximum theoretical potential or Nernst equation is [5]:

$$U_{rev} = -\frac{\Delta g^o}{n_e F} + \frac{RT}{n_e F} ln\left(\frac{p_{H_2}\left(p_{O_2}\right)^{0.5}}{p_{H_2O}}\right) \tag{1}$$

Overpotentials cause the irreversible losses. These are activation overpotential described as activation overpotential, concentration overpotential and ohmic overpotential are shown in eqs. (2-4) respectively [5].

$$U_{act} = \frac{RT}{\alpha n_e F} ln\left(\frac{j}{j_o}\right) \tag{2}$$

$$U_{con} = me^{(nj)} \tag{3}$$

$$U_{ohm} = j\frac{t_{ele}}{\kappa} \tag{4}$$

where j is operating current density, j_o is exchange current density, R is the universal gas constant, α is the charge transfer coefficient, n_e number of electrons and F is the Faraday's constant, T is the operating temperature of the fuel cell, m and n are two constant, t_{ele} is the thickness of the electrode and κ is the conductivity of phosphoric acid solution. The Output voltage of the PAFC is expressed as [5]:

$$U_f = U_{rev} - U_{act} - U_{con} - U_{ohm} \tag{5}$$

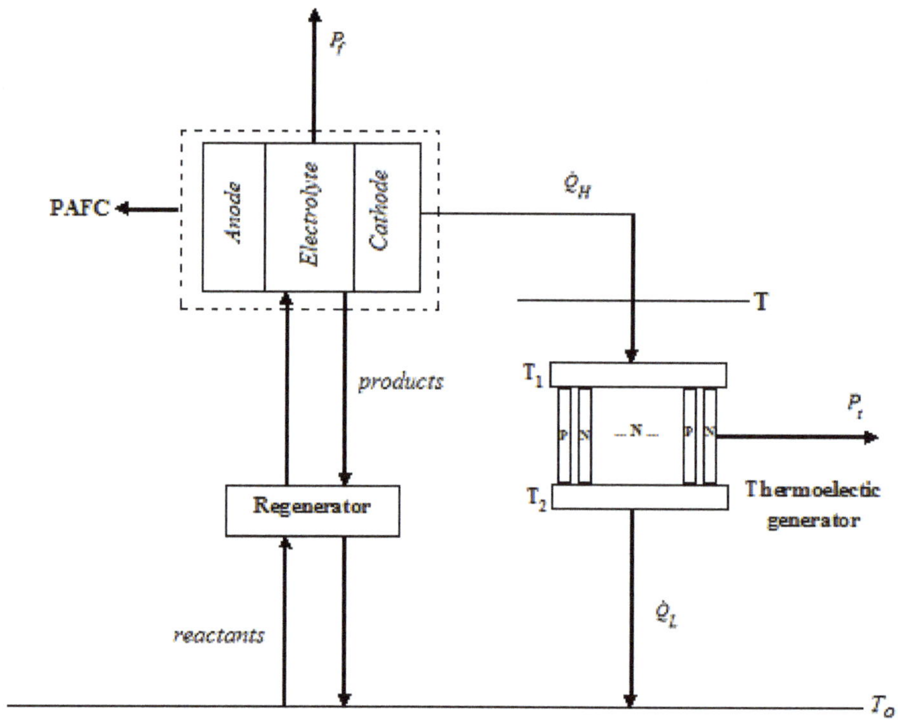

Figure 1. PAFC - thermoelectric generator hybrid system.

Power output and efficiency of the PAFC are calculated as eqs. (6,7) [5]:

$$P_f = U_f jA \tag{6}$$

$$\eta_f = \frac{P_f}{-\Delta\dot{H}} \tag{7}$$

where A is the polar plate area, $-\Delta\dot{H}$ is the total energy provided to fuel cell per unit time and Δh is the molar enthalpy change at the operating temperature. $-\Delta\dot{H}$ is described in eq. (8) [5]:

$$-\Delta\dot{H} = -\frac{jA\Delta h}{n_e F} \tag{8}$$

Exergy efficiency of the PAFC is:

$$\varphi_f = \frac{P_f}{-\Delta\dot{G}} \tag{9}$$

where $-\Delta\dot{G}$ Gibbs energy change per unit time and Δg is the molar Gibbs energy change. $-\Delta\dot{G}$ is defined as [1]:

$$-\Delta\dot{G} = -\frac{jA\Delta g}{n_e F} \tag{10}$$

Exergy destruction rate of the fuel cell is [3]:

$$Exd_f = \left(-\frac{\Delta h}{n_e Fa} - U_{cell} \right) jA \tag{11}$$

Heat transfer occurred in the regenerator is [6]:

$$\dot{Q}_r = \theta(1-\varepsilon_r)(T-T_o) \tag{12}$$

where θ heat conductance, ε_r, regenerator effectiveness and T_o is the ambient temperature. Heat released from the fuel cell is:

$$\dot{Q}_{H,f} = -\Delta\dot{H} - P_f - \dot{Q}_r \tag{13}$$

Secondly analysis of the thermoelectric generator must be performed. Heat input of thermoelectric generator can be defined in eqs. (14) and (16) [36]. These are equal as well as heat rejected from the PAFC in eq. (13). Heat rejected form the thermoelectric generator is shown in eqs. (15) and (17) [36].

$$\dot{Q}_{H,t} = N\left(\beta I T_1 - I^2 \frac{r}{2} + K(T_1 - T_2) \right) \tag{14}$$

$$\dot{Q}_{L,t} = N\left(\beta I T_2 + I^2 \frac{r}{2} + K(T_1 - T_2) \right) \tag{15}$$

$$\dot{Q}_{H,t} = k_H a_t (T - T_1) \tag{16}$$

$$\dot{Q}_{L,t} = k_L a_t (T_2 - T_o) \tag{17}$$

Power supplied from the thermoelectric generator is:

$$P_t = \dot{Q}_{H,t} - \dot{Q}_{L,t} \tag{18}$$

where β is Seebeck coefficient, T_1 is the temperature of the hot junction, T_2 is the temperature of the cold junction, N is the number of thermoelectric units, I is the electrical current generating in semi- conductor couple, r is the electrical resistance, K is the thermal conductance of the semi-conductance couple, k_H and k_L are the heat transfer coefficients and a_t is the heat transfer area of the heat exchangers. In this study, calculations are conducted for maximum power output of the thermoelectric generator. Optimum electrical current for maximum power can be defined as [36]:

$$I = \frac{\beta(T_1 - T_2)}{2r} \tag{19}$$

Number of modules required to heat input can be calculated by equalizing eqs. (14) and (16):

$$N = \frac{8a_t k_H r(T - T_1)}{(T_1 - T_2)\left(8Kr + \beta^2 (3T_1 + T_2)\right)} \tag{20}$$

Similarly T_2 is solved by equalizing eqs. (15) and (17) and result is:

$$T_2 = \frac{3\beta^2 k_H T - 8Kk_L r - 3\beta^2 k_H T_1 - 3\beta^2 k_L T_1 + \beta^2 k_L T_o}{2\alpha^2 k_L} + \frac{\sqrt{\left(8Kk_L r + \beta^2 (3k_H T_1 - 3k_H T + 3k_L T_1 - k_L T_o)\right)^2 + 4\beta^2 k_L \left(8Kr\left(k_H (T - T_1) + k_L T_o\right) + \beta^2 T_1 \left(k_H (T - T_1) + 3k_L T_o\right)\right)}}{2\beta^2 k_L} \tag{21}$$

Finally, T_1 is defined as using (13) and (16):

$$T_1 = -\frac{\left(-\frac{A\Delta hj}{Fn_e} - a_t k_H T - (1 - \varepsilon_r) k_r (T - T_o) - \frac{Aj\left(-\Delta g - e^{nj} Fmn_e - \frac{Fjn_e t_{ele}}{\kappa} - \frac{RT\, ln\left(\frac{j}{j_o}\right)}{\alpha} + RT\, ln\left(\frac{p_{H_2} (p_{o_2})^{0.5}}{p_{H_2O}}\right)\right)}{Fn_e} \right)}{a_t k_H} \tag{22}$$

Thermoelectric properties depending on temperature are shown in (23) - (28) [35]:

$$\beta = 2 \times (22224 + 930.6T_m - 0.9905T_m^{\ 2}) \times 10^{-9} \tag{23}$$

$$\rho = (5112 + 163.4T_m - 0.6279T_m^{\ 2}) \times 10^{-10} \tag{24}$$

$$\lambda = (62605 - 277.7T_m + 0.413T_m^{\ 2}) \times 10^{-4} \tag{25}$$

$$T_m = \left(\frac{T_1 + T_2}{2}\right) \tag{26}$$

$$r = \left(\frac{\rho L_1}{A_1} + \frac{\rho L_2}{A_2}\right) \tag{27}$$

$$K = \left(\frac{\lambda A_1}{L_1} + \frac{\lambda A_2}{L_2}\right) \tag{28}$$

where T_m is the mean temperature, L_1 is the length of p type material, L_2 is the length of n type material and, similarly, A_1 is the area of p type material, A_2 is the length of n type material. Efficiency and reversible power of the thermoelectric generator are shown in eqs. (29) and (30) respectively.

$$\eta_t = \frac{P_t}{\dot{Q}_{H,t}} \tag{29}$$

$$P_{rev,t} = Q_{H,t}\left(1-\frac{To}{T}\right)-Q_{L,t}\left(1-\frac{To}{T_L}\right) \tag{30}$$

Exergy efficiency and exergy destruction rate are expressed in eqs. (31) and (32) respectively.

$$\varphi_t = \frac{P_t}{P_{rev,t}} \tag{31}$$

$$Exd_t = T_o\left(\frac{\dot{Q}_{L,t}}{T_L}-\frac{\dot{Q}_{H,t}}{T}\right) \tag{32}$$

Finally, thermodynamic parameters for PAFC- thermoelectric generator is shown in (33) - (36). These are power output, efficiency, exergy efficiency and exergy destruction rate respectively.

$$P_{hy} = P_f + P_t \tag{33}$$

$$\eta_{hy} = \frac{P_{hy}}{-\Delta\dot{H}} \tag{34}$$

$$\varphi_t = \frac{P_{hy}}{-\Delta\dot{G}+P_{rev,t}} \tag{35}$$

$$Exd_{hy} = Exd_f + Exd_t \tag{36}$$

Firstly, hybrid system is investigated by using thermoeconomic method called EXCEM [54] where *EC* is ratio of thermodynamic loss rate (in this paper exergy loss rate, which is described as is ratio of rate, which is difference of exergy output from the exergy input, is considered) to capital cost [54].

$$EC_k = \frac{(Exergy\ loss\ rate)_k}{Z_k} \tag{37}$$

Where *Z* is the capital cost of the system and it is calculated by multiplying unit cost (*c*) for per kW and power output of the considered system:

$$Z_k = P_k c_k \tag{38}$$

Another economic evaluation criteria is the total cost of the system and it is defined as [55]:

$$\dot{C}_k = \dot{Z}_k + \dot{C}_{fuel} \tag{39}$$

where $\dot{Z}_k$ is capital cost to per unit of time [55]:

$$\dot{Z}_k = \frac{Z_k CRF \Phi}{N} \tag{40}$$

where *CRF* is the capital recovery factor, Φ is the maintenance factor and *N* is the operating hours in a year [55].

$$CRF = \frac{in(1+in)^{lt}}{(1+in)^{lt}-1} \tag{41}$$

In eq. (41), *in* represents interest rate and *lt* is the life time of the considered system. Fuel cost to per unit time is shown eq. (42), where c_{fuel} is unit cost of fuel, *LHV* lower heating value of the fuel and $\dot{m}_{fuel}$ is mass flow rate of the fuel [55]:

$$\dot{C}_{fuel} = c_{fuel}\left(\frac{LHV}{1000}\right)\dot{m}_{fuel} \tag{42}$$

mass flow rate is calculated as eq (43):

$$\dot{m}_{fuel} = \left(1.05\times10^{-8}\right)\frac{P_f}{U_f} \tag{43}$$

In calculations, parameters in eqs.(38)-(43), specific cost of the PAFC (c_f) is 4\$/W[56,57], specific cost of thermoelectric generator (c_t) is 11.92 \$/W [58], interest rate is 10%, operating house in a year is 8200 hour [55], Φ is 1.06 [55], specific cost of the fuel (c_{fuel}) is 10 \$/GJ [59], lower heating value of the fuel is 120.7 MJ/kg.

RESULTS AND DISCUSSION

Results obtained from calculations are shown in figs. 2-10. Parameters used in the calculations are listed in table 1 and 2 which is given from the refs. [5-7, 35-37]. Results are presented as power density *(p = P/A)* and exergy destruction density *(exd= Exd/A)* for power output and exergy rate.

One can be seen the change of power densities with current density in fig 2. As it is shown, all power outputs have an optimum (maximum) point. Power density of hybrid system reaches its maximum at *j*= 10200 Am^{-2} and it is equal to 8735.340 Wm^{-2}. When investigating power densities for PAFC and thermoelectric generator, it is seen that they have optimum points too. These points are provided at *j* = 10100 Am^{-2} with 8538.440 Wm^{-2} for PAFC and at *j* = 12500 Am^{-2} with 351.036 Wm^{-2} value for TEG. Changes of hybrid system and PAFC are parabolic, however, TEG increases nearly linear about *j*=8500 Am^{-2} after this point it changes logarithmically and reaches its optimum.

Table 1. Parameters used in PAFC [5-7, 37].

Parameter	Unit	Value
p_{H_2}, p_{H_2O}	*atm*	1. 0347
p_{O_2}	*atm*	0.21
ne	-	2
α	-	1
j_o	Am^{-2}	$6x10^{-2}$
t_{ele}	*m*	$1x10^{-6}$
κ	$mhom^{-1}$	3.63
m	-	$3x10^{-5}$
n	-	$8x10^{-4}$
F	*C/mol*	96485
R	*J/molK*	8.314
Δg	*J/mol*	-228610
Δh	*J/mol*	-243342
T	*K*	473.15
epr	-	0.85
To	*K*	293.15

Table 2. Parameters used in thermoelectric generator [35, 36].

Parameter	Unit	Value
k_H, k_L	Wm^{-2}	170
a_t	m^2	1
L_1, L_2	mm	6
A_1, A_2	m^2	0.0001

In figure 3 energy efficiencies are illustrated. As it is seen that they have no optimum point, their maximum values are obtained at minimum current density. Maximum efficiencies of hybrid system, PAFC and TEG are 81.35%, 81.28% and 5.51% respectively. Effect of TEG on the hybrid system increases at higher current densities, however, average effect of it on the hybrid system is about 2%.

Variation of exergy efficiencies are in fig. 4. Curves in the fig 4. is very similar in fig.3. Maximum exergy efficiencies are obtained at minimum current density and the bigger current density means the lower exergy efficiency. Maximum exergy efficiency of hybrid system is 86.6%, 86.5% for the PAFC and 14.9% for the TEG. Comparing results of energy and exergy efficiencies, it can be seen that exergy efficiencies bigger than energy efficiencies. However, effect of the TEG on the hybrid system in terms of the exergy efficiency is same with effect of the energy efficiency.

Exergy destruction densities changes are shown in fig.5. All exergy densities grow up with current density and this change nearly linear until j = 10000 Am^{-2}, after this point they increase dramatically. Their maximums, which are at j = 12900 Am^{-2}, 21473 Wm^{-2} for the hybrid system, 15925 Wm^{-2} for the PAFC and 5548 $Wm^{-2}$for the TEG.

Total cost rate of the hybrid system, PAFC and TEG are indicated in fig. 6. Total cost rates of hybrid system and PAFC have an critical (maximum point) contrast to TEG. Total cost of hybrid system and PAFC increase until their critical point and then they begin to decrease. Their critical point are obtained at j = 10900

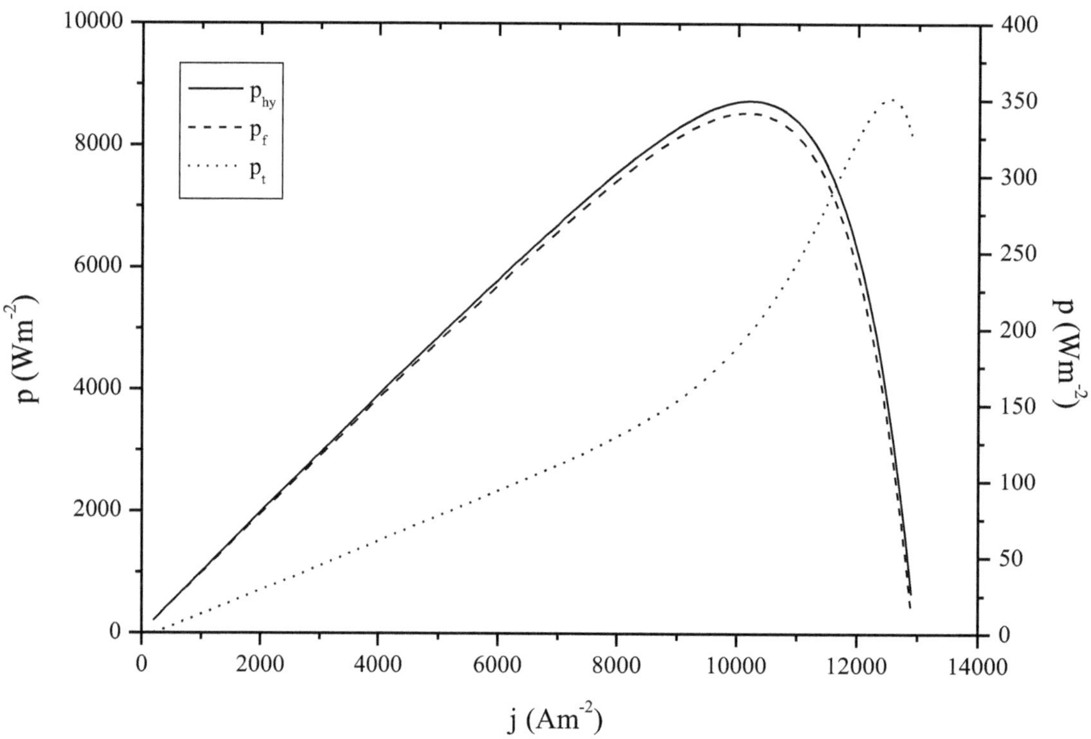

Figure 2. Variation of power density with current density.

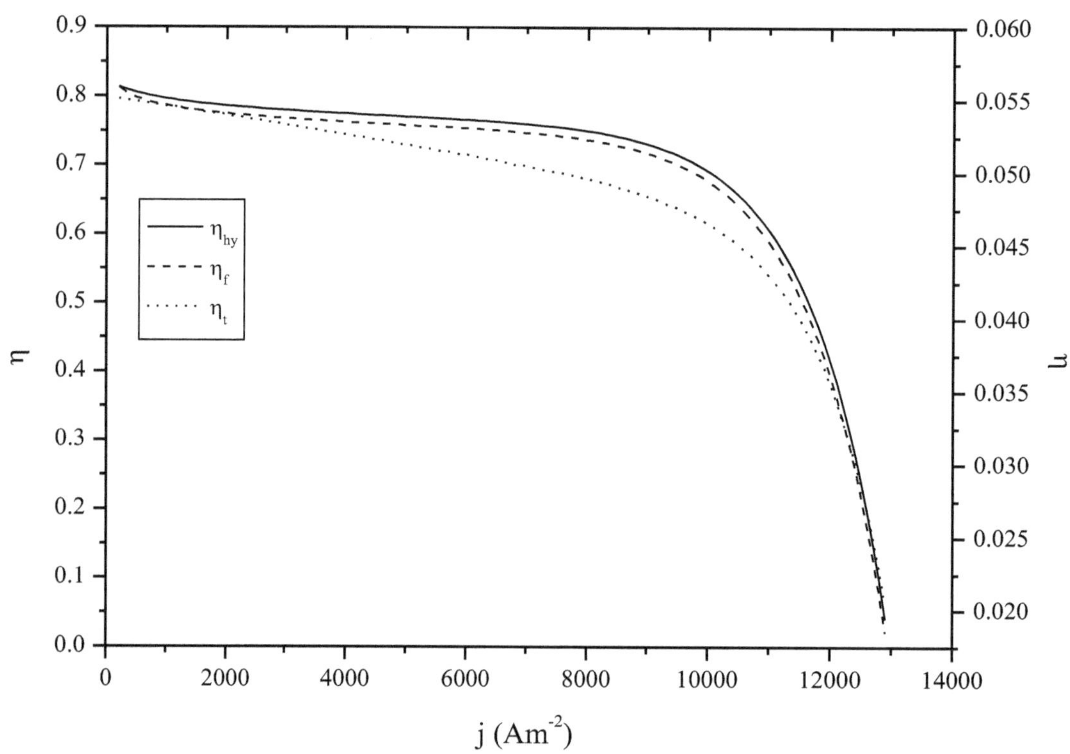

Figure 3. Variation of energy efficiency with current density.

Am^{-2} for the hybrid system and j = 10800 Am^{-2} for the PAFC and corresponding values are 0.633 $\$h^{-1}$ and 0.606 $\$h^{-1}$ respectively. The maximum total cost value of TEG is obtained at j = 12900 Am^{-2} with 0.389 $\$h^{-1}$.

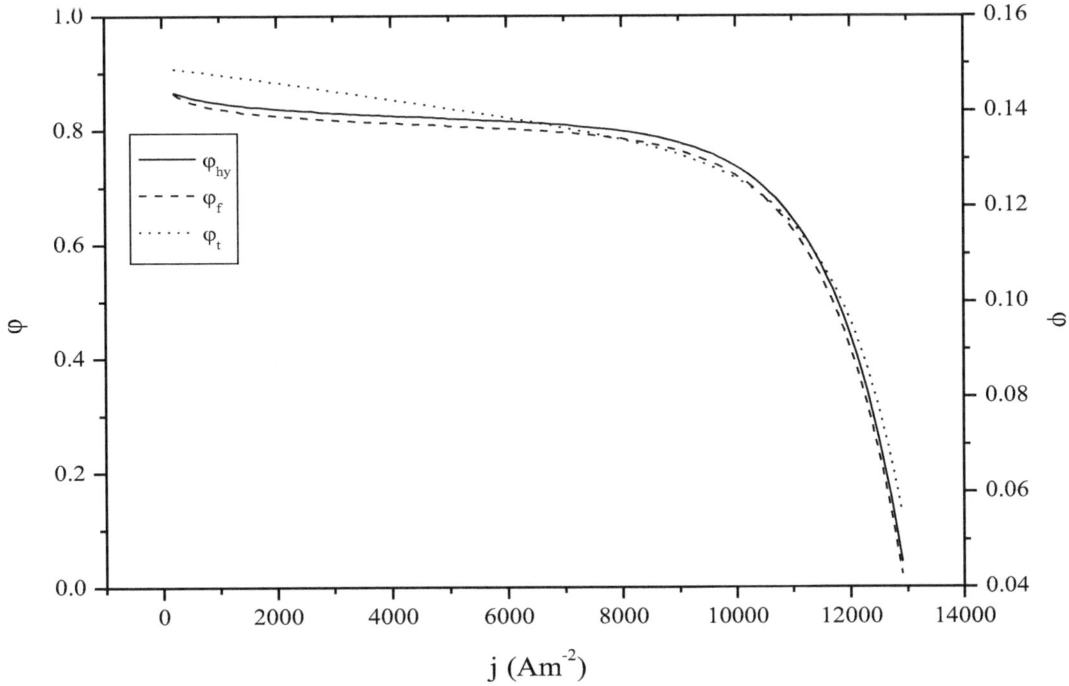

Figure 4.Variation of exergy efficiency with current density.

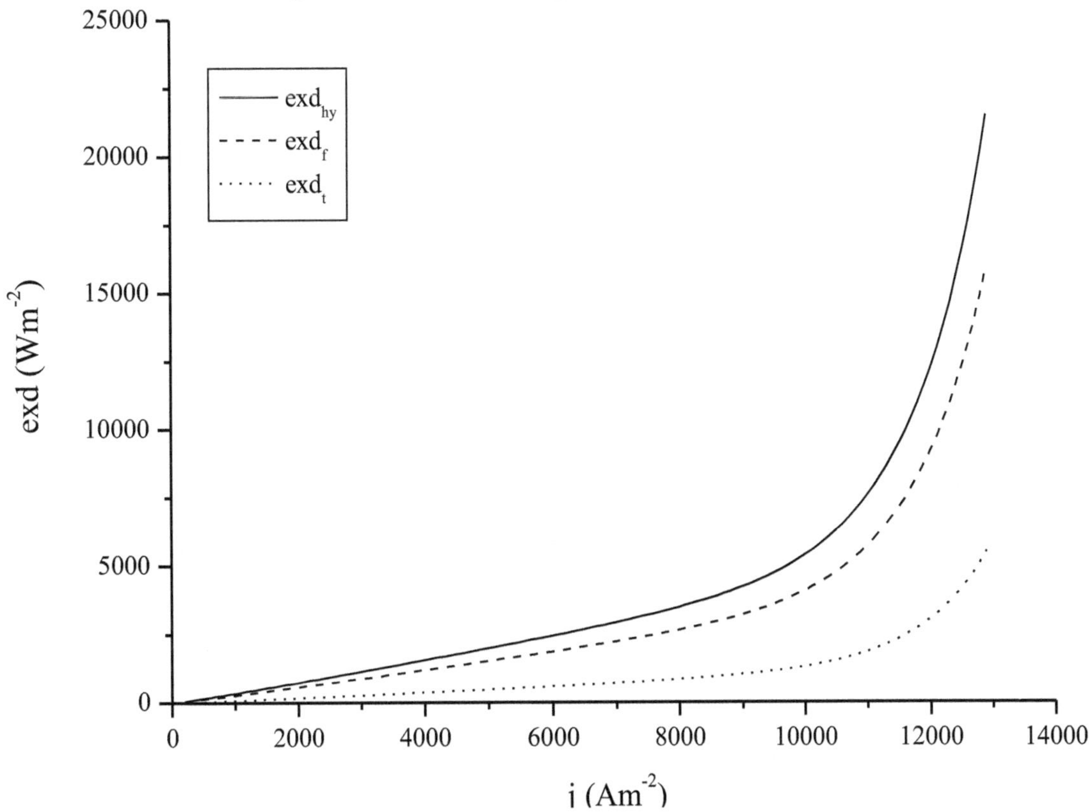

Figure 5. Variation of exergy destruction rate with current density.

EC of the considered systems can be shown in fig.7.*EC* of all considered systems show similar tendency. *EC* of hybrid system and the TEG rise up nearly linear about j =11000 Am^{-2} and this change is very slow, after this point they increase importantly. Similarly *EC* of PAFC increases nearly linear until j =12000 Am^{-2} after this point, it grows up dramatically. Maximum values of hybrid system, PAFC and TEG are 2.782 W\$$^{-1}$, 10.922 W\$$^{-1}$, 2.491 W\$$^{-1}$ respectively and change in them are 98.6%, 99.6% and 64.9% respectively.

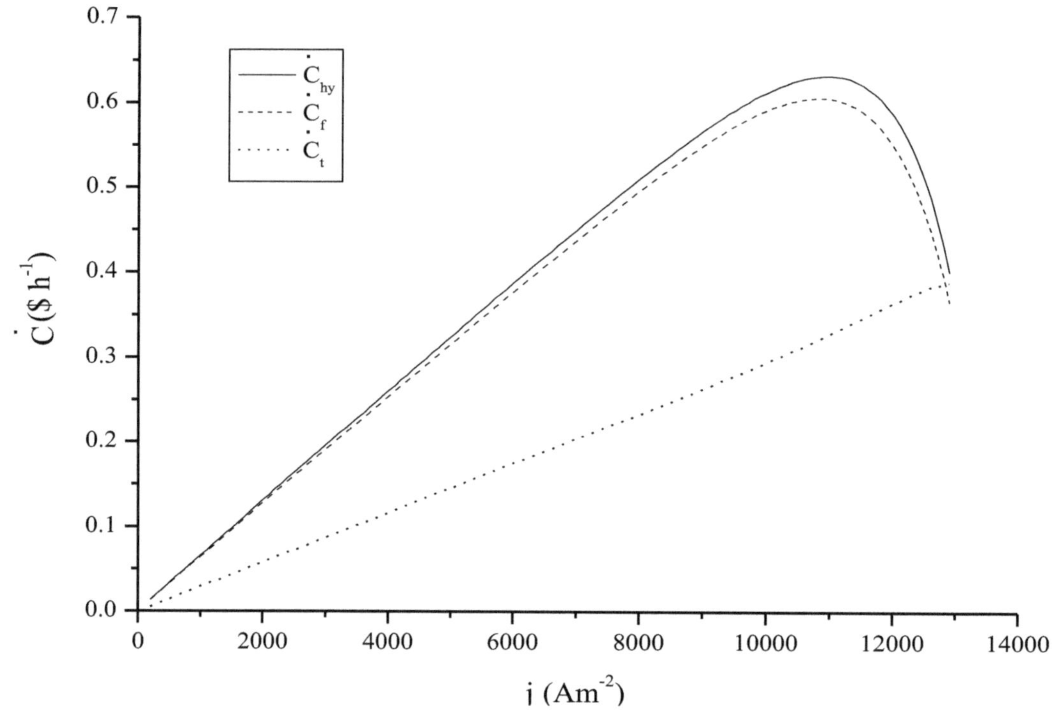

Figure 6. Variation of cost rate with current density.

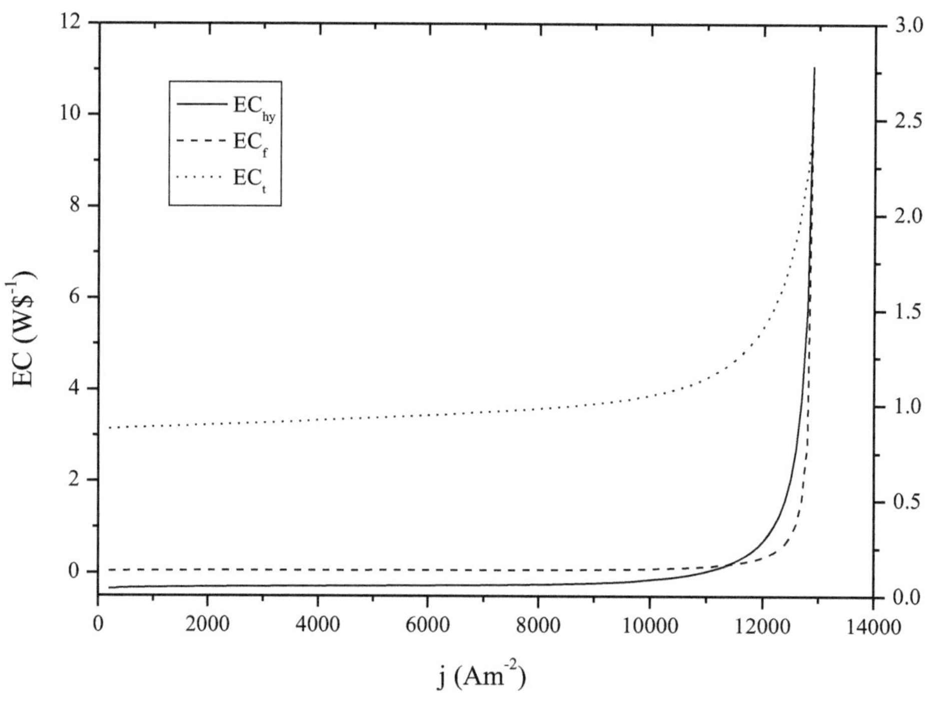

Figure 7. Variation of EC with current density.

p-η-φ curves is shown in fig. 8. This chart may be called as performance curve. In this graphic, energy and exergy efficiencies of hybrid system are investigated according to power density. In there, p_{max}, p_{η}, p_{φ}, η_{max}, η_{p}, φ_{max}, φ_{p} mean maximum power density, power density at the maximum energy efficiency, power density at maximum exergy efficiency, maximum energy efficiency, energy efficiency at maximum power density, maximum exergy efficiency and exergy efficiency at the maximum power density. η_p and φ_p are equal to 67.9%

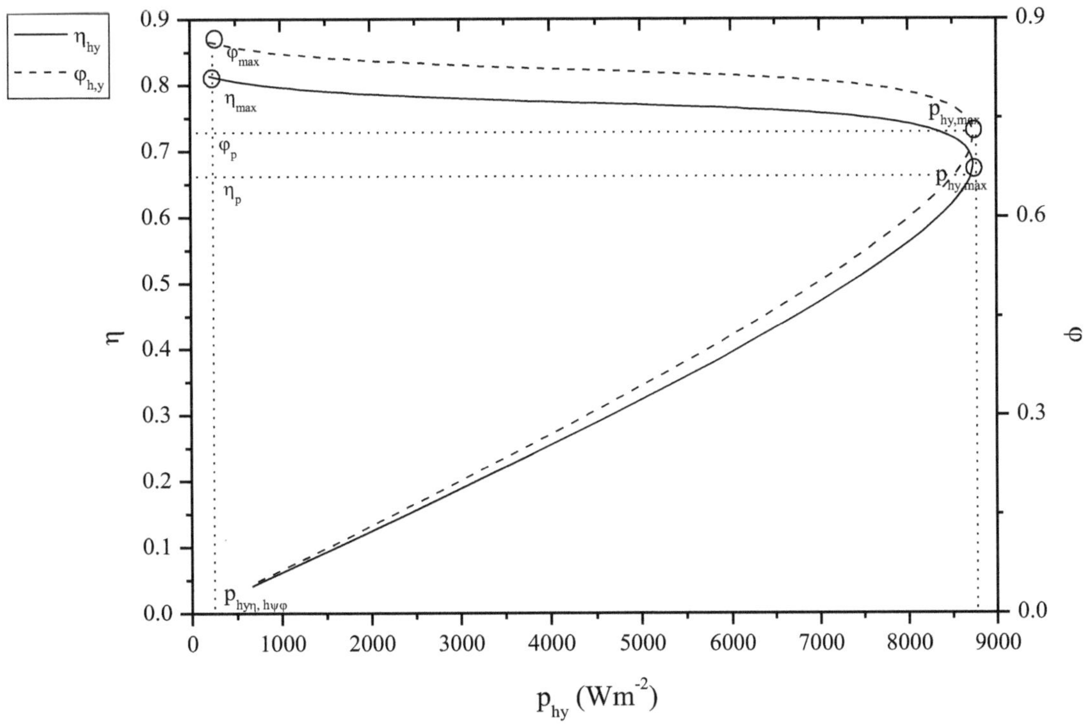

Figure 8. p-η-φ curve for the hybrid system.

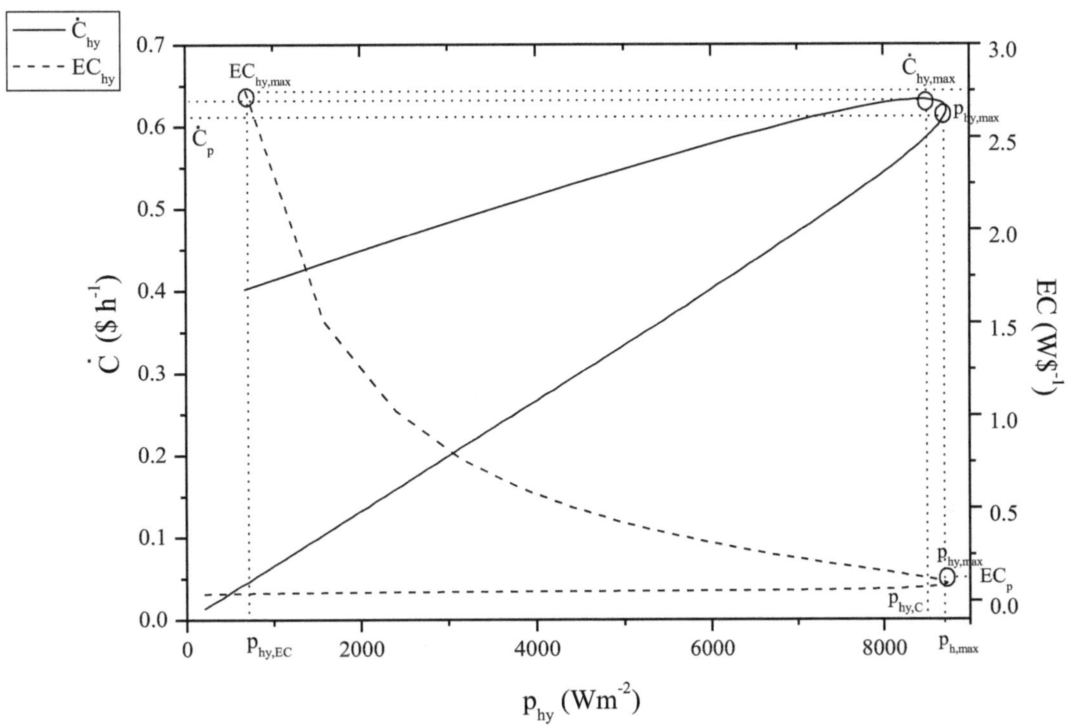

Figure 9. p-$\dot{C}$-EC curve for the hybrid system.

and 72.3%, these values correspond to 82.7% and 83.4% of their maximum respectively. p_η and p_φ are equal to each other and their values are 205.173Wm^{-2} and it is equal to only 2.3% of the maximum power density.

p-$\dot{C}$*-EC* curves can be shown in fig.9 where p_C, p_{EC}, $\dot{C}_{max}$, $\dot{C}_p$, EC_{max}, EC_p represent to power density at total cost rate, power density at *EC*, maximum total cost rate, total cost rate at power density, maximum exergy loss rate and *EC* at maximum power density respectively. p_C is equal to 8485.99 Wm^{-2} which is 97.1% of the maximum power density and p_{EC} is 8397.20 Wm^{-2} which is equal to 96.1% of the maximum power density.

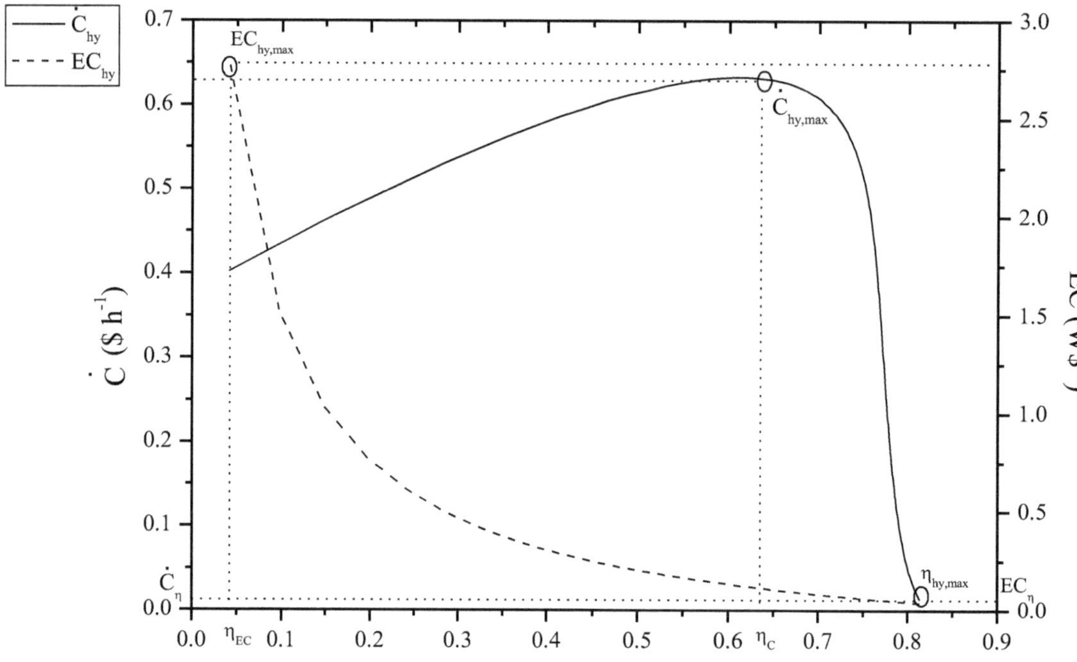

Figure 10. η-$\dot{C}$-EC curve for the hybrid system.

Similarly, $\dot{C}_p$, is 0.620 and it is 98% of the maximum total capital cost rate and EC_p is equal to 0.0917 W$\$^{-1}$ which is only 3.3% of the maximum *EC*.

Finally, η--$\dot{C}$-*EC* chart is indicated in fig.10 where η_C, η_{EC}, $\dot{C}_{max}$, $\dot{C}_\eta$, EC_{max}, EC_η are energy efficiency at the maximum total capital cost rate, energy efficiency at the maximum *EC*, maximum total capital cost rate, total capital cost rate at the maximum energy efficiency, maximum *EC* and *EC* the maximum energy efficiency respectively. η_C and η_{EC} are 61.7% and 4.1%, these values correspond to 75.8% and 5% of the maximum efficiency. In addition to that $\dot{C}_\eta$ and EC_η are 0.013 $\$h^{-1}$ and 0.039 W$\$^{-1}$, these values are 2% and 1.4% of their maximum.

CONCLUDING REMARKS

In this paper, a PAFC-TEG hybrid system is investigated in terms of its performance, economical and thermoeconomic aspects. Some important results are listed follows:

- Maximum power density of the hybrid system is 8735.340 Wm^{-2} at j= 10200 Am^{-2}

-Maximum energy efficiency of the hybrid system is 81.35% at low current densities, effect of TEG on the hybrid system 2% on average. Maximum exergy efficiency of the hybrid system is 86.6%, and it is obtained at low current densities similar to energy efficiency.

- The maximum total cost of the hybrid system is obtained j= 10900 Am^{-2} and this cost is equal to 0.633 $\$h^{-1}$

-The maximum *EC* of hybrid system is 10.922 W$\$^{-1}$.

- Exergy loss rate of the hybrid system increases dramatically at j= 11000 Am^{-2} and it is equal to 15925 Wm^{-2}.

As it is shown where the maximum power density, the total cost rate is maximum too, until 11000 Am^{-2} exergy lost rate does not change so importantly. In addition to these, energy efficiency, exergy efficiency and exergy destruction rate are high and the total cost rate, exergy loss rate have low values at the small current densities. According to these results, current densities should be chosen as low as possible in terms of economic aspects, economical and thermoeconomic analyses are recommended to be conducted for fuel cell hybrid systems.

ACKNOWLEDGEMENTS

Authors would like to thank the reviewers for their valuable comments, which have been utilized in improving the quality of the paper.

NOMECLATURE

A	polar rate area [m^2]
a	heat transfer area [m^2]
c	unit cost for per kW [$\$kW^{-1}$]
$\dot{C}$	heat capacity [$\$h^{-1}$]
CRF	Capital recovery factor
EC	ratio of thermodynamic loss rate to capital cost [$kW\$^{-1}$]
exd	exergy destruction density [Wm^{-2}]
Exd	exergy destruction rate [W]
F	Faraday constant [$Cmol^{-1}$]
Δg	change of the molar Gibbs free energy [$Jmol^{-1}$]
$\Delta\dot{G}$	change rate of the molar Gibbs free energy [W]
Δh	change of the molar enthalpy [$Jmol^{-1}$]
$\Delta\dot{H}$	change rate of the enthalpy [W]
I	electrical current generating in semi-conductance couple [A]
j	current density [Am^{-2}]
k	heat transfer coefficient [$W\ m^{-2}\ K^{-1}$]
K	heat conductance of the semi-conductance couple [$W\ m^{-1}\ K^{-1}$]
L	length of p and n type material [mm]
LHV	lower heating value of the fuel [$kJ\ kg^{-1}$]
m	a constant
$\dot{m}$	mass flow rate [kgs^{-1}]
n	number electrons and a constant
N	number of thermoelectic units and operation hours in a year [$hyear^{-1}$]
p	pressure [atm, kPa], power density [Wm^{-2}]
P	power [W]
$PAFC$	phosphoric acid fuel cell
$\dot{Q}$	heat rate [W]
r	electrical resistance of the semiconductor couple [Ω]
R	universal gas constant [$Jmol^{-1}K^{-1}$]
t	thickness [m]
T	temperature [K]
TEG	thermoelectric generator
U	potential [V]
Z	capital cost [$]
$\dot{Z}$	capital cost per hour [$\$h^{-1}$]

Subscripts

act	activation
C	heat capacity
con	concentration
EC	ratio of thermodynamic loss rate to capital cost
f	fuel cell
m	mean
e	electron

hy	hybrid
H	hot side
H_2	hydrogen
H_2O	water
k	kth component
L	cold side
max	maximum
o	environment condition and exchange current density
ohm	ohmic
O_2	oxygen
ohm	ohmic overpotential
p	power
r	regenerator
rev	reversible
t	thermoelectric module
1	hot junction and p type material
2	cold junction and n type material
η	efficiency

Greek letters

α	charge transfer coefficient
β	Seeback coefficient [VK^{-1}]
λ	temperature dependent thermoelectric property
ρ	temperature dependent thermoelectric property
κ	thermal conductivity of phosphoric acid solution [$mhom^{-1}$]
η	efficiency
θ	heat conductance [kWK^{-1}]
Φ	maintenance factor
φ	exergy efficiency
ε	effectiveness

REFERENCES

[1] Zhao Y, Ou C, Chen J. (2008). A new analytical approach to model and evaluate the performance of a class of irreversible fuel cells, International Journal of Hydrogen Energy, 33, 4161 – 4170.

[2] Zhang X, Guo J, Chen J. (2010) . The parametric optimum analysis of a proton exchange membrane (PEM) fuel cell and its load matching, Energy, 35, 5294-5299.

[3] Zhang H, Lin G, Chen J. (2011). Performance analysis and multi-objective optimization of a new molten carbonate fuel cell system, International Journal of Hydrogen Energy, 36, 4015 – 4021.

[4] Zhang H, Chen L, Zhang J, Chen J. (2014). Performance analysis of a direct carbon fuel cell with molten carbonate electrolyte, Energy, 68, 1-9.

[5] Zhang H, Lin G, Chen J. (2012). Multi-objective optimization analysis and load matching of a phosphoric acid fuel cell system, International Journal of Hydrogen Energy, 37, 3438-3446.

[6] Yang P, Zhang H, Hu Z. (2016). Parametric study of a hybrid system integrating a phosphoric acid fuel cell with an absorption refrigerator for cooling purposes, International Journal of Hydrogen Energy, 41, 3579 -3590.

[7] Chen X , Wang Y, Zahao Y, Zhou Y. (2016). A study of double functions and load matching of a phosphoric acid fuel cell/heat-driven refrigerator hybrid system, Energy 101, 359-365.

[8] Chen L, Zhang H,Gao S, Yan H. (2014)., Performance optimum analysis of an irreversible molten carbonate fuel cell - Stirling heat engine hybrid system, Energy 64, 923-930.

[9] Chen L, Gao S, Zhang H. (2013). Performance Analysis and Multi-Objective Optimization of an Irreversible Solid Oxide Fuel Cell-Stirling Heat Engine Hybrid System, Int. J. Electrochem. Sci., 8, 10772 - 10787.

[10]Açıkkalp E., Thermo-environmental performance analysis of irreversible solid oxide fuel cell – Stirling heat engine, International Journal of Ambient Energy, article in press DOI: 10.1080/01430750.2017.1345011.

[11] Yang P, Zhang H. (2015). Parametric analysis of an irreversible proton exchange membrane fuel cell/absorption refrigerator hybrid system, Energy, 85, 458-467.
[12] Zhao Y, Chen J. (2009). Modeling and optimization of a typical fuel cell–heat engine hybrid system and its parametric design criteria, Journal of Power Sources, 186, 96–103.
[13] Zhang H, Lin G, Chen J. (2011). Performance Evaluation and Parametric Optimum Criteria of an Irreversible Molten Carbonate Fuel Cell-Heat Engine Hybrid System, Int. J. Electrochem. Sci,. 6, 4714 - 4729.
[14] Zhang X, Chen J. (2010). Performance analysis and parametric optimum criteria of a class of irreversible fuel cell/heat engine hybrid systems, International Journal of Hydrogen Energy 35, 284 – 293.
[15] Zhang X,Wang Y, Guo J, Shih T-M, Chen J. (2014). A unified model of high-temperature fuel-cell heat engine hybrid systems and analyses of its optimum performances, International Journal of Hydrogen Energy, 39, 1811 –1825.
[16] Zhang X, Su S, Chen J, Zhao Y, Brandon N. (2011), A new analytical approach to evaluate and optimize the performance of an irreversible solid oxide fuel cell-gas turbine hybrid system, International Journal of Hydrogen Energy, 36, 15304 –15312.
[17] Haseli H, Dincer I, Naterer GF. (2008). Thermodynamic analysis of a combined gas turbine power system with a solid oxide fuel cell through exergy, Thermochimica Acta, 480, 1–9.
[18] Açıkkalp E. (2017). Ecologic and Sustainable Objective Thermodynamic Evaluation of Molten Carbonate Fuel Cell - Supercritical CO_2 Brayton Cycle Hybrid System, International Journal of Hydrogen Energy, 42, 6272-6280.
[19] Haseli H, Dincer I, Naterer GF. (2008). Thermodynamic modeling of a gas turbine cycle combined with a solid oxide fuel cell, International Journal of Hydrogen Energy, 33, 5811 –5822.
[20] Açıkkalp E. (2017). Performance analysis of irreversible solid oxide fuel cell – Brayton heat engine with ecological based thermo-environmental criterion, Energy Conversion and Management, 148, 279-286.
[21] Zhang X, Guo J, Chen J. (2012). Influence of multiple irreversible losses on the performance of a molten carbonate fuel cell-gas turbine hybrid system, International Journal of Hydrogen Energy 37, 8664 –8671.
[22] Sánchez D, Chacartegui R, Jiménez-Espadafor F, Sánchez T. (2009). A new concept for high temperature fuel cell hybrid systems using supercritical carbon dioxide, J. Fuel Cell Sci. Technol., 6, 021306.
[23] Sanchez D, Munoz de Escalona J.M, Chacartegui R, Munoz A., Sanchez T. (2011). A comparison between molten carbonate fuel cells based hybrid systems using air and supercritical carbon dioxide Brayton cycles with state of the art technology, Journal of Power Sources, 196, 4347–4354.
[24] Zhang X, Liu H, Ni M, Chen J. (2015). Performance evaluation and parametric optimum design of a syngas molten carbonate fuel cell and gas turbine hybrid system, Renewable Energy, 80, 407-414.
[25] Mehrpooya M ,Bahramian P, Pourfayaz F, Rosen M.A. Introducing and analysis of a hybrid molten carbonate fuel cell-supercritical carbon dioxide Brayton cycle system, Sustainable Energy Technologies and Assessments 18 (2016) 100–106.
[26] Zhang H, Su S, Lin G, Chen J. (2012). Performance Analysis and Multi-Objective Optimization of a Molten Carbonate Fuel Cell Braysson Heat Engine Hybrid System, Int. J. Electrochem. Sci., 7, 3420 - 3435.
[27] Açıkkalp E. (2017). Performance analysis of irreversible molten carbonate fuel cell – Braysson heat engine with ecological objective approach, Energy Conversion and Management, 13, 2432–2437
[28] Huang C, Pan Y, Wang Y, Su G, Chen J. (2016). An efficient hybrid system using a thermionic generator to harvest waste heat from a reforming molten carbonate fuel cell, Energy Conversion and Management 121, 186–193.
[29]Mahmoudi S.M.S., Ghavimi A.R. (2016). Thermoeconomic analysis and multi objective optimization of a molten carbonate fuel cell – Supercritical carbon dioxide – Organic Rankine cycle integrated power system using liquefied natural gas as heat sink, Applied Thermal Engineering, 107, 1219–1232.
[30] Chen L, Gong J, Sun F, Wu C. (2002). Effect of heat transfer on the performance of thermoelectric generators, Int J ThermSci, 41(1), 95–99.
[31] Meng F, Chen L, Sun F. (2010). Effects of heat reservoir temperatures on the performance of thermoelectric heat pump driven by thermoelectric generator, Int J Low-Carbon Technol,. 5(4),273–82.
[32] Meng F, Chen L, Sun F, Wu C. (2009), Thermodynamic analysis and optimisation of a new-type thermoelectric heat pump driven by a thermoelectric generator, Int J Ambient Energy, 30(2), 95–101.
[33]Chen L, Meng F, Sun F. Effect of heat transfer on the performance of thermoelectric generator-driven thermoelectric refrigerator system. Cryogenics 2012;52(1):58–65.
[34] Kaushik S, Manikandan S, Hans R. (2015). Energy and exergy analysis of thermoelectric heat pump system, Int J Heat Mass Transfer, 86, 843–52.
[35] Arora R, Kaushik SC, Arora R. (2015). Multi-objective and multi-parameter optimization of two-stage thermoelectric generator in electrically series and parallel configurations through NSGA-II, Energy, 30(91), 242–254.
[36] Manikandan., Kaushik S.C. (2015). Thermodynamic studies and maximum power point tracking in thermoelectric generator–thermoelectric cooler combined system, Cryogenics, 6752–6762.

[37] Meng F K, Chen L G, Sun F R. (2011). A numerical model and comparative investigation of a thermoelectric generator with multi-irreversibilities, Energy, 36(5), 3513-3522.
[38] Chen L G, Meng F K, Sun F R. (2012). Maximum power and efficiency of an irreversible thermoelectric generator with a generalized heat transfer law, Scientia Iranica, Transaction B: Mechanical Engineering, 19 (5), 1337-1345
[39] Chen L G, Meng F K, Sun F R. (2013). Internal and external simultaneous optimization of an irreversible multielement thermoelectric generator for maximum power output, International Journal of Low-Carbon Technologies, 8 (3), 188-196.
[40] Meng F K, Chen L G, Sun F R, Yang B. (2014). Thermoelectric power generation driven by blast furnace slag flushing water, Energy, 66, 965-972.
[41] Xiong B, Chen L G, Meng F K, Sun F R. (2014). Modeling and performance analysis of a two-stage thermoelectric energy harvesting system from blast furnace slag water waste heat, Energy, 77, 562-569.
[42] Chen L G, Meng F K, Sun F R. (2016). Thermodynamic analyses and optimizations for thermoelectric devices: the state of the arts, Science China: Technological Sciences, 59(3), 442-455.
[43] Meng F K, Chen L G, Sun F R. (2016). Effects of thermocouples physical dimension on the performance of TEG-TEH system, International Journal of Low-Carbon Technologies, 11(3), 375-382.
[44] Meng F K, Chen L G, Feng Y L, Xiong B. (2017). Thermoelectric generator for industrial gas phase waste heat recovery, Energy, 135, 83-90.
[45] Meng F K, Chen L G, Xie Z H, Ge Y L. (2017). Thermoelectric generator with air-cooling heat recovery device from wastewater, Thermal Science and Engineering Progress, 4, 106-112.
[46] Chen X , Wang Y, Cai L, Zhou Y. (2015), Maximum power output and load matching of a phosphoric acid fuel cell-thermoelectric generator hybrid system, Journal of Power Sources 294, 430-436.
[47] Zhao M , Zhang H, Hua Z, Zhang Z, Zhang J. (2015). Performance characteristics of a direct carbon fuel cell/thermoelectric generator hybrid system, Energy Conversion and Management 89, 683–689.
[48] Chen X, Chen L, Guo J, Chen J. (2011). An available method exploiting the waste heat in a proton exchange membrane fuel cell system, International Journal of Hydrogen Energy 36, 6099 – 6104.
[49] Feng H J, Chen L G, Xie Z H, Sun F R. (2015). Constructal optimization for a single tubular solid oxide fuel cell, Journal of Power Sources, 286, 406-413.
[50] Abbas S. S., Ahmadi M.H., Ahmadi M.A. (2015). Optimization performance and thermodynamic analysis of an irreversible nano scale Brayton cycle operating with Maxwell–Boltzmann gas, Energy Conversion and Management 101, 592-605.
[51] Ahmadi M H., Ahmad M.A., Pourfayaz F., Bidi M. (2016). Thermodynamic analysis and optimization for an irreversible heat pump working on reversed Brayton cycle. Energy Conversion and Management, 110, 260-267.
[52] Ahmadi M.H, Ahmadi M.A,. (2016). Multi objective optimization of performance of three-heat-source irreversible refrigerators based algorithm NSGAII, Renewable and Sustainable Energy Reviews, 60, 784-794.
[53] Ahmadi M.H., Sayyaadi H, Hosseinzadeh H. (2015). Optimization of Output Power and Thermal Efficiency of Solar-Dish Stirling Engine Using Finite Time Thermodynamic Analysis, Heat Transfer—Asian Research 44, 347-376.
[54] Dincer I , Rosen M.A, Exergy: Energy, Environment and Sustainable Development, Elsevier Science, 2 edition.
[55] Aminyavari M., Haghighat A, Mamaghani A. S., Najafi B, Rinaldi F. (2016). Exergetic, economic, and environmental evaluations and multi-objective optimization of an internal-reforming SOFC-gas turbine cycle coupled with a Rankine cycle, Applied Thermal Engineering 108, 833–846.
[56] Kwak H-Y, Lee H-S, Jung J-Y, Jeon J-S, Park D-R (2004) . Exergetic and thermoeconomic analysis of a 200-kW phosphoric acid fuel cell plant, Fuel 83, 2087–2094.
[57] Staffell I, Green R. (2013). The cost of domestic fuel cell micro-CHP systems, International Journal of Hydrogen Energy, 38, 1088-1102.
[58]https://energy.gov/sites/prod/files/2015/02/f19/QTR%20Ch8%20%20Thermoelectic%20Materials%20TA%20Feb-13-2015.pdf (access date, 07.04.2017).
[59] Kazim A., (2005). Exergoeconomic analysis of a PEM fuel cell at various operating conditions, Energy Conversion and Management, 46, 1073–1081.

17

THERMODYNAMIC PERFORMANCE ANALYSIS OF DEDICATED MECHANICALLY SUBCOOLED VAPOUR COMPRESSION REFRIGERATION SYSTEM

S. Agarwal[1], A. Arora[1,*], B. B. Arora[1]

ABSTRACT

In this work, the thermodynamic analysis of dedicated mechanically subcooled vapour compression refrigeration system is presented. A software based computer program in EES has been formulated for computation of the performance results. The effect of varation of degree of subcooling (5-30°C), evaporator temperature (-20 to 10°C) and condenser temperature (30-50°C) has been investigated for energetic and exergetic performance of the system. The analysis of the system has been carried out using zero ODP and very low GWP (1 to 4) refrigerants viz.HFO-R1234ze and R1234yf to compare the performance of HFC-R134a. The results depicts that the COP and exergetic efficiency of dedicated subcooled VCR cycle are better than that of simple VCR cycle. Refrigerant R1234ze performs better than R1234yf and comparable to R134a.

Keywords: *VCR, Subcooling, COP, R1234ze, R1234yf, Exergetic Efficiency, Dedicated, R134a*

INTRODUCTION

Demand of energy, economy and eco-sustainability are the major considerations of a refrigeration system design. The demand of high grade energy is growing continuously in the developing countries viz. India. The cost of electricity is also increasing with its demand. The distortion of surrounding ecosystem with the use of high GWP refrigerants (HFCs) in the refrigeration and air-conditioning systems is a current issue. The high GWP HFCs have to be phased out as per the European Union (EU) regulation (Regulation (EU) No 517/2014 [1]). Many researchers have proved that the mechanically sub-cooling of vapour compression refrigeration (VCR) cycle improves its performance. Zubair [2, 3] and Zubair et al. [4] have investigated a dedicated and an integrated mechanical sub-cooling system from the thermodynamic standpoint. In these studies, it was found that the system performance had improved when operating in situations where the difference between the condensing and evaporating temperatures is large. Couvillion et al. [5] developed a mathematical model of a dedicated mechanical sub-cooling system by considering the individual component models of the equipment involved in the system. They found an improvement of 6-8% in the coefficient of performance (COP) and 20-170% in the capacity over a conventional simple cycle. Qureshi et al. [6] analysed that the load carrying capacity of the evaporator increased by approximately 0.5kW when R22 was subcooled in the main cycle by 5-8°C. They noted that using the sub-cooling, the second-law efficiency of the cycle increased by an average 21% and this percentage increase is direcltly proportional to the ambient temperature. Khan and Zubair [7] have investigated finite time thermodynamic models of simple vapour compression refrigeration, two stage VCR system (TSS), dedicated mechanical sub-cool VCR system (DMSS), integrated mechanical sub-cooling VCR system (IMSS). They predicted that the optimum value of COP is obtained by varying the parameters of heat exchanger and reduced sub-cooler saturation temperature. Thornton et al. [8] have used dedicated mechanical sub-cooling VCR cycle to sub-cool the condenser exit of main VCR cycle. They predicted the optimum value of sub-cooling evaporator temperature which established the design rule for optimum distribution of heat exchanger area. Arora et al. [9, 10, 11], Arora and Kaushik [12] and Arora [13] have analysed vapour compression and absorption refrigeration systems for energetic and exergetic performance improvement. Llopis et al. [14, 15] have theoretically analysed the energy performance enhancement of CO_2 transcritical refrigeration system using dedicated mechanical sub-cooling cycle. They observed 20% increase in COP and 28.8% increases in cooling capacity of combined and reviewed sub-cooling method viz. internal heat exchanger, economizers, thermoelectric systems, dedicated sub-cooling methods of CO_2 refrigeration cycle. They concluded that the sub-cooling is a worth method to increase the performance of refrigeration cycle and deserve future development. Many researchers have suggested the replacement of R134a with R1234ze and R1234yf in refrigeration

This paper was recommended for publication in revised form by Regional Editor Sandip Kale
[1]Department of Mechanical Engineering, Delhi Technological University, Delhi-110042, India.
**E-mail address: profakhilesharora@gmail.com*

systems. Mota-Babiloni et al. [16] reviewed the recent developments in commercial refrigeration and focused on system modifications (as dedicated sub-cooling or the implementation of ejectors), tri-generation technologies (electrical, heating and cooling demand) and better evaporation conditions control. They concluded that the hydrocarbon, HFO and CO_2 are HFC replacements. Kalla et al. [17, 18] reviewed alternative refrigerants for vapour compression refrigeration system and investigated the performance of R22 and its substitutes in air-conditioners. Yataganbaba et al. [19] performed exergy analysis on a two evaporator vapour compression refrigeration system using R1234yf, R1234ze and R134a as refrigerants. The two refrigerants R1234yf and R1234ze were good alternatives to R134a regarding their environment friendly properties. Dixit et al. [24-26] carried out energy and exergy analysis of absorption-compression cascade and waste heat driven triple effect refrigeration cycles.

The literature survey gathered from different established data sources viz. Google Scholar, Research gate, Science direct etc. reveals that the extensive research have been done on simple VCR cycle. Out of these, some studies based on dedicated mechanical sub-cool cycle have been considered to enhance the performance of the cycle. It is observed that limited work has been carried out to explore the performance of dedicated mechanical subcool cycle using HFO refrigerants viz. R1234yf and R1234ze.

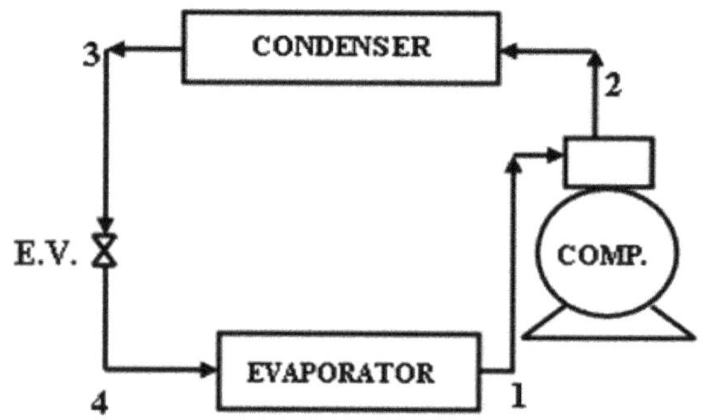

(a)

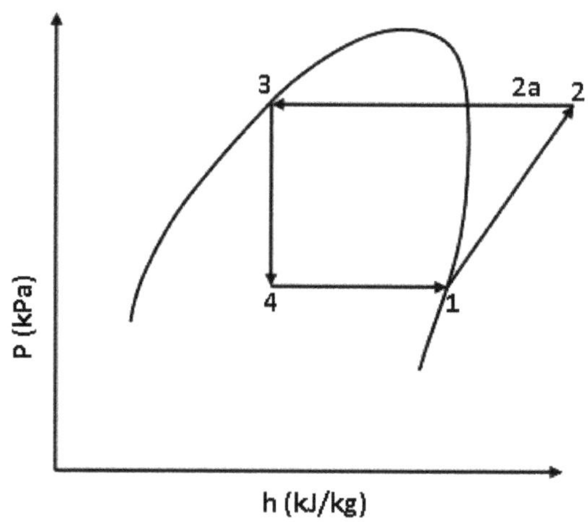

(b)

Figure 1. (a) Saturated Vapour Compression Refrigeration Cycle, (b) P-h diagram of saturated vapour compression refrigeration cycle

The present work investigates the thermodynamic performance of dedicated mechanically subcooled VCR cycle using R1234yf, R1234ze and R134a. The comparison of considered refrigerant has also been explored for drop-in replacement of R134a to R1234ze and R1234yf. The first and second law based performance parameters viz. COP and exergetic efficiency have been computed with the variation in degree of sub-cooling (DOS), evaporator temperature and condenser temperature.

DESCRIPTION OF SYSTEM AND MODELLING

Description of System

The current work presents first and second law based thermodynamic analysis of dedicated mechanically subcooled vapour compression refrigeration system (DSC) using R1234ze, R1234yf and R134a. The mechanical sub-cooling of main VCR cycle has been carried out through the subcooler or evaporator of another simple VCR cycle designated as subcooler VCR cycle.

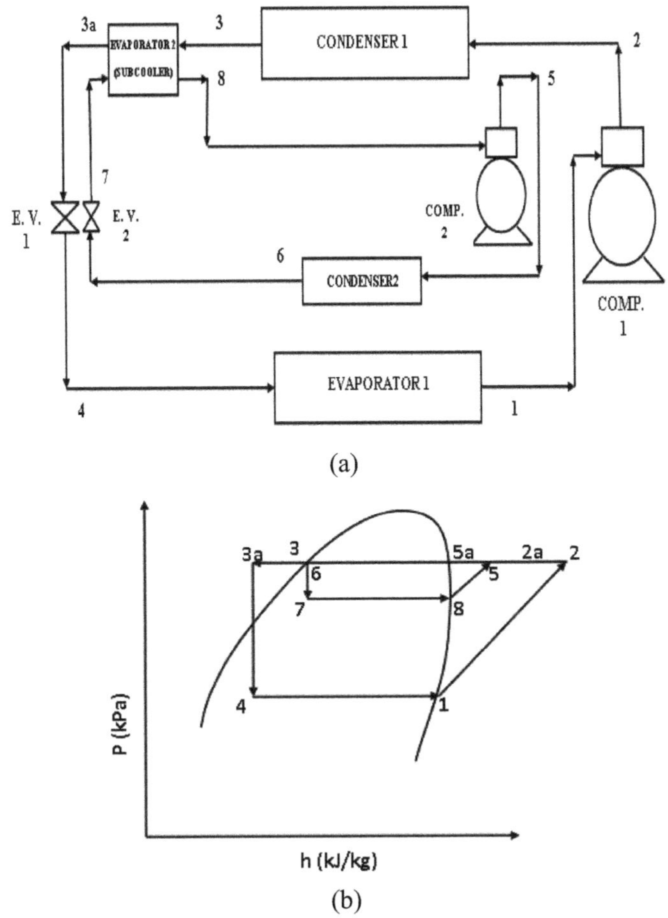

Figure 2. (a) Schematic diagram of dedicated mechanically subcooled vapour compression refrigeration cycle, (b) P-h diagram of dedicated sub-cooled vapour compression cycle

Figure 2(a) and 2(b) illustrate the schematic and P-h diagrams of dedicated mechanically subcooled vapour compression refrigeration system respectively. It comprises of two sections i.e. subcooled and subcooler vapour compression refrigeration cycles. The evaporator of subcooler VCR cycle receives the heat rejected by the condensate liquid refrigerant of subcooled VCR cycle and acts as subcooler. In this way, the two cycles are coupled together through subcooler or evaporator 2 in order to achieve low temperature in evaporator1. The refrigerants considered for the energy and exergy analysis of the system are R134a, R1234yf and R1234ze. The same refrigerant has been considered in the two sections simultaneously.

The subcooled VCR cycle is in lower section which consists of compressor1, evaporator1, expansion valve1, subcooler and condenser1 and the subcooler VCR cycle is upper section which consists of compressor2, condenser2, expansion valve2 and evaporator2 or subcooler. The vapour refrigerant leaving the evaporator1 of subcooled VCR cycle at state point1 enters into the compressor1 and is being compressed. The refrigerant at high temperature and pressure leaving the compressor1 at state point 2 and enters in to the condenser1. The liquid condensate has been subcooled from state point 3-3a by the evaporator2 of the subcooler cycle. The subcooled liquid refrigerant is throttled by the expansion valve1 from state point 3a-4. The low temperature liquid-vapour mixture of refrigerant enters in to the evaporator1 and low temperature is achieved in the evaporator1.

Similarly, the saturated vapour of the refrigerant leaving the evaporator2 or subcooler at state point 5 enters in to the compressor2. The compressed high pressure, high temperature vapour refrigerant enters into the condenser2 at state point 6. The vapour phase of the refrigerant is converted in to liquid in the condenser2. The liquid condensate refrigerant leaving the condenser2 at state point7 is expanded freely by the expansion valve2 from state 7-8. The low temperature liquid-vapour mixture of refrigerant enters into the evaporator2 or subcooler and absorbs heat from the condensate of condenser1 from state 8-5 and transformed into saturated vapour.

In this way, the mechanical sub-cooling is produced by the subcooler VCR cycle in the subcooler from states 3-3a. The evaporator temperature ($T_{e,SC}$) of the subcooler cycle is more than the evaporator temperature of subcooled cycle. The sub-cooling of liquid condensate refrigerant of subcooled cycle enhances the net specific refrigerating effect as shown in Figure 2 from states 4-1. The increase in refrigerating effect enhances the coefficient of performance (COP) of the subcooled cycle. However, the compressor work of the subcooler cycle reduces the COP of the overall cycle.

Thermodynamic Modelling

In the present work, the energy and exery analysis of dedicated mechanically subcooled vapour compression cycle have been carried out. In which the subcooler VCR cycle produces mechanical sub-cooling in the subcooled VCR cycle through subcooler using considered refrigerants R134a, R1234yf and R1234ze. The various performance parameters viz. compressor work, COP, exergetic efficiency, total exergy destruction rate and exergy destruction ratio have been explored for different variables viz. degree of sub-cooling, evaporator temperature, effectiveness of subcooler, isentropic efficiency of compressor and condenser temperature.

A set of computer code has been formulated in Engineering Equation Solver (EES) (Klein and Alvarado [22]) to perform the analysis. The steady state governing equations, based on first and second law of thermodynamics have been developed.

Energy Analysis

The first law based work, energy and mass conservation principles are presented in equations (1)-(2).

$$\sum \dot{Q} - \sum \dot{W} = \sum \dot{m}_o h_o - \sum \dot{m}_i h_i \tag{1}$$

$$\sum \dot{m}_i - \sum \dot{m}_o = 0 \tag{2}$$

where $\dot{Q}$, $\dot{W}$ and $\dot{m}$ are the rate of heat, work and mass transfer crossing the boundary of the system respectively.

The coefficient of performance (COP) of simple, dedicated mechanically subcooled, subcooler and overall VCR cycles are given by the equation (3)-(9).

$$COP_{SVCR} = \frac{\dot{Q}_e}{\dot{W}_{comp}} \tag{3}$$

where $\dot{Q}_e$ is the net refrigeration effect produced and $\dot{W}_{comp}$ is the actual work of the compressor.

$$COP_{DSC} = \frac{\dot{Q}_{e,DSC}}{\dot{W}_{comp1}} \tag{4}$$

$$\dot{Q}_{e,DSC} = \dot{m}_{r,DSC}(h_1 - h_4) \tag{5}$$

$$\dot{W}_{comp1} = \dot{m}_{r,DSC}(h_2 - h_1) \tag{6}$$

$$COP_{SC} = \frac{\dot{Q}_{e,SC}}{\dot{W}_{comp2}} \tag{7}$$

$$\dot{W}_{comp2} = \dot{m}_{r,\,SC}(h_6 - h_5) \tag{8}$$

$$COP_{\,OC} = \frac{\dot{Q}_{e,DSC}}{\dot{W}_{comp1}+\dot{W}_{comp2}} \tag{9}$$

where $\dot{Q}_{e,DSC}$, $\dot{Q}_{e,SC}$, $\dot{W}_{comp1}$, $\dot{W}_{comp2}$, $\dot{m}_{r,\,DSC}$ and $\dot{m}_{r,\,SC}$ are net refrigerant effect, net compressor work and mass flow rate of dedicated mechanically subcooled and subcooler VCR cycles respectively.

The effectiveness of subcooler ε_{SC} is the ratio of minimum heat transfer to the maximum heat transfer.

$$\varepsilon_{SC} = \frac{T_5\text{-}T_8}{T_3\text{-}T_8} \tag{10}$$

where ε_{SC} is the effectiveness of subcooler and T_3,T_5 and T_8 are temperatures at respective state points.

The energy balance in the subcooler is given by equation (11)

$$\dot{m}_{r,DSC}\,(h_3 - h_{3a}) = \dot{m}_{r,SC}\,(h_5 - h_8) \tag{11}$$

Exergy Analysis

The second law of thermodynamics expresses the concept of exergy. The exergy is defined as the degree measure of quality or usefulness of energy to impact the environment. The maximum useful work which can be extracted from a system as it reversibly comes into equilibrium with its environment. (Bejan et al. (1996), Arora and Kaushik [12]).

For the refrigerant flowing in a refrigerating system, the exergy is defined as follows (Bejan et al. [20]):

$$\dot{E} = \dot{m}_r[(h - h_0) - T_0(s - s_0)] \tag{12}$$

where h_0 and s_0 are the enthalpy and entropy values of the refrigerant at dead state pressure P_0 and temperature T_0.

Exergy destruction($\dot{E}_D$) or internal exergy destruction losses which are caused by irreversibilites of the system is the algebric sum of total exergy at the inlet and outlet of the system. General exergy balance is given by (Dincer and Kanoglu [21]):

$$\dot{E}_D = \dot{E}_{in} - \dot{E}_{out} \tag{13}$$

where $\dot{E}_D$ is the rate of exergy destruction and $\dot{E}_{in}$ and $\dot{E}_{out}$ are the total exergy transferred by heat, work and mass.

Exergy Destruction ($\dot{E}_D$) in the Components of Dedicated Mechanically Subcooled Vapour Compression Refrigeration Cycle (DSC)

The exergy destruction in each component of the dedicated mechanically subcooled vapour compression refrigeration cycle is given by the following equations:

Evaporator 1

$$\left(\dot{E}_D\right)_{e,DSC} = \dot{E}_{X4} + \dot{Q}_{e,DSC}\left(1 - \frac{T_0}{T_{b,DSC}}\right) - \dot{E}_{X1} = \dot{m}_{r,DSC}(h_4 - T_0 s_4) + \dot{Q}_{e,DSC}\left(1 - \frac{T_0}{T_b}\right) - \dot{m}_{r,DSC}(h_1 - T_0 s_1) \tag{14}$$

Compressor 1

$$(\dot{E}_D)_{comp1} = \dot{E}_{X_1} + \dot{W}_{comp1} - \dot{E}_{X_2} = \dot{m}_{r,DSC}(\mathrm{h}_1 - T_0 s_1) + \dot{W}_{comp1} - \dot{m}_{r,DSC}(\mathrm{h}_2 - T_0 s_2) \tag{15}$$

Condenser 1

$$(\dot{E}_D)_{cond1} = \dot{E}_{X_2} - \dot{E}_{X_3} = \dot{m}_{r,DSC}(h_2 - T_0 s_2) - \dot{m}_{r,DSC}(h_3 - T_0 s_3) \tag{16}$$

Expansion Valve 1

$$(\dot{E}_D)_{Ex.\ Valve1} = \dot{E}_{X_3} - \dot{E}_{X_{3a}} = \dot{m}_{r,DSC}(h_{3a} - T_0 s_{3a}) - \dot{m}_{r,DSC}(h_4 - T_0 s_4) = \dot{m}_{r,DSC} T_0 (s_4 - s_{3a}) \tag{17}$$

Subcooler

$$(\dot{E}_D)_{SC} = (\dot{E}_{X_8} - \dot{E}_{X_5}) + (\dot{E}_{X_3} - \dot{E}_{X_{3a}}) = \dot{m}_{r,SC}\{(h_8 - h_5) - T_0(s_8 - s_5)\} + \dot{m}_{r,DSC}\{(h_3 - h_{3a}) - T_0(s_3 - s_{3a})\} \tag{18}$$

Compressor 2

$$(\dot{E}_D)_{comp2} = \dot{E}_{X_5} + \dot{W}_{comp2} - \dot{E}_{X_6} = \dot{m}_{r,SC}(h_5 - T_0 s_5) + \dot{W}_{comp2} - \dot{m}_{r,SC}(h_6 - T_0 s_6) \tag{19}$$

Condenser 2

$$(\dot{E}_D)_{cond2} = \dot{E}_{X_6} - \dot{E}_{X_7} = \dot{m}_{r,SC}(h_6 - T_0 s_6) - \dot{m}_{r,SC}(h_7 - T_0 s_7) \tag{20}$$

Expansion Valve 2

$$(\dot{E}_D)_{Ex.\ valve2} = \dot{E}_{X_7} - \dot{E}_{X_8} = \dot{m}_{r,SC}(h_7 - T_0 s_7) - \dot{m}_{r,SC}(h_8 - T_0 s_8) = \dot{m}_{r,SC} T_0 (s_8 - s_7) \tag{21}$$

Evaporator 2

$$(\dot{E}_D)_{e,SC} = \dot{E}_{X_8} + \dot{Q}_{e,SC}\left(1 - \frac{T_0}{T_{b,SC}}\right) - \dot{E}_{X_5} = \dot{m}_{r,SC}(h_8 - T_0 s_8) + \dot{Q}_{e,SC}\left(1 - \frac{T_0}{T_{b,SC}}\right) - \dot{m}_{r,SC}(h_5 - T_0 s_5) \tag{22}$$

where $\dot{E}_D$, $\dot{E}_X$, $\dot{m}_r$ and T_b are the exergy destruction rate, exergy rate, mass flow rate and temperature of space to be cooled for the dedicated mechanically subcooled and subcooler cycle respectively. T_0, h_0 and s_0 are the temperature, enthalpy and entropy of dead state.

Total Exergy Destruction

Total exergy destruction of the system is the sum of exergy destruction in each component of the system. The total exergy destruction in dedicated mechanically subcooled vapour compression refrigeration cycle (DSC) is given by:

$$\Sigma(\dot{E}_D)_{DSC} = (\dot{E}_D)_{e,DSc} + (\dot{E}_D)_{comp1} + (\dot{E}_D)_{cond1} + (\dot{E}_D)_{Ex.\ valve1} \tag{23}$$

$$\Sigma(\dot{E}_D)_{SC} = (\dot{E}_D)_{e,Sc} + (\dot{E}_D)_{comp2} + (\dot{E}_D)_{cond2} + (\dot{E}_D)_{Ex.\ valve2} \tag{24}$$

$$\sum(\dot{E}_D)_{OC} = (\dot{E}_D)_{e,DSc} + (\dot{E}_D)_{comp1} + (\dot{E}_D)_{cond1} + (\dot{E}_D)_{Ex.\ valve1} + (\dot{E}_D)_{SC} + (\dot{E}_D)_{comp2} + (\dot{E}_D)_{cond2} + (\dot{E}_D)_{Ex.\ valve2} \tag{25}$$

Exergetic Efficiency

The exergetic efficiency is the ratio of total exergy recovered to the total exery supplied and is given by (Dincer and Kanoglu (2010)):

$$\eta_{ex} = \frac{\sum \dot{E}_{out}}{\sum \dot{E}_{in}} = 1 - \frac{\sum \dot{E}_D}{\sum \dot{E}_{in}} \tag{26}$$

where η_{ex} is the exergetic efficiency of the cycle. $\sum \dot{E}_{out}$, $\sum \dot{E}_{in}$ and $\sum \dot{E}_D$ are thc total exergy recovered, supplied and destructed respectively.

The exergetic efficiency of vapour compression refrigeration system is defined as the ratio of the exergy of heat absorbed in the evaporator from the space to be cooled at temperature T_b to the actual compressor work input ($\dot{W}_{comp}$) (Arora and Kaushik [12]). Exergy efficiency can also be defined as The exergetic efficiency is the ratio of exery in products to the exergy of fuel. The exergetic efficiency of dedicated mechanically subcooled, subcooler and overall VCR cycles are given by the equations (27)-(29).

$$\eta_{ex\ (DSC)} = \frac{\left|\dot{Q}_{e,DSC}\left(1 - \frac{T_0}{T_{b,DSC}}\right)\right|}{\dot{W}_{comp1}} \tag{27}$$

$$\eta_{ex\ (SC)} = \frac{\left|\dot{Q}_{e,SC}\left(1 - \frac{T_0}{T_{b,SC}}\right)\right|}{\dot{W}_{comp2}} \tag{28}$$

$$\eta_{ex\ (OC)} = \frac{\left|\dot{Q}_{e,OC}\left(1 - \frac{T_0}{T_{b,DSC}}\right)\right|}{\dot{W}_{comp1} + \dot{W}_{comp2}} \tag{29}$$

where T_0 is the ambient or dead state temperature.

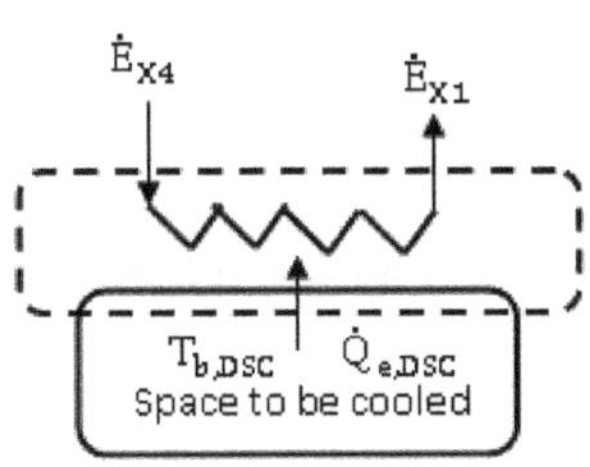

Figure 3. Exergy balance in evaporator of DSC and the space to be cooled [23]

Exergy Destruction Ratio (EDR)

The EDR is the total exergy destruction in the system to the exergy in the products (Arora and Kaushik [12]) and is given by equation (30).

$$EDR = \frac{\dot{E}_{D,total}}{\dot{E}_P} = \frac{1}{\eta_{ex}} - 1 \tag{30}$$

The exergy destruction ratio for the dedicated mechanically subcooled, subcooler and overall VCR cycles are given in following (31)-(33).

$$EDR_{DSC} = \frac{\Sigma(\dot{E}_D)_{DSC}}{\dot{E}_{P,DSC}} = \frac{\Sigma(\dot{E}_D)_{DSC}}{\left|\dot{Q}_{e,DSC}\left(1-\frac{T_0}{T_{b,DSC}}\right)\right|} \tag{31}$$

$$EDR_{SC} = \frac{\Sigma(\dot{E}_D)_{SC}}{\dot{E}_{P,SC}} = \frac{\Sigma(\dot{E}_D)_{SC}}{\left|\dot{Q}_{e,SC}\left(1-\frac{T_0}{T_{b,SC}}\right)\right|} \tag{32}$$

$$EDR_{OC} = \frac{\Sigma(\dot{E}_D)_{OC}}{\dot{E}_{P,DSC}} = \frac{\Sigma(\dot{E}_D)_{OC}}{\left|\dot{Q}_{e,DSC}\left(1-\frac{T_0}{T_{b,DSC}}\right)\right|} \tag{33}$$

Mathematical Modelling

The present work investigates the performance of dedicated mechanically subcooled vapour compression refrigeration cycle (DSC) on the bassis of first and second law of thermodynamics. The mathematical fomulation of thermodynamic relations have been carried out considering mass, energy and work conservation principles for system and system components.

Table 1. Input variable considered except where the variation of these system variables involved (Arora and Kaushik [12])

S.No.	Input variables	Values
1.	Evaporator1 temperature ($T_{e, DSC}$)	-10^{0}C
2.	Degree of subcooling ($(\Delta T)_{sc}$)(3-3a)	5^{0}C
3.	Condenser1,2 temperature ($T_{c,DSC}$ and $T_{c,SC}$), assuming summer conditions in a country near the equator line.	50^{0}C
4.	Isentropic efficiency of compressor ($\eta_{comp1}, \eta_{comp2}$)	80%
5.	Effectiveness of subcooler (ε_{SC})	0.8
6.	Net refrigerating effect ($\dot{Q}_{e,DSC}$)	3.5167 kW
7.	Ambient or dead state temperature (T_0) and pressure (P_0)	25^{0}C and 101.325 kPa

The following assumptions have been considered to formulate the complex thermodynamic model of dedicated mechanically subcooled vapour compression refrigeration cycle (DSC) except where the variation of physical parameters involved.

- It is assumed that the state of refrigerant is dry and saturated at the entry of compressor at state points 1 and 5.
- Difference between evaporator temperature and the temperature of space to be cooled ($T_{b, DSC}$ -$T_{e, DSC}$, $T_{b,SC}$ -$T_{e,SC}$) is 5 ^{0}C.
- The range of variation of evaportor1 temperature ($T_{e, DSC}$) is -20 ^{0}C to 10 ^{0}C.
- The range of variation of condenser1 temperature ($T_{e, SC}$) is 30 ^{0}C to 50 ^{0}C.

- The heat losses and pressure losses from the system and system components are negligible. The whole system operates in steady state condition.
- The range of degree of subcooling $(\Delta T)_{SC}$ is 5 to 30 ^{0}C.
- The pressure drop in evaporator (δ_e) and condenser (δ_c) is assumed to be zero.

RESULTS AND DISCUSSION

Energy and exergy analysis of dedicated mechanically subcooled vapour compression refrigeration cycle (DSC) have been performed using a computer software based program in Engineering Equation Solver (EES) (Klein and Alvarado [22]) which has been shown in Figure 1(a) and 2(b). The considered refrigerants for the analysis are R134a, R1234yf and R1234ze. The computation of the various performance parameters viz. compressor work, COP, exergetic efficiency, total exergy destruction rate and exergy destruction ratio (EDR) have been done by calling builtin functions (i.e. specific entropy, specific enthalpy, temperature and pressure etc.) of the EES library.

The results from the present analysis have been compared with the results of Arora et al. [12]. It is observed that the difference in the value of COP and exergetic efficiency is within a range of ±1%. The difference in results may be due to the pressure drop in evaporator and condenser is assumed to be zero.

Effect of Subcooling

The effect of degree of sub-cooling $(\Delta T)_{sc}$ on various performance parameters viz. compressor work $(\dot{W}_C)$, COP, exergetic efficiency (η_{ex}), total exergy destruction rate $(\dot{E}_{D,\ Total})$ and exergy destruction ratio (EDR) has been illustrated in Figure 4(a) to 4(d).

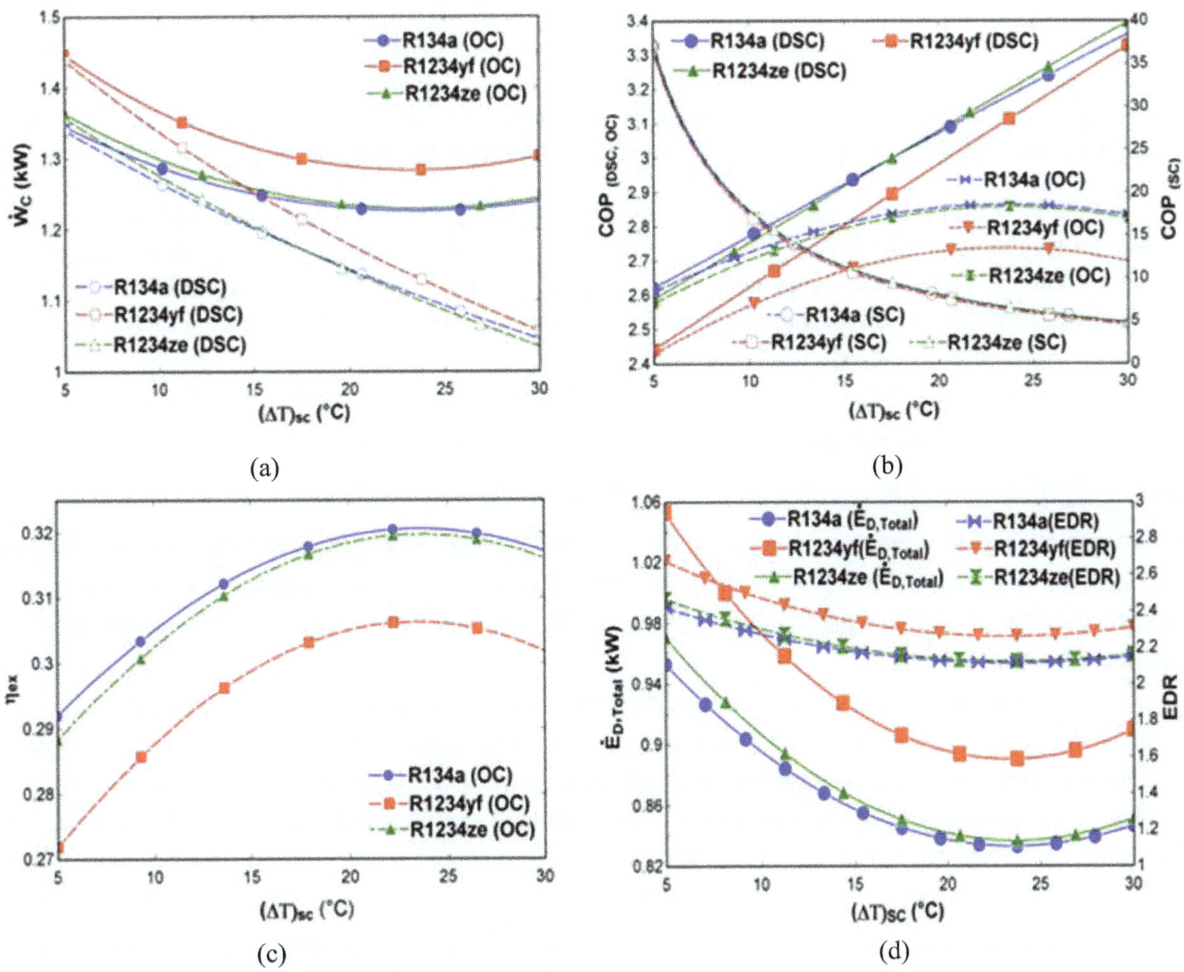

Figure 4. (a) Variation in compressor work $(\dot{W}_C)$ with degree of subcooling (ΔT_{sc}), (b) effect of degree of subcooling (ΔT_{sc}) on COP, (c) Effect of degree of subcooling (ΔT_{sc}) on exergetic efficiency (ηex), (d) variation in total exergy destruction rate (E˙_(D,Total)) and exergy destruction ratio (EDR) with subcooling (ΔT_{sc}) of OC

Figure 4(a) represents the effect of sub-cooling $(\Delta T)_{sc}$ on compressor work $(\dot{W}_C)$. For 1 Ton refrigeration, compressor work of dedicated mechanically subcooled (DSC) $(\dot{W}_{Comp1})$ decreases with increase in degree of subcooling (DOS) because the mass flow rate of refrigerant decreases with increase in degree of subcooling for constant refrigeration load of 3.5167kW. However, the compressor work first decreases and then increases for overall cycle (OC) with increase in degree of sub-cooling for the considered refrigerants. This is due to that the compressor work of OC $(\dot{W}_{Comp1} + \dot{W}_{Comp2})$ decreases for low values of DOS as the value of $\dot{W}_{Comp1}$ decreases and the values of $\dot{W}_{Comp2}$ are very small. For the higher values of DOS, the compressor work of subcooler cycle $(\dot{W}_{Comp2})$ increases while $\dot{W}_{Comp1}$ decreases. However the compressor work for the simple VCR cycle is is higher than that of the DSC. The optimize value of degree of sub-cooling lies within 20-25^{0}C for overall cycle. It has been observed that the compressor work is minimum for R1234ze and maximum for R1234yf for DSC and OC.

Figure 4(b) shows the variation of COP with degree of sub-cooling for the considered refrigerants R134a, R1234yf and R1234ze. The COP of the dedicated mechanically subcooled cycle increases and the COP of subcooler cycle decreases with increase in DOS for the considered refrigerants. However, the COP of the overall cycle first increases, maximise for optimum value of DOS which lies within 20-25^{0}C and then decreases with increase in DOS. In order to increase DOS, The temperature and pressure of evaporator of subcooler cycle has to achieved lower values which reduces the COP of the subcooler cycle while for the constant refrigerating load of $\dot{Q}_{e,DSC}$ = 3.5167kW, the compressor work $(\dot{W}_{Comp1})$ of DSC decreases linearly and the COP of DSC is $\frac{\dot{Q}_{e,DSC}}{\dot{W}_{comp1}}$ ("Eq.(4)"). Thus the COP of the DSC increases linearly with increase in DOS. The COP of OC is $\frac{\dot{Q}_{e,DSC}}{\dot{W}_{comp1}+\dot{W}_{comp2}}$ ("Eq.(9)") which increases with increase in COP of DSC and decreases with decrease in COP of subcooler cycle due to increase in compressor work of SC with increase in DOS. Thus the COP of OC first increases (as COP of DSC increases faster) reaches to maximum value and then decreases (as COP of SC cycle decreases faster) for the refrigerant considered. It is observed that the COP is maximum for R1234ze and minimum for R1234yf. The COP of simple VCR cycle is lower than the DSC and OC.

Figure 4(c) illustrates the effect of degree of sub-cooling (ΔT_{sc}) on exergetic efficiency $(\eta_{ex\,(OC)})$ of overall cycle. The exergetic efficiency of the overall cycle first increases reaches to a maximum value and then decreases with increase in DOS for the considered refrigerants. The optimum value of degree of sub-cooling lies between 20 to 25^{0}C for which the value of exergetic efficiency is maximum. The exergetic efficiency of overall cycle is given by $\eta_{ex\,(OC)} = \frac{\left|\dot{Q}_{e,OC}\left(1-\frac{T_0}{T_{b,DSC}}\right)\right|}{\dot{W}_{comp1}+\dot{W}_{comp2}}$ (eq. (29)). In which the term $\frac{|\dot{Q}_{e,OC}|}{\dot{W}_{comp1}+\dot{W}_{comp2}}$ is the $COP_{(OC)}$ which first increases, reaches to a maximum value and then decreases with increase in DOS (Figure 4(b)) and the term $\left|\left(1-\frac{T_0}{T_{b,DSC}}\right)\right|$ remains constant with increase in DOS because the value of T_0 and $T_{b,DSC}$ remains constant with increase in DOS. Hence the value of $\eta_{ex\,(OC)}$ first increases, reaches to a maximum value and then decreases. The optimum value of DOS lies between 20 to 25^{0}C for the considered refrigerants. Hence the exergetic efficiency of the OC follows the trend of COP. It can be observed that $\eta_{ex\,(OC)}$ is maximum for R134a and minimum for R1234yf.

Figure 4(d) shows the variation in total exergy destruction rate $(\dot{E}_{D,\ Total})$ and exergy destruction ratio (EDR) with degree of subcooling (ΔT_{sc}) for overall cycle. The value of $\dot{E}_{D,\ Total}$ and EDR first decreases, reaches to a minimum value and then increases with increase in DOS for the refrigerants considered. The value of $\dot{Q}_{e,OC}$ first increases reaches to a maximum value and then decreases with increase in DOS (Figure 4(b)). Therefore, the exergy recovered from the system and system components first increases, reaches to a maximum value and then decreases with increase in DOS. Consequently, the $\dot{E}_{D,\ Total}$ first decreases, reaches to a minimum value and then increases. EDR is given by $\frac{\Sigma(\dot{E}_D)_{OC}}{\left|\dot{Q}_{e,DSC}\left(1-\frac{T_0}{T_{b,DSC}}\right)\right|}$ (eq. (33)) which is proportional to $\dot{E}_{D,\ Total}$ and reciprocal to $\left|\dot{Q}_{e,DSC}\left(1-\frac{T_0}{T_{b,DSC}}\right)\right|$. Hence EDR first decreases, reaches to a minimum value and then increases with increase in DOS (ΔT_{sc}). It can also be observed that the values of $\dot{E}_{D,\ Total}$ and EDR are minimum for R134a and maximum for R1234yf.

Effect of Evaporator Temperature

The effect of evaporator1 temperature ($T_{e,\ DSC}$) on various performance parameters viz. compressor work $(\dot{W}_C)$, degree of subcooling $(\Delta T)_{sc}$, COP, exergetic efficiency (η_{ex}), total exergy destruction rate $(\dot{E}_{D,\ Total})$ and exergy destruction ratio (EDR) has been shown in Figure 5(a) to 5(d).

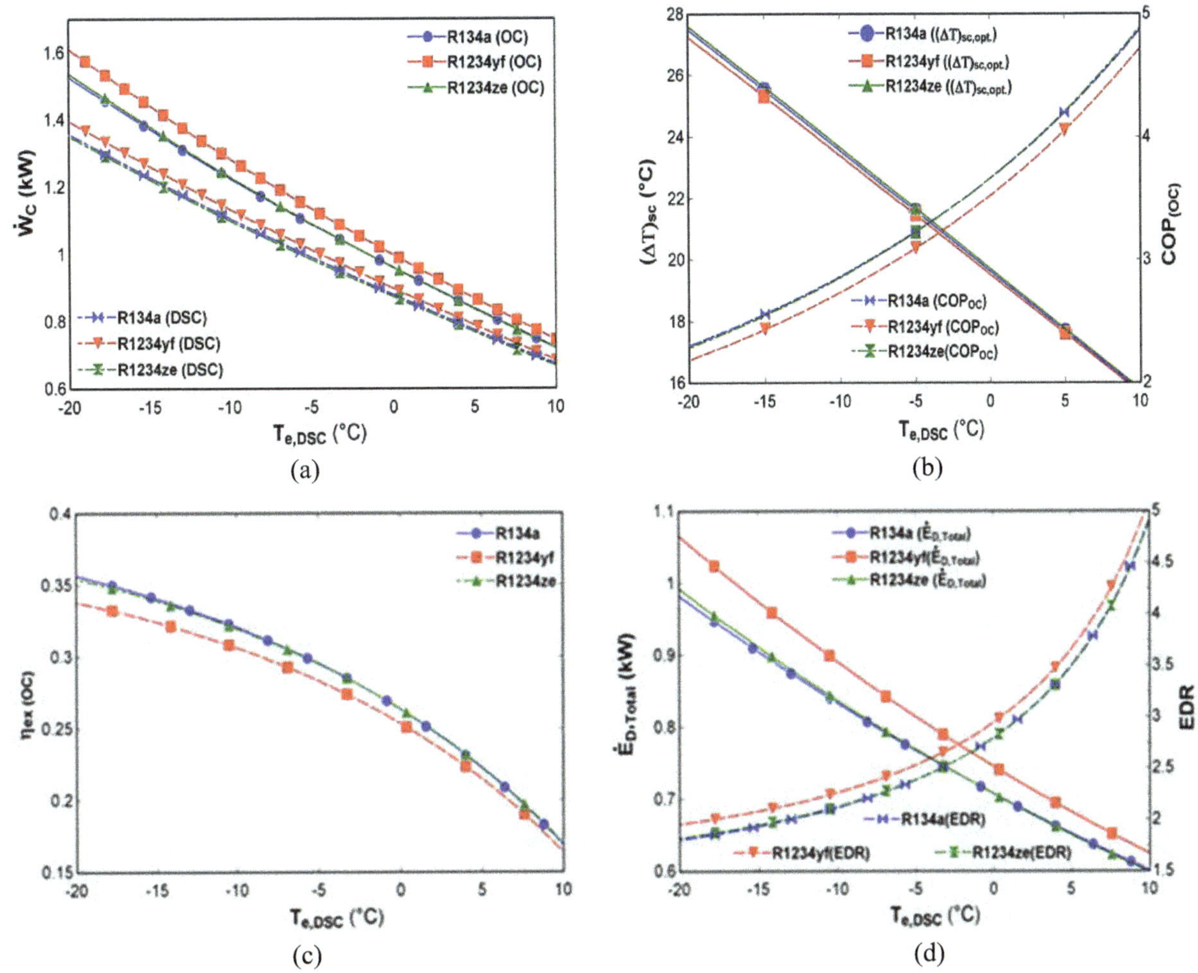

Figure 5. (a) Variation in compressor work$(\dot{W}_C)$ with evaporator1 Temperature ($T_{e,\ DSC}$), (b) effect of evaporator1 temperature ($T_{e,DSC}$) on degree of subcooling (ΔT_{sc}) and COP, (c) effect of evaporator1 temperature ($T_{e,DCS}$) on exergetic efficiency (η_{ex}) of overall cycle, (d) variation in total exergy destruction rate $(\dot{E}_{D,Total})$ and exergy destruction ratio (EDR) with evaporator1 temperature ($T_{e,DCS}$) of overall cycle

Figure 5(a) shows the varition in compressor work $(\dot{W}_C)$ with evaporator1 temperature ($T_{e,\ DSC}$) for dedicated mechanically subcooled and overall VCR cycle. The compressor work of DSC ($\dot{W}_{Comp1}$) and OC ($\dot{W}_{Comp1} + \dot{W}_{Comp2}$) decreases with increase in evaporator1 temperature ($T_{e,\ DSC}$) for considered refrigerants. As the evaporator1 temperature increases, the pressure ratio between evaporator and condenser decreases and for constant refrigeration load of 3.5167kW, the mass flow rate of refrigerant also decreases. Moreover, the evaporator2 temperature of sub-cooler cycle also increases. Hence the compressor work of DSC and OC decreases with increase in evaporator1 temperature. It is observed that the value of compressor work is minimum for R1234ze and maximum for R1234yf.

Figure 5(b) illustrates the effect of evaporator1 temperature ($T_{e,\ DSC}$) on DOS $(\Delta T)_{sc}$ and COP for OC. The value of DOS decreases and COP increases with increase in evaporator1 temperature for the refrigerant considered. For the constant refrigeration load of 3.5167kW, lower amount of sub-cooling is required in the subcooler to the refrigerant of DSC. Thus the DOS decreases with increase in $T_{e,\ DSC}$. However, the compressor work decreases as shown in Figure 5(a), the value of COP increases. It has been observed that the value of DOS and COP are maximum for R1234ze and minimum for R1234yf.

Figure 5(c) depicts the effect of evaporator1 temperature ($T_{e,\ DSC}$) on exergetic efficiency (η_{ex}) of OC for considered refrigerants. The exergetic efficiency is given by $COP_{(OC)} \times \left|\left(1-\frac{T_0}{T_{b,DSC}}\right)\right|$ in which the value of $COP_{(OC)}$ increases (Figure 5(b)) and the value of term $\left|\left(1-\frac{T_0}{T_{b,DSC}}\right)\right|$ decreases with increase in evaporator1 temperatue as ($T_{b,\ DSC}$ -$T_{e,\ DSC}$ =5^{0}C). `Hence, the value of $\eta_{ex\ (OC)}$ decreases with increase in evaporator1 temperature. It can be observed that the value of $\eta_{ex\ (OC)}$ is maximum for R134a and minimum for R1234yf.

Figure 5(d) illustrates the variation in total exergy destruction rate ($\dot{E}_{D,\ Total}$) and exergy destruction ratio (EDR) with evaporator1 temperature ($T_{e,\ DSC}$) of OC. The decreases with increase in $T_{e,\ DSC}$. It is observed that the value of $\dot{E}_{D,\ Total}$ and EDR are minimum for R1234ze and maximum for R1234yf. $\dot{E}_{D,\ Total}$ increases and EDR decreases with increase in $T_{e,\ DSC}$ for refrigerant considered. As the exergetic efficiency decreases with increase in evaporator1 temperature (Figure 5(c)), the rate of exergy recovered of system component decreases. By which the $\dot{E}_{D,\ Total}$ decreases with increase in $T_{e,\ DSC}$. However, the exergy of product decreases with increase in $T_{e,\ DSC}$ because the term $\left|\left(1-\frac{T_0}{T_{b,DSC}}\right)\right|$ decreases due to which the EDR of the overall cycle.

Effect of Condenser Temperature

The effect of degree of condenser1,2 temperature ($T_{c,\ DSC} = T_{c,\ SC}$) on various performance parameters viz. compressor work ($\dot{W}_C$), COP, exergetic efficiency (η_{ex}), total exergy destruction rate ($\dot{E}_{D,\ Total}$) and exergy destruction ratio (EDR) has been depicted in Figure 6(a) to 6(d).

Figure 6(a) illustrates the effect of condenser1,2 temperature ($T_{c,\ DSC} = T_{c,\ SC}$) on compressor work ($\dot{W}_{C,SVCR}$, $\dot{W}_{Comp1}$, $\dot{W}_{Comp2}$and($\dot{W}_{Comp1} + \dot{W}_{Comp2}$)) of simple vapour compression refrigeration cycle (SVCR), DSC, SC and OC for refrigerants considered. The values of compressor work increases with increase in condenser temperature. As the condenser temperature increases, the pressure ratio between evaporator and condenser increases for which more work has been done by the compressor. Therefore, the value of $\dot{W}_C$ increases with increase in T_c. It is also observed that the value of $\dot{W}_C$ is minimum for R1234ze and maximum for R1234yf in DSC, OC and SVCR.

Figure 6(b) shows the effect of condenser1,2 temperature ($T_{c,\ DSC} = T_{c,\ SC}$) on COP of DSC, OC, SC and SVCR for considered refrigerants. The COP of DSC, OC and SVCR decreases with increase in $T_{c,\ DSC} = T_{c,\ SC}$. This is due to that for constant refrigeration load of 3.5167kW, the compressor work increases with increase in condenser temperature. Thus the COP of DSC, OC and SVCR decreases with increase in condenser temperature. However, the COP of sub-cooler VCR cycle (SC) increases with increase in condenser temperature. The rate of heat of heat transfer to evaporator2 increases in the sub-cooler, consequently the refrigeration load of sub-cooler cycle increases and hence COP of sub-cooler cycle increases with increase in condenser temperature. It has also been observed that the COP is maximum for R1234ze and minimum for R1234yf.

Figure 6(c) presents the variation in exergetic efficiency (η_{ex}) with condenser temperature ($T_{c,\ DSC} = T_{c,\ SC}$) of DSC, OC and SVCR cycles for considered refrigerants. The value of η_{ex} decreases with increase in $T_{c,\ DSC} = T_{c,\ SC}$. Exergetic efficiency is given by the relation, $\eta_{ex} = COP \times \left|\left(1-\frac{T_0}{T_b}\right)\right|$ in which the term $\left|\left(1-\frac{T_0}{T_b}\right)\right|$ remains constant and the value of COP decreases with increase in condenser temperature. Hence the value of η_{ex} decreases with increase in $T_{c,\ DSC} = T_{c,\ SC}$. It can also be observed that the value of η_{ex} is maximum for R1234ze and minimum for R1234yf.

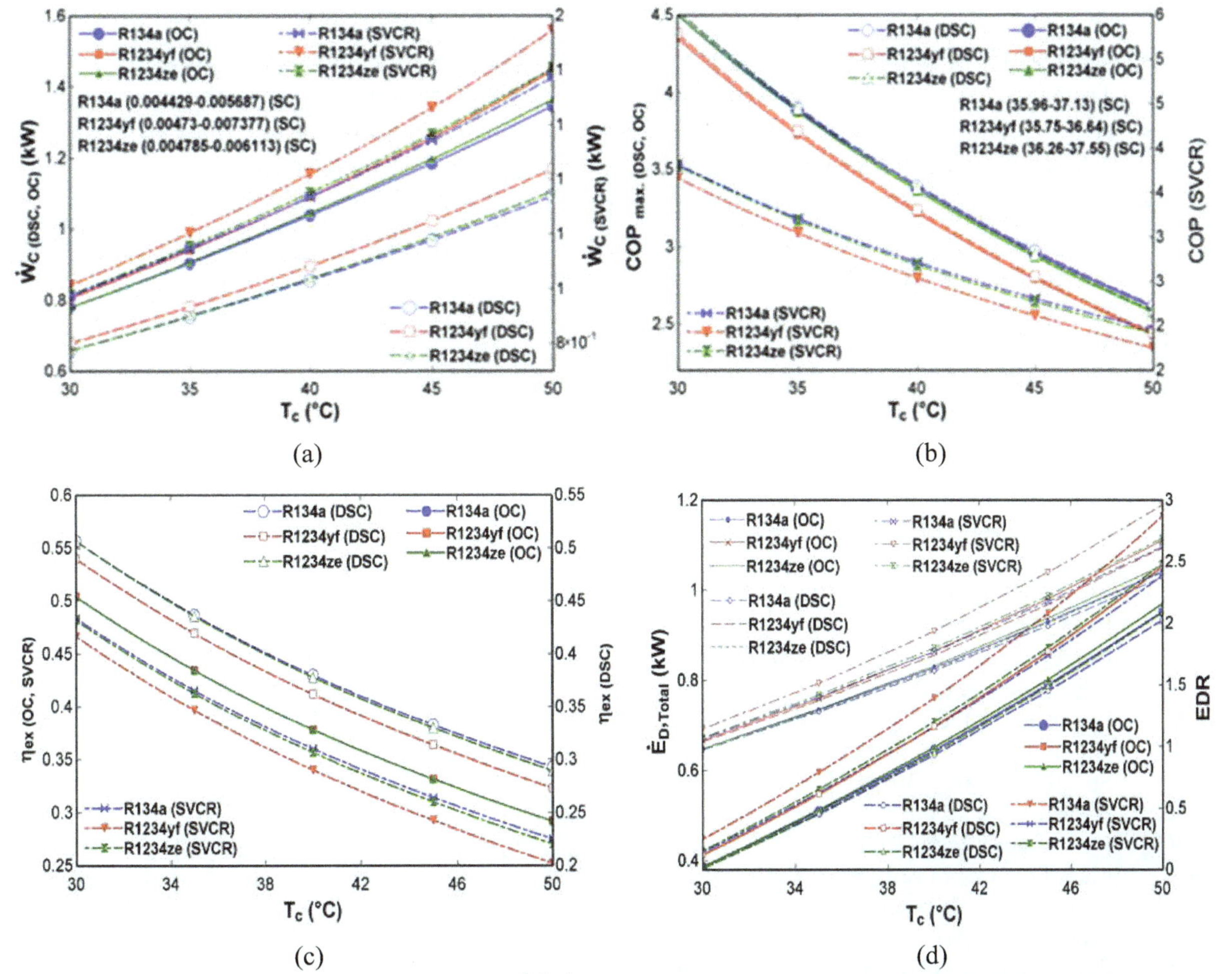

Figure 6. (a) Variation in compressor work $(\dot{W}_C)$ with condenser temperature (T_c), (b) effect of condenser temperature ($T_{e,DSC}$) on COP, (c) Effect of condenser temperature (T_c) on exergetic efficiency (η_{ex}), (d) Variation in total exergy destruction rate $(\dot{E}_{D,Total})$ and exergy destruction ratio (EDR) with condenser temperature (T_c)

Figure 6(d) depicts the effect of condenser temperature ($T_{c,\,DSC} = T_{c,\,SC}$) on total exergy destruction rate $(\dot{E}_{D,Total})$ and exergy destruction ratio (EDR) of DSC, OC and SVCR cycles. The value of $\dot{E}_{D,\ Total}$ and EDR increases with increase in $T_{c,\,DSC} = T_{c,\,SC}$. As the condenser temperature increases, the exergy destruction rate in system components increases as well as the exergy of products $\left(\left|\dot{Q}_{e,DSC}\left(1 - \frac{T_0}{T_{b,DSC}}\right)\right|\right)$ decreases because the value of COP decreases (Figure 6(b)). Hence the total exergy destruction rate and EDR increase with increase in $T_{c,\,DSC} = T_{c,\,SC}$. It has been seen that the value of $\dot{E}_{D,\ Total}$ and EDR is minimum for R134a and maximum for R1234yf.

CONCLUSIONS

The current work presents the energy and exergy analysis of dedicated mechanically sub-cooled vapour compression refrigeration cycle (DSC). The comparison of DSC, overall cycle (OC), sub-cooler cycle (SC) and simple vapour compression refrigeration cycle (SVCR) has also been considered for R134a, R1234ya and R1234ze using a computer software based program in EES. The theoretical performance analysis involves the variation of DOS, evaporator temperature and condenser temperature to check the performance parameters viz. compressor work, COP, exergetic efficiency, and total exergy destruction rate and exergy destruction ratio. The main concluding points of the current work are as follows:

- The COP of dedicated mechanically sub-cooled VCR cycle (DSC) and overall cycle (OC) is higher than the simple VCR cycle.

- The optimum value of DOS lies between 20^0C to 25^0C for which $\dot{W}_C$, COP and η_{ex} are maximum in OC.
- The COP of OC improves while η_{ex} goes down with increase in evaporator temperature.
- The high condenser temperature reduces the performance (COP andη_{ex}) of the cycle.
- Higher condenser temperature promotes exergy destruction while exergy destruction reduces for lower values of evaporator temperature. The optimum value of DOS lies between 20^0C to 25^0C.
- The performance of R1234ze is better than R1234yf and comparable to R134a.
- Finally, it is inferred that the dedicated mechanically sub-cooled VCR cycle performs better than that of simple VCR cycle and the performance of HFO R1234ze is better than R1234yf.

NOMENCLATURE

COMP	Compressor
Cond	Condenser
COP	Coefficient of performance
DOS	Degree of sub-cooling
DSC	Dedicated mechanically sub-cooled VCR cycle
$(\Delta T)_{sc}$	Degree of subcooling
$\dot{E}_D$	Exergy destruction rate (kW)
ED	Exergy destruction
EDR	Exergy destruction ratio
EES	Energy equation solver
$\dot{E}$	Rate of exergy (kW)
E.V.	Expansion valve
GWP	Global warming potential
h	Specific enthalpy (kJ/kg)
HFC	Hydrofluorocarbon
HFO	Hydrofluoroolefin
$\dot{m}_r$	Mass flow rate of refrigerant (kg/s)
ODP	Ozone depleting potential
P	Pressure (kappa)
SVCR	Simple vapour compression refrigeration
$\dot{Q}_e$	Rate of net refrigerating effect (kW)
s	Specific entropy (kJ/kg^{0}C)
T	Temperature (^{0}C)
T_b	Boundary temperature (^{0}C)
T_e	Evaporator temperature (^{0}C)
T_0	Dead state temperature (^{0}C)
VCR	Vapour compression refrigeration
$\dot{W}$	Work rate (kW)
ε	Effectiveness
η	Efficiency
Σ	represents summation
0	Dead state
C	Condenser
Comp	Compressor
Cond	Condenser
e	Evaporator
ex	Exergetic
SC	Su-cooling cycle
OC	Overall cycle
i	Input
o	Output

REFERENCES

[1] Parliament, E., and CotE, U. (2014). Regulation (EU) No 517/2014 of The European Parliament and of the council of 16 April 2014 on fluorinated greenhouse gases and repealing Regulation (EC) No 842/2006. Official J. Eur. Union, 57, 195-230.

[2] Zubair, S. M. (1990). Improvement of refrigeration/air-conditioning performance with mechanical sub-cooling. Energy, 15(5), 427-433.

[3] Zubair, S. M. (1994). Thermodynamics of a vapor-compression refrigeration cycle with mechanical subcooling. Energy, 19(6), 707-715.

[4] Zubair, S. M., Yaqub, M., and Khan, S. H. (1996). Second-law-based thermodynamic analysis of two-stage and mechanical-subcooling refrigeration cycles. International Journal of Refrigeration, 19(8), 506-516.

[5] Couvillion, R. J. (1988). Analysis of a Vapour-compression Refrigeration System with mechanical subcooling. ASHRAE Trans., 94(2), 641-660.

[6] Qureshi, B. A., Inam, M., Antar, M. A., and Zubair, S. M. (2013). Experimental energetic analysis of a vapor compression refrigeration system with dedicated mechanical sub-cooling. Applied Energy, 102, 1035-1041.

[7] Zubair, S. M. (2001). Thermodynamic optimization of finite time vapor compression refrigeration systems. Energy conversion and management, 42(12), 1457-1475.

[8] Thornton, J. W., Klein, S. A., and Mitchell, J. W. (1994). Dedicated mechanical subcooling design strategies for supermarket applications. International Journal of Refrigeration, 17(8), 508-515.

[9] Arora, A., Arora, B. B., Pathak, B. D., and Sachdev, H. L. (2007). Exergy analysis of a vapour compression refrigeration system with R-22, R-407C and R-410A. International journal of Exergy, 4(4), 441.

[10] Arora, A., Dixit, M., and Kaushik, S. C. (2016). Computation of optimum parameters of a half effect water lithium bromide vapour absorption refrigeration system. Journal of Thermal Engineering, 2(2), 683-692.

[11] Arora, A., Dixit, M., and Kaushik, S. C. (2016). Energy and exergy analysis of a double effect parallel flow LiBr/H2O absorption refrigeration system. Journal of Thermal Engineering, 2(1), 541-549.

[12] Arora, A., and Kaushik, S. C. (2008). Theoretical analysis of a vapour compression refrigeration system with R502, R404A and R507A. International journal of refrigeration, 31(6), 998-1005.

[13] Arora, A., Arora, B. B., Pathak, B. D., and Sachdev, H. L. (2007). Exergy analysis of a vapour compression refrigeration system with R-22, R-407C and R-410A. International journal of Exergy, 4(4), 441.

[14] Llopis, R., Cabello, R., Sánchez, D., and Torrella, E. (2015). Energy improvements of CO2 transcritical refrigeration cycles using dedicated mechanical subcooling. International Journal of Refrigeration, 55, 129-141.

[15] Llopis, R., Nebot-Andrés, L., Sánchez, D., Catalán-Gil, J., and Cabello, R. (2018). Subcooling methods for CO 2 refrigeration cycles. A Review. International Journal of Refrigeration.

[16] Mota-Babiloni, A., Navarro-Esbrí, J., Barragán-Cervera, Á., Molés, F., Peris, B., and Verdú, G. (2015). Commercial refrigeration–an overview of current status. International journal of refrigeration, 57, 186-196.

[17] Kalla, S. K., Arora, B. B., and Usmani, J. A. (2018). Alternative refrigerants for HCFC 22 - A review. Journal of Thermal Engineering, 4(3), 1998-2017.

[18] Kalla, S. K., Arora, B. B., and Usmani, J. A. (2018). Performance analysis of R22 and its substitutes in air conditioners. Journal of Thermal Engineering, 4(1), 1724-1736.

[19] Yataganbaba, A., Kilicarslan, A., and Kurtbaş, İ. (2015). Exergy analysis of R1234yf and R1234ze as R134a replacements in a two evaporator vapour compression refrigeration system. International journal of refrigeration, 60, 26-37.

[20] Bejan, A., Tsatsaronis, G., Moran, M., and Moran, M. J. (1996). Thermal design and optimization. John Wiley and Sons.

[21] Dincer, I., and Kanoglu, M. (2010). Refrigeration systems and applications (Vol. 2). New York: Wiley.

[22] Klein, S. A., and Alvarado, F. (2012) Engineering Equation Solver. F-Chart Software, Middleton, WI. Version 9, 224-3D.

[23] Kotas, T. J. (1995) The exergy method of thermal plant analysis, Reprint ed., Krieger Publishing Company, Krieger Drive, Malabar, Florida-32950.

[24] Dixit, M., Kaushik, S. C., and Arora, A. (2017) Energy and exergy analysis of absoption-compression cascade refrigeration system. Journal of Thermal Engineering, 3(5), 1466-1477.

[25] Dixit, M., Arora, A., and Kaushik, S. C. (2016) Energy and exergy analysis of a waste heat driven cycle for triple effect refrigeration. Journal of Thermal Engineering, 2(5), 954-961.

[26] Dixit, M., Kaushik, S. C., and Arora, A. (2017) Energy and exergy analysis of absoption-compression cascade refrigeration system. Journal of Thermal Engineering, 3(5), 1466-1477.

A CFD INVESTIGATION OF Al_2O_3/WATER FLOW IN A DUCT HAVING BACKWARD-FACING STEP

R. Ekiciler[1,*], E. Aydeniz[2], K. Arslan[3]

ABSTRACT

Al_2O_3/water forced convection nanofluid flow was numerically studied in a duct with backward-facing step. Nanoparticle volume fraction was changed between 1%-5%. Diameter of nanoparticle was constant (d_p=40 nm). The Reynolds number was increased from 100 to 500. The step and total height of the duct were 4.8 mm and 9.6 mm, respectively. The bottom wall, which was positioned after the step, was heated with 2000 W/m^2 and the rest of the walls were adiabatic. Nusselt number, velocity profiles and friction factor were investigated in detail. It was obtained that Nusselt number increases with increasing nanoparticle volume fraction and Reynolds number.

Keywords: *Forced Convection, Nanofluid, Backward-facing Step, Laminar Flow, Diameter of Nanoparticle*

INTRODUCTION

Flow separation and reattachment of the fluid caused by sudden expansion is important in many engineering applications. Heat transfer applications of backward facing step (BFS) are mainly used in avionics systems, high performance heat exchangers, cooling of nuclear reactors, wide-angle diffusers, airfoils, electronic cooling equipment, cooling of turbine blades, combustion chambers, environmental control systems, axial and centrifugal compressors blade and is also used in many other heat transfer devices.

In the literature, the separation and reattachment flow over a backward-facing step was firstly carried out in the 1950's. Studies on separated flow have been researched more extensively by increasing popularity over in the last decade. Goldstein et al. [1] performed an experimental study to investigate the reattachment point under laminar and subsonic flow conditions over a step. They noticed that the reattachment point and boundary displacement thickness change with Reynolds number. Denham and Patrick [2] investigated 2D laminar flow over the BFS. Water was used as working fluid. The channel's expansion and aspect ratios are equal to 3 and 20, respectively. The Reynolds number was varied from 50 to 250. They observed that flow characteristics of BFS approximates to other 2D geometries having sudden expansion. Yet, both recirculation length and mass flow rate circulating were smaller. At Re=229, they noticed a fluctuation, which shows the beginning of transition to turbulent separated boundary layer. Rinoie et al. [3] experimentally investigated turbulent flow over the backward-facing step to clarify behavior of turbulent flow inside the reattachment shear layer using turbulent energy balance assisting in understanding of turbulent flow of BFS. Step height and aspect ratio were 0.02 m and 10, respectively. They observed that turbulent structures can be classified in three categories (the dead air, reversed flow and separated shear layer zone). In the turbulent energy balance, transverse diffusion assisting to balance the dissipation term was positive. Nie and Armally [4] examined numerically the effects of step height on heat transfer and flow characteristics in a three dimensional BFS channel. Flow was thought as laminar and forced convection conditions. Air was passed through the channel as working fluid. Thermo-physical properties of air were assumed constant along the channel. The study was performed at constant Reynold number (Re=343). Constant and uniform heat flux (q"w=50 W/m2) was subjected to downward of the step wall while the other walls were insulated. Three different step height was performed in this study. They observed at the end of the study that Nusselt number, general three- dimensional flow characteristic features, gradient locating in recirculation region increases with increasing step height. Also, they noticed that step height effects the friction coefficient distribution. Lan et al. [5] conducted numerical simulation of 3D turbulent forced convection flow in backward-facing step channel. Range of Reynolds number was at 20000-50000. The channel's step height and

This paper was recommended for publication in revised form by Regional Editor Tolga Taner
[1] Department of Mechanical Engineering, Gazi University, Ankara, Turkey, recepekiciler@gazi.edu.tr
[2] Department of Mechanical Engineering, Karabük University, Karabük, Turkey, eaydeniz@gmail.com
[3] Department of Mechanical Engineering, Karabük University, Karabük, Turkey, kamilarslan@karabuk.edu.tr
**E-mail address: recepekiciler@gazi.edu.tr*

expansion ratio was 4.8 and 1.48 mm, respectively. This study was done to understand the effects of Reynolds number and aspect ratio on heat transfer and fluid flow features. The Nusselt number, temperature, and general flow features distributions were presented. They concluded that Reynolds number and aspect ratio influence slightly flow separation and reattachment point but heat transfer is substantially influenced.

Nowadays, more and more energy is needed to supply energy of the World due to the developing technologies. Therefore, diversified heat transfer technologies have been set out to be developed. One of them is initiated by using nanofluids that are used for increasing heat transfer. Nanofluids are novel heat transfer fluids having nanoparticles whose sizes change from 1 nm to 100 nm suspended in the base fluids. In recent years nanofluids have been progressively studied. Abu- Nada [6] is the first researcher who conducted a numerical study concerning heat transfer by using nanofluids in ducts with backward-facing step. In this study, five different nanoparticles (CuO, Cu, TiO_2, Ag, and Al_2O_3) were used with volume fractions from 0.05% to 2.0%. Water was the base fluid. Duct was thought 2D and its expansion ratio was 2. The range of Reynolds number was 200-600. He reported that Nusselt number increases with increasing volume fractions. Al-aswadi et al. [7] studied numerically 2D laminar forced convection flow of nanofluids in channel with BFS. The base fluid was water. They conducted this study by using Al_2O_3, Au, Ag, CuO, TiO_2, Cu, SiO_2 and diamond nanoparticles with 0.05 volume fraction. The study was performed for Re=50, 100 and 175. Expansion ratio and step height and of the devised channel were 2 mm and 4.8 mm, respectively. All the walls were insulated. They presented the effects of Reynolds number and nanofluids types on pressure drop, velocity distribution and friction coefficient. They noticed that the highest wall shear stress is observed for SiO_2/water nanofluid. They found that nanofluids prepared nanoparticle with low density increase velocity much more compared to nanofluids prepared nanoparticle with high density. Also, they obtained that Reynolds number and skin friction coefficient are inversely proportional. Ekiciler and Arslan [8] numerically conducted a study of heat transfer and flow characterictics in forced convection laminar flow conditions by using 2D duct with BFS. SiO_2/water nanofluid was forced. The nanoparticle volume fraction was ranged from 1% to 4%. Reynolds number was selected from 75 to 225. Both of the step and inlet heights of the duct were 4.8 mm. As the downstream wall was subjected to constant and uniform heat flux of 2000 W/m^2, the other walls were insulated. It was found from the results of numerical simulation that the Nusselt number increases with increasing the nanoparticle volume fraction; however, there was no significant change on Darcy friction factor.

It is obvious from the above literature review that the case of forced convective heat transfer over BFS utilizing different nanoparticle volume fraction of Al_2O_3/water nanofluid seem have not been investigated in the past and this has motivated the present study. The present study deals with 2D laminar forced convective nanofluid flow over a BFS. Results of interests such as velocity distributions, Nusselt number, and friction factor for laminar forced convection over a horizontal BFS were reported to illustrate the effect of types of nanofluids on these parameters.

MATHEMATICAL MODELLING

Laminar forced heat transfer and flow of nanofluids in a 2D duct with BFS, which is presented in Figure 1, has been numerically investigated. The length of downstream and upstream wall were 0.05 m and 0.1 m, respectively. Total duct height was 9.6 mm. While downstream stepped wall was heated with uniform and constant heat flux of 2000 W/m2, the other walls were insulated. Solid particles with size under 100 nm were considered to be able to use single phase approach, thus single phase approach was practiced for nanofluid modeling and the phase of nanofluid was assumed to be as a continuous [9]. The following assumptions were adopted for this numerical study: (i) Both heat transfer and fluid flow in duct were in 2D and steady-state; (ii) fluid flow was incompressible and laminar flow; (iii) the physical properties of nanofluid, such as specific heat, density, thermal conductivity were taken as temperature independent; (iv) negligible viscous dissipation and radiation heat transfer; and (v) the base fluid and the nanoparticles were in thermal equilibrium.

With assumptions mentioned above, governing equations; continuity, momentum and energy are given below, respectively:

$$\vec{\nabla}.\vec{V} = 0 \tag{1}$$

$$\rho \frac{D\vec{V}}{Dt} = -\Delta p + \mu \nabla^2 \vec{V} \tag{2}$$

$$\rho c_p \frac{DT}{Dt} = k \nabla^2 T \tag{3}$$

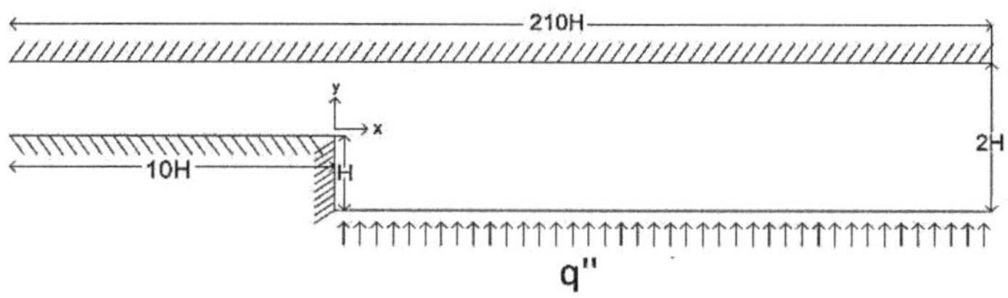

Figure 1. Schematic diagram for backward facing step

Table 2. Thermo-physical properties of nanoparticles and pure water

Types of Nanofluids	ρ(kg/m^3)	μ(N/ms)	k(W/mK)	Cp(kJ/kgK)
Water	996.6	0.000855	0.613	4179
Al_2O_3 [10]	3970	-	40	765

Vajjha and Das [11] offered the equation of effective thermal conductivity for nanofluid as:

$$k_{eff} = k_{static} + k_{brownian} \tag{4}$$

Ghasemi and Aminossadati [12] presented the k_{static} as:

$$k_{static} = k_f \left[\frac{(k_s + 2k_f) - 2\varphi(k_f - k_s)}{(k_s + 2k_f) + \varphi(k_f - k_s)} \right] \tag{5}$$

where k_s and k_f are the thermal conductivies of the solid particles and the base fluid, respectively.

Vajjha and Das [11] presented thermal conductivity originating from Brownian motion as:

$$k_{brownian} = 5x10^4 \beta \varphi \rho_f c_{pf} \sqrt{\frac{KT}{\rho_s d_p}} f(T,\varphi) \tag{6}$$

where;

$$f(T,\varphi) = (2.8217x10^{-2}\varphi + 3.917x10^{-3})\frac{T}{T_r} + (-3.0669x10^{-2}x10^{-2}\varphi - 3.91123x10^{-3}) \tag{7}$$

where T is the fluid temperature and taking constant at 300 K, K is the Boltzmann constant and T_r is the reference temperature.

Corcione [13] suggested the equation of effective thermal expansion for nanofluid as:

$$\beta_{eff} = \frac{(1-\varphi)(\rho\beta)_f + \varphi(\rho\beta)_s}{(1-\varphi)\rho_f + \varphi\rho_s} \quad (8)$$

where β_s and β_f are the thermal expansion of nanoparticles and base fluid, respectively.

Table 3. β value of different nanoparticles and its boundary conditions

Parameters	Al_2O_3
β	$8.4407(100\ \varphi\ )^{-1.07304}$ [11]
Concentration	$1\% \le \varphi \le 4\%$
Temperature	298 K≤T≤363 K

Corcione [13] presented viscosity of nanofluid as:

$$\frac{\mu_{eff}}{\mu_f} = \frac{1}{1-34.87(d_p/d_f)^{-0.3}\varphi^{1.03}} \quad (9)$$

where;

$$d_f = \left[\frac{6M}{N\pi\rho_{fo}}\right]^{1/3} \quad (10)$$

where μ_f and μ_{eff} are viscosity of base fluid and nanofluid, respectively. φ is volume fraction of nanofluid, d_f is the base fluid equivalent diameter and d_p is diameter of nanoparticle., M is the base fluid's moleculer weight, N is the Avogadro number and ρ_{fo} is the mass density of the base fluid calculated at T=293 K.

Corcione [13] proposed the effective density of nanofluid as:

$$\rho_{eff} = (1-\varphi)\rho_f + \varphi\rho_s \quad (11)$$

where ρ_s, ρ_f and ρ_{eff} are densities of nanoparticle, base fluid and nanofluid, respectively.

Corcione [13] offered the effective specific heat of nanofluid as:

$$\left(c_p\right)_{eff} = \frac{(1-\varphi)(\rho c_p)_f + \varphi(\rho c_p)_s}{(1-\varphi)\rho_f + \varphi\rho_s} \quad (12)$$

where c_{pf} and c_{ps} are the specific heat of base fluid and nanoparticles, respectively.

Average Nusselt number, Darcy friction factor and heat transfer coefficient are calculated as:

$$Nu = \frac{hD_h}{k} \quad (13)$$

$$f = \frac{\Delta P(D_h / L)}{\rho u_i^2 / 2} \tag{14}$$

$$h = \rho u_i A_c c_p (T_o - T_i) / A_s (T_w - T_m) \tag{15}$$

NUMERICAL PROCEDURE AND MESH INDEPENDENCE TEST

In the numerical computations, the finite-volume method based commercial CFD software Ansys Fluent 17.0 was used to perform the numerical calculations by solving the governing equations along with the boundary conditions. The convection terms in mass, momentum and energy equations were discretized using a second order upwind scheme. The standard scheme was employed for discretization of pressure and the SIMPLE algorithm was used to resolve the coupling between velocity and pressure. The Green-Gauss cell based method was applied for discretization of the momentum and energy equations. To obtain convergence, each equation for mass, momentum, and energy were iterated until the residual falls below $1x10^{-6}$. No convergence problems were observed during the calculations. The hexahedral mesh distribution was used for the duct with BFS. The numbers of mesh points or control volumes were increased close to the wall of the duct and near to the BFS to enhance the resolution and accuracy as given in Figure 2.

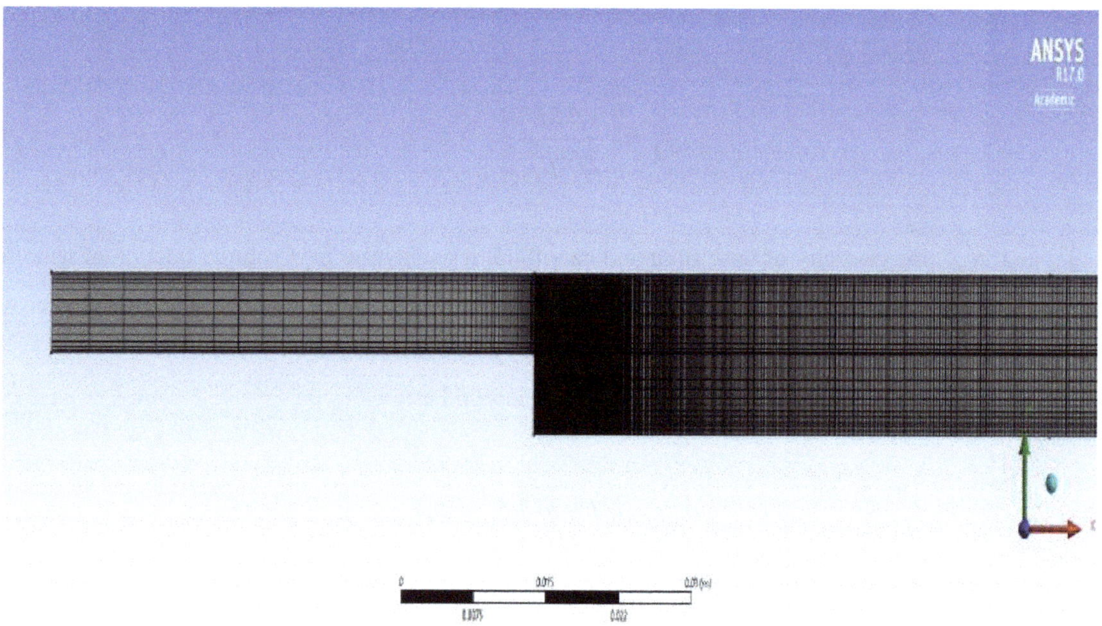

Figure 2. Mesh distribution of the computational domain

The mesh independence study was performed by refining the mesh number until the variation in both average Nusselt number and average friction factor were less than 0.1%. To obtain the optimum mesh number, a grid independence study was conducted using fifteen different mesh numbers changing from $6.2x10^2$ to $1.5\ x10^5$ for Re=500. Changing of Nusselt number and average friction factor values with mesh number for pure water flow is given in Figure 3 as an illustration. It was observed that a further refinement of mesh number from $9x10^3$ to $1.5x10^5$, the changing of average Nusselt number and average friction factor is negligible. Hence, optimum mesh number with minimum computational time and maximum accuracy approximately can be seen at $2.6x10^4$ for this study.

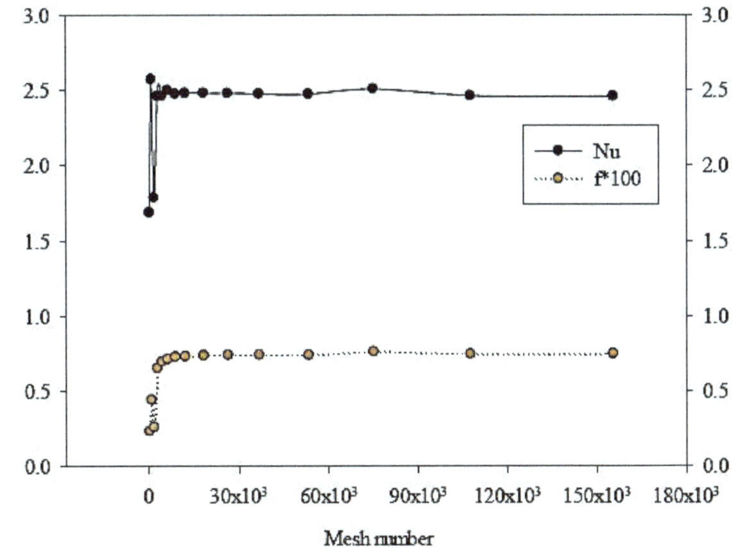

Figure 3. Mesh independence study for Nusselt number and friction factor at Re=500

To understand the accuracy of the numerical simulation, values of average pressure was compared with a study conducted by Al-aswadi et al. [7] obtained from literature in Figure 4. The results obtained by present study are good agreement with the study conducted by Al-aswadi et al. [7].

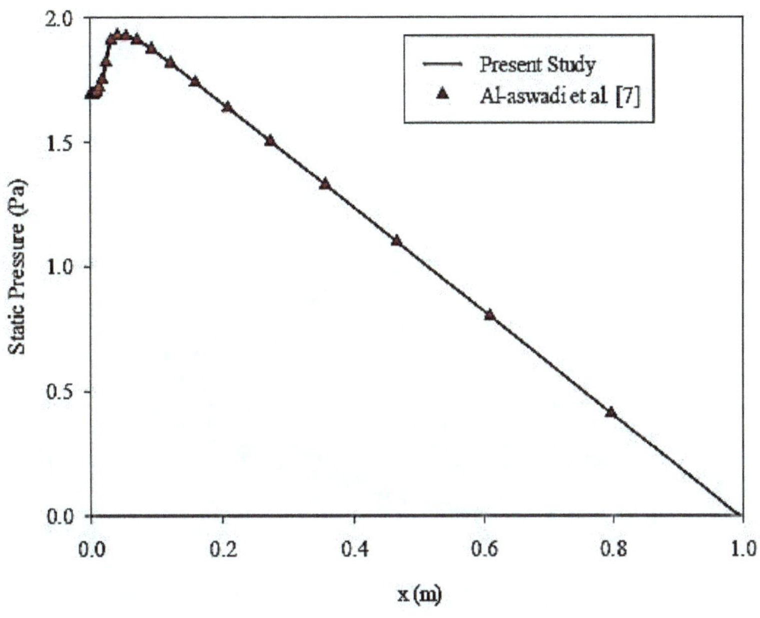

Figure 4. Comparison of pressure drop of the present results for pure water at Re=175

RESULTS AND DISCUSSION

Velocity distribution of Al_2O_3/water nanofluid for different nanoparticle volume fraction is presented at Re=400 in Figure 5. It can be seen from the figure that nanoparticle volume fraction has an active role in velocity distribution. Velocity magnitude increases with increasing the nanoparticle volume fraction. The highest velocity distribution of Al_2O_3/water nanofluid is at 5% volume fraction. Figure 5(a) and Figure 5(b) testimony that there is a recirculation region occurs after the step. Figure 5(c), Figure 5(d), and Figure 5(e) show that the nanofluid flow is independent of effect of sudden expansion.

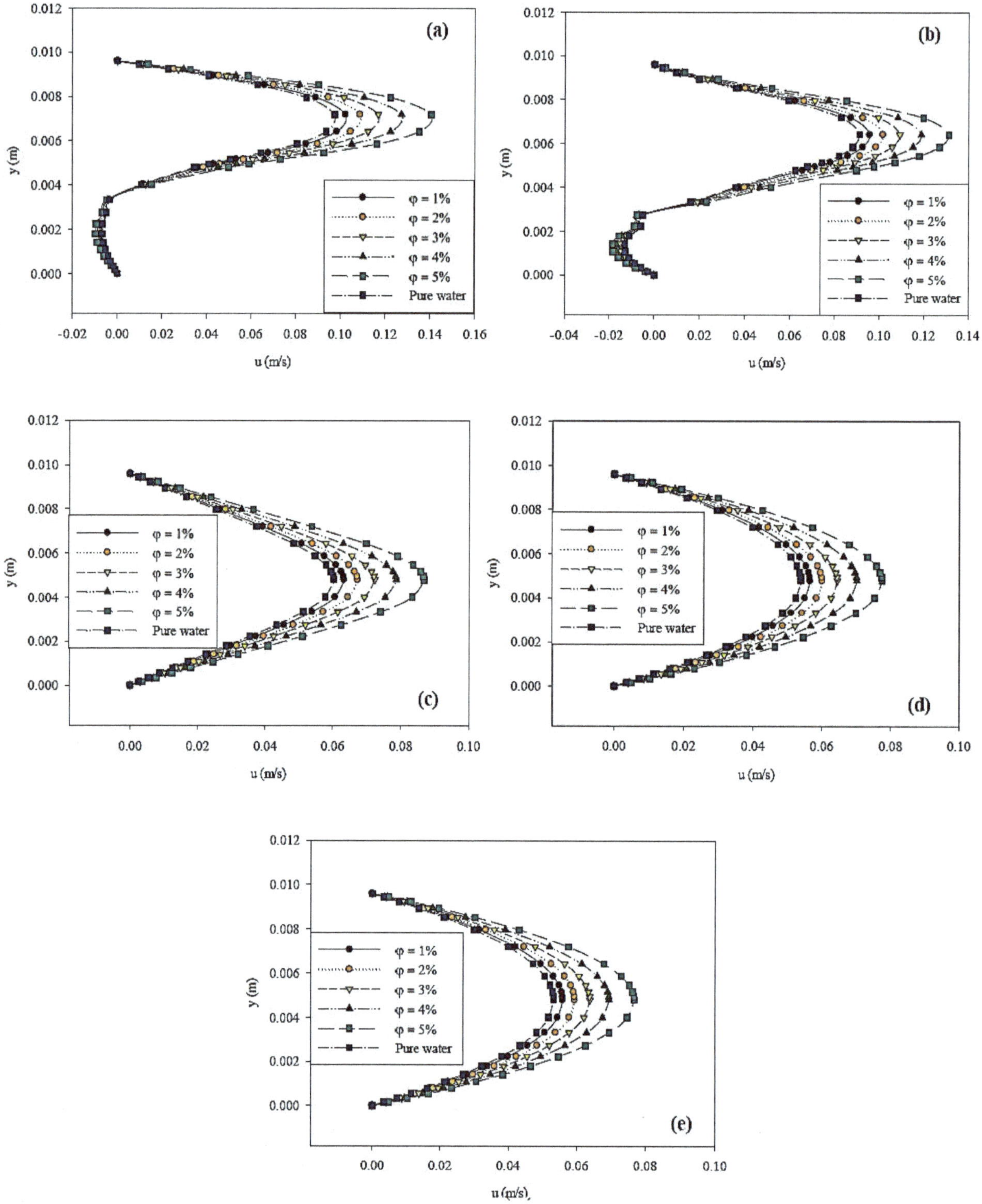

Figure 5. The effects of nanoparticle volume fraction on velocity distribution using Al_2O_3/water at Re=400 for different locations; (a) x=0.01 m, (b) x=0.03 m, (c) x=0.13 m, (d) x=0.35 m, (e) x=1.0 m

Figure 6 shows average Nusselt number distribution of Al_2O_3/water nanofluid and pure water as a function of Reynolds number. The nanoparticle volume fraction is changed from 1% to 5%. It realized that Nusselt number increases with increasing Reynolds number. Al_2O_3/water nanofluid having 5% volume fraction has the highest Nusselt number while pure water has the lowest one. This result demonstrates that Al_2O_3/water nanofluid flow enhances the heat transfer.

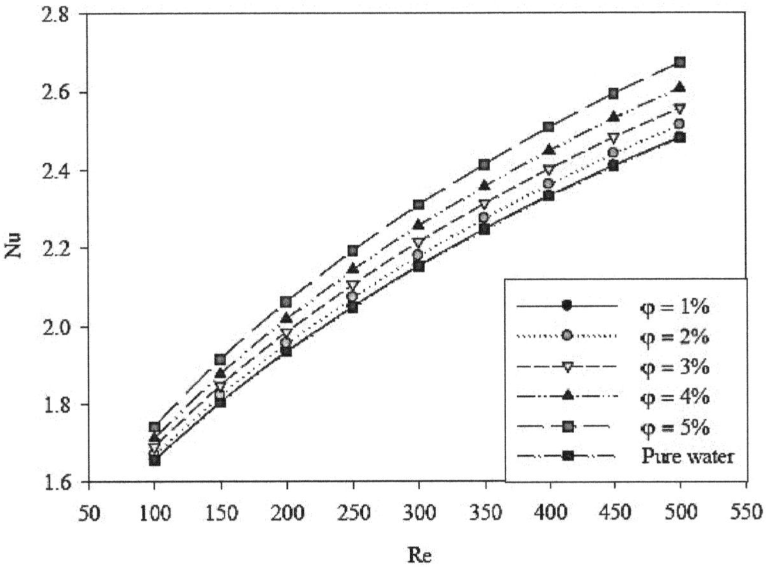

Figure 6. Average Nusselt number distribution using Al_2O_3/water nanofluid and pure water for different nanoparticle volume fractions

The effect of nanoparticle volume fraction on average friction factor distribution of Al_2O_3/water nanofluid and pure water is given in Figure 7. It is revealed that average friction factor decreases with Reynolds number for all nanoparticle volume fractions. In addition, average friction factor does not almost influence by nanoparticle volume fractions, which is the results with Ref. [14].

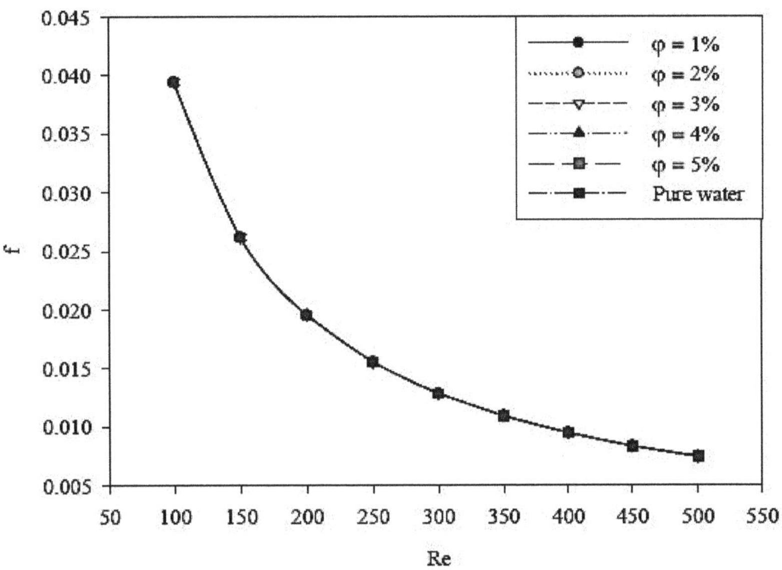

Figure 7. Average friction factor distribution using Al_2O_3/water nanofluid and pure water for different nanoparticle volume fractions

The local Nusselt number distribution is presented along the heated wall for Re=400 in Figure 8. It is noticed that the local Nusselt number increases substantially to its maximum value which is obtained at the separation point. After that it decreases until the reattachment point and then it takes almost a fixed shape, when the flow is being thermally fully developed condition. Moreover, local Nusselt number distribution presents the same characteristic behavior for all nanoparticle volume fractions. Local Nusselt number increases with increasing CuO nanoparticle volume fraction, also. Maximum local Nusselt number can be obtained at 4.0% nanoparticles volume fraction for Re=400.

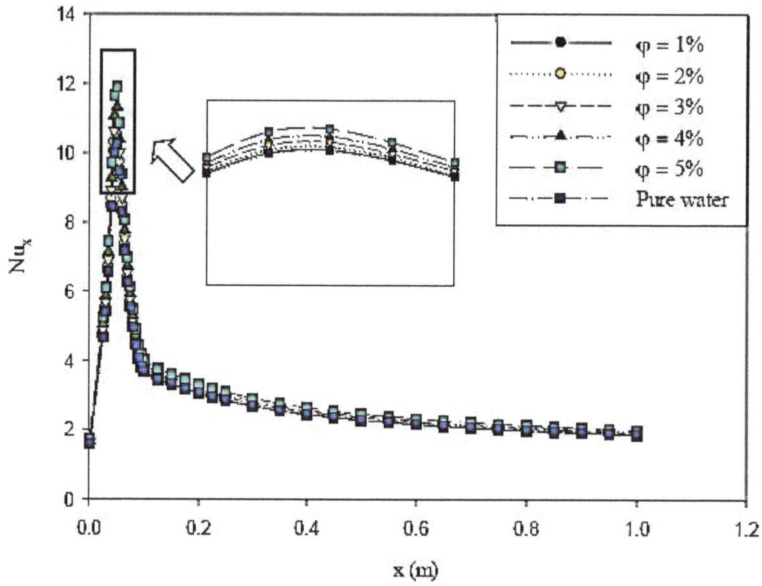

Figure 8. Local Nusselt number distribution using Al_2O_3/water nanofluid and pure water for different nanoparticle volume fractions at Re=400

The local Darcy friction factor distribution is illustrated along the heated wall different nanoparticle volume fractions and Re=400 in Figure 9. It is noticed that the local Darcy friction factor decreases and then increases until it reaches a point where remains constant along the rest of heated wall. This shows that the local Darcy friction factor approaches the hydrodynamically fully developed condition. Also, it is realized that the average Darcy friction factor is independent from nanoparticle volume fraction.

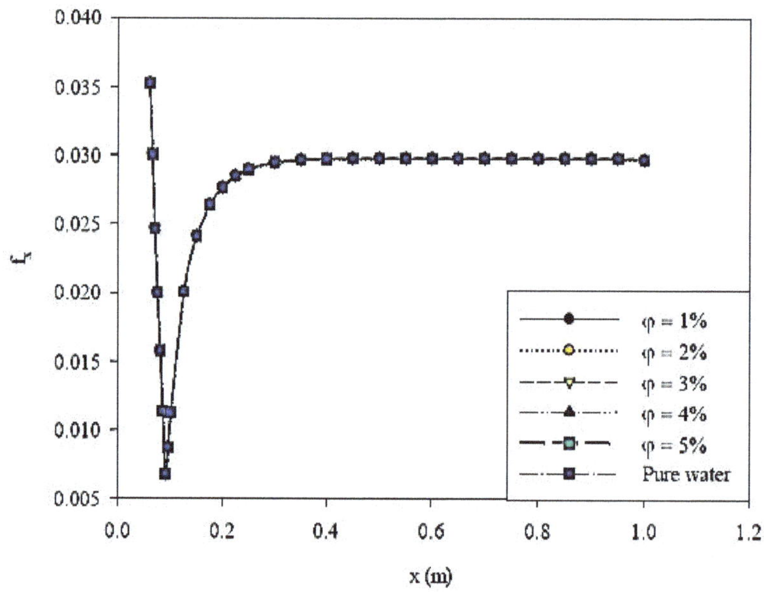

Figure 9. Local friction factor distribution using Al_2O_3/water nanofluid and pure water for different nanoparticle volume fractions at Re=400

CONCLUDING REMARKS

Al_2O_3/water nanofluid over two-dimensional backward- facing step is simulated under laminar forced convection condition. As a result, Al_2O_3/water nanofluid enhances heat transfer. The average Nusselt number increases with increasing the Reynolds number and nanoparticle volume fraction. Also, Darcy friction factor decreases with increasing Reynolds number. There is no effect of changing of nanoparticle volume fractions on average Darcy friction factor.

NOMENCLATURE

c_p	Specific heat [J/kg K]
D_h	Hydraulic diameter, 2h [m]
d_p	Nanoparticles diameter [nm]
ER	Expansion ratio, ER= h/H
F	Average friction factor
h	Inlet channel height [m]
H	Total channel height [m]
K	Boltzmann constant
K	Thermal conductivity [W/m K]
L	Length of the heated section [m]
L_S	The length of step [m]
M	Molecular weight [kg/kmol]
N	Avogadro number
Nu	Average Nusselt number, hD_h/k
q”	Heat flux [W/m^2]
P	Pressure [Pa]
Re	Reynolds number, $\rho u_i D_h/\mu$
S	Step height, [nm]
T	Fluid temperature [K
u	Velocity component on x direction [m/s]
v	Velocity component on y direction [m/s]
w	Velocity component on z direction [m/s]
Φ	Nanoparticle volume fraction
B	Thermal expansion coefficient [1/K]
ρ	Density [kg/m^3]
μ	Dynamic viscosity [N s/m^2]
τ_W	Wall shear stress [N/m^2]
υ	Kinematic viscosity [m^2/s]
eff	Effective
f	Fluid
s	Solid
nf	Nanofluid
w	Wall
i	Inlet condition
o	Outlet condition
x	Local
c	Cross-section

REFERENCES

[1] Goldstein, R. J., Eriksen, V. L., Olson, R. M., & Eckert, E. R. G. (1970). Laminar separation, reattachment, and transition of the flow over a downstream-facing step. Journal of Basic Engineering, 92(4), 732-739.

[2] Denham, M. K., Patrick, M. A. (1974). Laminar flow over a downstream-facing step in a two-dimensional flow channel. Transactions of the American Institute of Chemical Engineers 52(4), 361-367.

[3] Rinoie, K., Shirai, Y., Saitao, Y., Sunada, Y. (1998) Behaviours of separated and reattaching flow formed over backward-facing step. 21st ICAS Congress.

[4] Nie, J. H., Armaly, B. F. (2002). Three-dimensional convective flow adjacent to backward-facing step-effects of step height. International journal of heat and mass transfer, 45(12), 2431-2438.

[5] Lan, H., Armaly, B. F., Drallmeier, J. A. (2009). Three-dimensional simulation of turbulent forced convection in a duct with backward-facing step. International Journal of Heat and Mass Transfer, 52(7-8), 1690-1700.

[6] Abu-Nada, E. (2008). Application of nanofluids for heat transfer enhancement of separated flows encountered in a backward facing step. International Journal of Heat and Fluid Flow, 29(1), 242-249.

[7] Al-Aswadi, A. A., Mohammed, H. A., Shuaib, N. H., Campo, A. (2010). Laminar forced convection flow over a backward facing step using nanofluids. International Communications in Heat and Mass Transfer, 37(8), 950-957.

[8] Ekiciler, R., Arslan, K. (2016) Numerical analysis of SiO_2/water nanofluid flow over backward facing step. International Conference of Engineering and Natural Sciences pp 331–339.

[9] Moraveji, M. K., Darabi, M., Haddad, S. M. H., Davarnejad, R. (2011). Modeling of convective heat transfer of a nanofluid in the developing region of tube flow with computational fluid dynamics. International communications in heat and mass transfer, 38(9), 1291-1295.

[10] Salman, B. H., Mohammed, H. A., Kherbeet, A. S. (2012). Heat transfer enhancement of nanofluids flow in microtube with constant heat flux. International Communications in Heat and Mass Transfer, 39(8), 1195-1204.

[11] Vajjha, R. S., Das, D. K. (2009). Experimental determination of thermal conductivity of three nanofluids and development of new correlations. International Journal of Heat and Mass Transfer, 52(21-22), 4675-4682.

[12] Ghasemi, B., & Aminossadati, S. M. (2010). Brownian motion of nanoparticles in a triangular enclosure with natural convection. International Journal of Thermal Sciences, 49(6), 931-940.

[13] Corcione, M. (2010). Heat transfer features of buoyancy-driven nanofluids inside rectangular enclosures differentially heated at the sidewalls. International Journal of Thermal Sciences, 49(9), 1536-1546.

[14] Ekiciler, R., & Arslan, K. (2018). CuO/Water Nanofluid Flow over Microscale Backward-Facing Step and Analysis of Heat Transfer Performance. Heat Transfer Research, 49(15).

Permissions

All chapters in this book were first published in JTE, by Yildiz Technical University; hereby published with permission under the Creative Commons Attribution License or equivalent. Every chapter published in this book has been scrutinized by our experts. Their significance has been extensively debated. The topics covered herein carry significant findings which will fuel the growth of the discipline. They may even be implemented as practical applications or may be referred to as a beginning point for another development.

The contributors of this book come from diverse backgrounds, making this book a truly international effort. This book will bring forth new frontiers with its revolutionizing research information and detailed analysis of the nascent developments around the world.

We would like to thank all the contributing authors for lending their expertise to make the book truly unique. They have played a crucial role in the development of this book. Without their invaluable contributions this book wouldn't have been possible. They have made vital efforts to compile up to date information on the varied aspects of this subject to make this book a valuable addition to the collection of many professionals and students.

This book was conceptualized with the vision of imparting up-to-date information and advanced data in this field. To ensure the same, a matchless editorial board was set up. Every individual on the board went through rigorous rounds of assessment to prove their worth. After which they invested a large part of their time researching and compiling the most relevant data for our readers.

The editorial board has been involved in producing this book since its inception. They have spent rigorous hours researching and exploring the diverse topics which have resulted in the successful publishing of this book. They have passed on their knowledge of decades through this book. To expedite this challenging task, the publisher supported the team at every step. A small team of assistant editors was also appointed to further simplify the editing procedure and attain best results for the readers.

Apart from the editorial board, the designing team has also invested a significant amount of their time in understanding the subject and creating the most relevant covers. They scrutinized every image to scout for the most suitable representation of the subject and create an appropriate cover for the book.

The publishing team has been an ardent support to the editorial, designing and production team. Their endless efforts to recruit the best for this project, has resulted in the accomplishment of this book. They are a veteran in the field of academics and their pool of knowledge is as vast as their experience in printing. Their expertise and guidance has proved useful at every step. Their uncompromising quality standards have made this book an exceptional effort. Their encouragement from time to time has been an inspiration for everyone.

The publisher and the editorial board hope that this book will prove to be a valuable piece of knowledge for researchers, students, practitioners and scholars across the globe.

List of Contributors

N. Kurtulmuş
Adana Science and Technology University, Faculty of Engineering and Natural Science, Mech. Eng. Dept., Adana/Turkey

M. Bilgili
Cukurova University, Ceyhan Engineering Faculty, Mech. Eng. Dept., Adana, Turkey

B. Şahin
Cukurova University, Faculty of Engineering and Architecture, Mech. Eng. Dept., Adana, Turkey

A. H. Bademlioglu
Department of Energy Systems Engineering, Bursa Technical University, Bursa, TURKEY

R. Yamankaradeniz and O. Kaynakli
Department of Mechanical Engineering, Uludag University, Bursa, TURKEY

N.Doseva and D. Chakyrova
Department of Thermal Engineering, Varna Technical University, Varna, BULGARIA

Yasin Karagöz
Yildiz Technical University, Mechanical Engineering Department, Automotive Division, İstanbul, Turkey

Hasan Köten
Medeniyet University, Mechanical Engineering Department, Istanbul, Turkey

H. Kaya
Department of Mechanical Engineering, Bartın University, Bartın, Turkey

R. Ekiciler
Department of Mechanical Engineering, Gazi University, Ankara, Turkey

K. Arslan
Department of Mechanical Engineering, Karabük University, Karabük, Turkey

A. Ahmadi Nadooshan, A. Jahanbakhshi and M. Bayareh
Department of Mechanical Engineering, Shahrekord University, Shahrekord, IRAN

B. Aksu
Gas and Gas Installation Program, TOBB Vocational School of Technical Sciences, Karabuk Uni., 78050, Karabuk, TURKEY

C. Uysal
Automotive Technologies Program, TOBB Vocational School of Technical Sciences, Karabuk Uni., 78050, Karabuk, TURKEY

H. Kurt
Mechanical Engineering Department, Faculty of Engineering & Architecture, Necmettin Erbakan Uni., 42140, Konya, TURKEY

Mohammad H. Ahmadi
Faculty of Mechanical Engineering, Shahrood University of Technology, Shahrood, Iran

Fathollah Pourfayaz and Mohammad Hossein Jahangir
Department of Renewable Energies and Environmental, Faculty of New Sciences and Tech., University of Tehran, Tehran, Iran

M. Ozdogan, L. Namli and A. Durmus
Department of Mechanical Engineering, Ondokuz Mayıs University, Samsun, TURKEY

N. S. Bondareva and M. A. Sheremet
Department of Theoretical Mechanics, Tomsk State University, Tomsk, Russia

Y. Durmusoglu and G. Kocak
Istanbul Technical University, Department of Marine Engineering, 34940, Tuzla, Istanbul, Turkey

Yu. Baidak and I. Vereitina
Department of Thermodynamics and Renewable Energetics, Odessa national Academy of Food Tech., Odessa, UKRAINE

C. Yilmaz
Department of Mechanical Engineering, Afyon Kocatepe University, Afyonkarahisar, Turkey

T. H. Cetin, B. Ozturkmen and M. Kanoglu
Department of Mechanical Engineering, University of Gaziantep, Gaziantep, Turkey

M. M. Keshtkar and P. Talebizadeh
Department of Mechanical Engineering, Kerman Branch, Islamic Azad University, Kerman, Iran

T. Mohapatra and S. K. Rout
Department of Mechanical Engineering,C. V. Raman College of Engineering, Bhubaneswar, Odisha, India

Fatma Gül Çelik, Emin Açıkkalp and Hasan Yamik
Department of Mechanical Engineering, Engineering Faculty, Bilecik S.E. University, Bilecik, Turkey

S. Agarwal, A. Arora and B. B. Arora
Department of Mechanical Engineering, Delhi Technological University, Delhi-110042, India

R. Ekiciler
Department of Mechanical Engineering, Gazi University, Ankara, Turkey

E. Aydeniz
Department of Mechanical Engineering, Karabük University, Karabük, Turkey

Index

CPSIA information can be obtained
at www.ICGtesting.com
Printed in the USA
LVHW062010100122
708212LV00002B/149

9 781639 891337